T0211601

Data Science for Economics and Finance

.

Sergio Consoli • Diego Reforgiato Recupero •
Michaela Saisana

Editors

Data Science for Economics and Finance

Methodologies and Applications

 Springer

Editors
Sergio Consoli
European Commission
Joint Research Centre
Ispra (VA), Italy

Diego Reforgiato Recupero
Department of Mathematics and Computer
Science
University of Cagliari
Cagliari, Italy

Michaela Saisana
European Commission
Joint Research Centre
Ispra (VA), Italy

ISBN 978-3-030-66893-8   ISBN 978-3-030-66891-4   (eBook)
https://doi.org/10.1007/978-3-030-66891-4

This Springer imprint is published by the registered company Springer Nature Switzerland AG.
The registered company address is: Gewerbestrasse 11, 6330 Cham, Switzerland

Foreword

To help repair the economic and social damage wrought by the coronavirus pandemic, a transformational recovery is needed. The social and economic situation in the world was already shaken by the fall of 2019, when one fourth of the world's developed nations were suffering from social unrest, and in more than half the threat of populism was as real as it has ever been. The coronavirus accelerated those trends and I expect the aftermath to be in much worse shape. The urgency to reform our societies is going to be at its highest. Artificial intelligence and data science will be key enablers of such transformation. They have the potential to revolutionize our way of life and create new opportunities.

The use of data science and artificial intelligence for economics and finance is providing benefits for scientists, professionals, and policy-makers by improving the available data analysis methodologies for economic forecasting and therefore making our societies better prepared for the challenges of tomorrow.

This book is a good example of how combining expertise from the European Commission, universities in the USA and Europe, financial and economic institutions, and multilateral organizations can bring forward a shared vision on the benefits of data science applied to economics and finance, from the research point of view to the evaluation of policies. It showcases how data science is reshaping the business sector. It includes examples of novel big data sources and some successful applications on the use of advanced machine learning, natural language processing, networks analysis, and time series analysis and forecasting, among others, in the economic and financial sectors. At the same time, the book is making an appeal for a further adoption of these novel applications in the field of economics and finance so that they can reach their full potential and support policy-makers and the related stakeholders in the transformational recovery of our societies.

We are not just repairing the damage to our economies and societies, the aim is to build better for the next generation. The problems are inherently interdisciplinary and global, hence they require international cooperation and the investment in collaborative work. We better learn what each other is doing, and we better learn the tools and language that each discipline brings to the table, and we better start now. This book is a good place to kick off.

Society of Sloan Fellows Professor of Management   Roberto Rigobon
Professor, Applied Economics
Massachusetts Institute of Technology
Cambridge, MA, USA

Preface

Economic and fiscal policies conceived by international organizations, governments, and central banks heavily depend on economic forecasts, in particular during times of economic and societal turmoil like the one we have recently experienced with the coronavirus spreading worldwide. The accuracy of economic forecasting and nowcasting models is however still problematic since modern economies are subject to numerous shocks that make the forecasting and nowcasting tasks extremely hard, both in the short and medium-long runs.

In this context, the use of recent *Data Science* technologies for improving forecasting and nowcasting for several types of economic and financial applications has high potential. The vast amount of data available in current times, referred to as the *Big Data* era, opens a huge amount of opportunities to economists and scientists, with a condition that data are opportunately handled, processed, linked, and analyzed. From forecasting economic indexes with little observations and only a few variables, we now have millions of observations and hundreds of variables. Questions that previously could only be answered with a delay of several months or even years can now be addressed nearly in real time. Big data, related analysis performed through (Deep) Machine Learning technologies, and the availability of more and more performing hardware (Cloud Computing infrastructures, GPUs, etc.) can integrate and augment the information carried out by publicly available aggregated variables produced by national and international statistical agencies. By lowering the level of granularity, Data Science technologies can uncover economic relationships that are often not evident when variables are in an aggregated form over many products, individuals, or time periods. Strictly linked to that, the evolution of ICT has contributed to the development of several decision-making instruments that help investors in taking decisions. This evolution also brought about the development of *FinTech*, a newly coined abbreviation for Financial Technology, whose aim is to leverage cutting-edge technologies to compete with traditional financial methods for the delivery of financial services.

This book is inspired by the desire for stimulating the adoption of Data Science solutions for Economics and Finance, giving a comprehensive picture on the use of Data Science as a new scientific and technological paradigm for boosting these sectors. As a result, the book explores a wide spectrum of essential aspects of Data Science, spanning from its main concepts, evolution, technical challenges, and infrastructures to its role and vast opportunities it offers in the economic and financial areas. In addition, the book shows some successful applications on advanced Data Science solutions used to extract new knowledge from data in order to improve economic forecasting and nowcasting models. The theme of the book is at the frontier of economic research in academia, statistical agencies, and central banks. Also, in the last couple of years, several master's programs in Data Science and Economics have appeared in top European and international institutions and universities. Therefore, considering the number of recent initiatives that are now pushing towards the use of data analysis within the economic field, we are pursuing with the present book at highlighting successful applications of Data Science and Artificial Intelligence into the economic and financial sectors. The book follows up a recently published Springer volume titled: "*Data Science for Healthcare: Methodologies and Applications*," which was co-edited by Dr. Sergio Consoli, Prof. Diego Reforgiato Recupero, and Prof. Milan Petkovic, that tackles the healthcare domain under different data analysis angles.

How This Book Is Organized

The book covers the use of Data Science, including Advanced Machine Learning, Big Data Analytics, Semantic Web technologies, Natural Language Processing, Social Media Analysis, and Time Series Analysis, among others, for applications in Economics and Finance. Particular care on model interpretability is also highlighted. This book is ideal for some educational sessions to be used in international organizations, research institutions, and enterprises. The book starts with an introduction on the use of Data Science technologies in Economics and Finance and is followed by 13 chapters showing successful stories on the application of the specific Data Science technologies into these sectors, touching in particular topics related to: novel big data sources and technologies for economic analysis (e.g., Social Media and News); Big Data models leveraging on supervised/unsupervised (Deep) Machine Learning; Natural Language Processing to build economic and financial indicators (e.g., Sentiment Analysis, Information Retrieval, Knowledge Engineering); Forecasting and Nowcasting of economic variables (e.g., Time Series Analysis and Robo-Trading).

Target Audience

The book is relevant to all the stakeholders involved in digital and data-intensive research in Economics and Finance, helping them to understand the main opportunities and challenges, become familiar with the latest methodological findings in

(Deep) Machine Learning, and learn how to use and evaluate the performances of novel Data Science and Artificial Intelligence tools and frameworks. This book is primarily intended for data scientists, business analytics managers, policy-makers, analysts, educators, and practitioners involved in Data Science technologies for Economics and Finance. It can also be a useful resource to research students in disciplines and courses related to these topics. Interested readers will be able to learn modern and effective Data Science solutions to create tangible innovations for Economics and Finance. Prior knowledge on the basic concepts behind Data Science, Economics, and Finance is recommended to potential readers in order to have a smooth understanding of this book.

Ispra (VA), Italy   Sergio Consoli
Cagliari, Italy   Diego Reforgiato Recupero
Ispra (VA), Italy   Michaela Saisana

Acknowledgments

We are grateful to Ralf Gerstner and his entire team from Springer for having strongly supported us throughout the publication process.

Furthermore, special thanks to the Scientific Committee members for their efforts to carefully revise their assigned chapter (each chapter has been reviewed by three or four of them), thus leading us to largely improve the quality of the book. They are, in alphabetical order: Arianna Agosto, Daniela Alderuccio, Luca Alfieri, David Ardia, Argimiro Arratia, Andres Azqueta-Gavaldon, Luca Barbaglia, Keven Bluteau, Ludovico Boratto, Ilaria Bordino, Kris Boudt, Michael Bräuning, Francesca Cabiddu, Cem Cakmakli, Ludovic Calès, Francesca Campolongo, Annalina Caputo, Alberto Caruso, Michele Catalano, Thomas Cook, Jacopo De Stefani, Wouter Duivesteijn, Svitlana Galeshchuk, Massimo Guidolin, Sumru Guler-Altug, Francesco Gullo, Stephen Hansen, Dragi Kocev, Nicolas Kourtellis, Athanasios Lapatinas, Matteo Manca, Sebastiano Manzan, Elona Marku, Rossana Merola, Claudio Morana, Vincenzo Moscato, Kei Nakagawa, Andrea Pagano, Manuela Pedio, Filippo Pericoli, Luca Tiozzo Pezzoli, Antonio Picariello, Giovanni Ponti, Riccardo Puglisi, Mubashir Qasim, Ju Qiu, Luca Rossini, Armando Rungi, Antonio Jesus Sanchez-Fuentes, Olivier Scaillet, Wim Schoutens, Gustavo Schwenkler, Tatevik Sekhposyan, Simon Smith, Paul Soto, Giancarlo Sperlì, Ali Caner Türkmen, Eryk Walczak, Reinhard Weisser, Nicolas Woloszko, Yucheong Yeung, and Wang Yiru.

A particular mention to Antonio Picariello, estimated colleague and friend, who suddenly passed away at the time of this writing and cannot see this book published.

Ispra (VA), Italy   Sergio Consoli
Cagliari, Italy   Diego Reforgiato Recupero
Ispra (VA), Italy   Michaela Saisana

Contents

Data Science Technologies in Economics and Finance: A Gentle Walk-In

Luca Barbaglia, Sergio Consoli, Sebastiano Manzan, Diego Reforgiato Recupero, Michaela Saisana, and Luca Tiozzo Pezzoli

Abstract This chapter is an introduction to the use of data science technologies in the fields of economics and finance. The recent explosion in computation and information technology in the past decade has made available vast amounts of data in various domains, which has been referred to as *Big Data*. In economics and finance, in particular, tapping into these data brings research and business closer together, as data generated in ordinary economic activity can be used towards effective and personalized models. In this context, the recent use of data science technologies for economics and finance provides mutual benefits to both scientists and professionals, improving forecasting and nowcasting for several kinds of applications. This chapter introduces the subject through underlying technical challenges such as data handling and protection, modeling, integration, and interpretation. It also outlines some of the common issues in economic modeling with data science technologies and surveys the relevant big data management and analytics solutions, motivating the use of data science methods in economics and finance.

1 Introduction

The rapid advances in information and communications technology experienced in the last two decades have produced an explosive growth in the amount of information collected, leading to the new era of big data [31]. According to [26], approximately three billion bytes of data are produced every day from sensors, mobile devices, online transactions, and social networks, with 90% of the data in

Authors are listed in alphabetic order since their contributions have been equally distributed.

L. Barbaglia · S. Consoli (✉) · S. Manzan · M. Saisana · L. Tiozzo Pezzoli
European Commission, Joint Research Centre, Ispra (VA), Italy
e-mail: sergio.consoli@ec.europa.eu

D. Reforgiato Recupero
Department of Mathematics and Computer Science, University of Cagliari, Cagliari, Italy

the world having been created in the last 3 years alone. The challenges in storage, organization, and understanding of such a huge amount of information led to the development of new technologies across different fields of statistics, machine learning, and data mining, interacting also with areas of engineering and artificial intelligence (AI), among others. This enormous effort led to the birth of the new cross-disciplinary field called "Data Science," whose principles and techniques aim at the automatic extraction of potentially useful information and knowledge from the data. Although data science technologies have been successfully applied in many different domains (e.g., healthcare [15], predictive maintenance [16], and supply chain management [39], among others), their potentials have been little explored in economics and finance. In this context, devising efficient forecasting and nowcasting models is essential for designing suitable monetary and fiscal policies, and their accuracy is particularly relevant during times of economic turmoil. Monitoring the current and the future state of the economy is of fundamental importance for governments, international organizations, and central banks worldwide. Policy-makers require readily available macroeconomic information in order to design effective policies which can foster economic growth and preserve societal well-being. However, key economic indicators, on which they rely upon during their decision-making process, are produced at low frequency and released with considerable lags—for instance, around 45 days for the Gross Domestic Product (GDP) in Europe—and are often subject to revisions that could be substantial. Indeed, with such an incomplete set of information, economists can only approximately gauge the actual, the future, and even the very recent past economic conditions, making the nowcasting and forecasting of the economy extremely challenging tasks. In addition, in a global interconnected world, shocks and changes originating in one economy move quickly to other economies affecting productivity levels, job creation, and welfare in different geographic areas. In sum, policy-makers are confronted with a twofold problem: timeliness in the evaluation of the economy as well as prompt impact assessment of external shocks.

Traditional forecasting models adopt a mixed frequency approach which bridges information from high-frequency economic and financial indexes (e.g., industrial production or stock prices) as well as economic surveys with the targeted low-frequency variable, such as the GDP [28]. An alternative could be dynamic factor models which, instead, resume large information in few factors and account of missing data by the use of Kalman filtering techniques in the estimation. These approaches allow the use of impulse-responses to assess the reaction of the economy to external shocks, providing general guidelines to policy-makers for actual and forward-looking policies fully considering the information coming from abroad. However, there are two main drawbacks to these traditional methods. First, they cannot directly handle huge amount of unstructured data since they are tailored to structured sources. Second, even if these classical models are augmented with new predictors obtained from alternative big data sets, the relationship across variables is assumed to be linear, which is not the case for the majority of the real-world cases [21, 1].

Data science technologies allow economists to deal with all these issues. On the one hand, new big data sources can integrate and augment the information carried by publicly available aggregated variables produced by national and international statistical agencies. On the other hand, machine learning algorithms can extract new insights from those unstructured information and properly take into consideration nonlinear dynamics across economic and financial variables. As far as big data is concerned, the higher level of granularity embodied on new, available data sources constitutes a strong potential to uncover economic relationships that are often not evident when variables are aggregated over many products, individuals, or time periods. Some examples of novel big data sources that can potentially be useful for economic forecasting and nowcasting are: retail consumer scanner price data, credit/debit card transactions, smart energy meters, smart traffic sensors, satellite images, real-time news, and social media data. Scanner price data, card transactions, and smart meters provide information about consumers, which, in turn, offers the possibility of better understanding the actual behavior of macro aggregates such as GDP or the inflation subcomponents. Satellite images and traffic sensors can be used to monitor commercial vehicles, ships, and factory tracks, making them potential candidate data to nowcast industrial production. Real-time news and social media can be employed to proxy the mood of economic and financial agents and can be considered as a measure of perception of the actual state of the economy.

In addition to new data, alternative methods such as machine learning algorithms can help economists in modeling complex and interconnected dynamic systems. They are able to grasp hidden knowledge even when the number of features under analysis is larger than the available observations, which often occurs in economic environments. Differently from traditional time-series techniques, machine learning methods have no "a priori" assumptions about the stochastic process underlying the state of the economy. For instance, deep learning [29], a very popular data science methodology nowadays, is useful in modeling highly nonlinear data because the order of nonlinearity is derived or learned directly from the data and not assumed as is the case in many traditional econometric models. Data science models are able to uncover complex relationships, which might be useful to forecast and nowcast the economy during normal time but also to spot early signals of distress in markets before financial crises.

Even though such methodologies may provide accurate predictions, understanding the economic insights behind such promising outcomes is a hard task. These methods are black boxes in nature, developed with a single goal of maximizing predictive performance. The entire field of data science is calibrated against out-of-sample experiments that evaluate how well a model trained on one data set will predict new data. On the contrary, economists need to know how models may impact in the real world and they have often focused not only on predictions but also on model inference, i.e., on understanding the parameters of their models (e.g., testing on individual coefficients in a regression). Policy-makers have to support their decisions and provide a set of possible explanations of an action taken; hence, they are interested on the economic implication involved in model predictions. Impulse response functions are a well-known instruments to assess the impact of a shock in one variable on an outcome of interest, but machine learning algorithms do not support this functionality. This could prevent, e.g., the evaluation of stabilization policies for protecting internal demand when an external shock hits the economy. In order to fill this gap, the data science community has recently tried to increase the transparency of machine learning models in the literature about *interpretable AI* [22]. Machine learning applications in economics and finance can now benefit from new tools such as Partial Dependence plots or Shapley values, which allow policy-makers to assess the marginal effect of model variables on the predicted outcome. In summary, data science can enhance economic forecasting models by:

- Integrating and complementing official key statistic indicators by using new real-time unstructured big data sources
- Assessing the current and future economic and financial conditions by allowing complex nonlinear relationships among predictors
- Maximizing revenues of algorithmic trading, a completely data-driven task
- Furnishing adequate support to decisions by making the output of machine learning algorithms understandable

This chapter emphasizes that data science has the potential to unlock vast productivity bottlenecks and radically improve the quality and accessibility of economic forecasting models, and discuss the challenges and the steps that need to be taken into account to guarantee a large and in-depth adoption.

2 Technical Challenges

In recent years, technological advances have largely increased the number of devices generating information about human and economic activity (e.g., sensors, monitoring, IoT devices, social networks). These new data sources provide a rich, frequent, and diversified amount of information, from which the state of the economy could be estimated with accuracy and timeliness. Obtaining and analyzing such kinds of data is a challenging task due to their size and variety. However, if properly exploited, these new data sources could bring additional predictive power than standard regressors used in traditional economic and financial analysis.

As the data size and variety augmented, the need for more powerful machines and more efficient algorithms became clearer. The analysis of such kinds of data can be highly computationally intensive and has brought an increasing demand for efficient hardware and computing environments. For instance, Graphical Processing Units (GPUs) and cloud computing systems in recent years have become more affordable and are used by a larger audience. GPUs have a highly data parallel architecture that can be programmed using frameworks such as CUDA[1] and OpenCL.[2] They

---

[1]NVIDIA CUDA: https://developer.nvidia.com/cuda-zone.
[2]OpenCL: https://www.khronos.org/opencl/.

consist of a number of cores, each with a number of functional units. One or more of these functional units (known as *thread processors*) process each thread of execution. All thread processors in a core of a GPU perform the same instructions, as they share the same control unit. Cloud computing represents the distribution of services such as servers, databases, and software through the Internet. Basically, a provider supplies users with on-demand access to services of storage, processing, and data transmission. Examples of cloud computing solutions are the Google Cloud Platform,[3] Microsoft Azure,[4] and Amazon Web Services (AWS).[5]

Sufficient computing power is a necessary condition to analyze new big data sources; however, it is not sufficient unless data are properly stored, transformed, and combined. Nowadays, economic and financial data sets are still stored in individual silos, and researchers and practitioners are often confronted with the difficulty of easily combining them across multiple providers, other economic institutions, and even consumer-generated data. These disparate economic data sets might differ in terms of data granularity, quality, and type, for instance, ranging from free text, images, and (streaming) sensor data to structured data sets; their integration poses major legal, business, and technical challenges. Big data and data science technologies aim at efficiently addressing such kinds of challenges.

The term "big data" has its origin in computer engineering. Although several definitions for big data exist in the literature [31, 43], we can intuitively refer to data that are so large that they cannot be loaded into memory or even stored on a single machine. In addition to their large *volume*, there are other dimensions that characterize big data, i.e., *variety* (handling with a multiplicity of types, sources and format), *veracity* (related to the quality and validity of these data), and *velocity* (availability of data in real time). Other than the four big data features described above, we should also consider relevant issues as data trustworthiness, data protection, and data privacy. In this chapter we will explore the major challenges posed by the exploitation of new and alternative data sources, and the associated responses elaborated by the data science community.

2.1 Stewardship and Protection

Accessibility is a major condition for a fruitful exploitation of new data sources for economic and financial analysis. However, in practice, it is often restricted in order to protect sensitive information. Finding a sensible balance between accessibility and protection is often referred to as *data stewardship*, a concept that ranges from properly collecting, annotating, and archiving information to taking a "long-term care" of data, considered as valuable digital assets that might be reused in

---

[3] Google Cloud: https://cloud.google.com/.

[4] Microsoft Azure: https://azure.microsoft.com/en-us/.

[5] Amazon Web Services (AWS): https://aws.amazon.com/.

future applications and combined with new data [42]. Organizations like the World Wide Web Consortium (W3C)[6] have worked on the development of interoperability guidelines among the realm of open data sets available in different domains to ensure that the data are FAIR (*Findable*, *Accessible*, *Interoperable*, and *Reusable*).

Data protection is a key aspect to be considered when dealing with economic and financial data. Trustworthiness is a main concern of individuals and organizations when faced with the usage of their financial-related data: it is crucial that such data are stored in secure and privacy-respecting databases. Currently, various privacy-preserving approaches exist for analyzing a specific data source or for connecting different databases across domains or repositories. Still several challenges and risks have to be accommodated in order to combine private databases by new anonymization and pseudo-anonymization approaches that guarantee privacy. Data analysis techniques need to be adapted to work with encrypted or distributed data. The close collaboration between domain experts and data analysts along all steps of the data science chain is of extreme importance.

Individual-level data about credit performance is a clear example of sensitive data that might be very useful in economic and financial analysis, but whose access is often restricted for data protection reasons. The proper exploitation of such data could bring large improvements in numerous aspects: financial institutions could benefit from better credit risk models that identify more accurately risky borrowers and reduce the potential losses associated with a default; consumers could have easier access to credit thanks to the efficient allocation of resources to reliable borrowers, and governments and central banks could monitor in real time the status of their economy by checking the health of their credit markets. Numerous are the data sets with anonymized individual-level information available online. For instance, mortgage data for the USA are provided by the Federal National Mortgage Association (Fannie Mae)[7] and by the Federal Home Loan Mortgage Corporation (Freddie Mac):[8] they report loan-level information for millions of individual mortgages, with numerous associated features, e.g., repayment status, borrower's main characteristics, and granting location of the loan (we refer to [2, 35] for two examples of mortgage-level analysis in the US). A similar level of detail is found in the European Datawarehouse,[9] which provides loan-level data of European assets about residential mortgages, credit cards, car leasing, and consumer finance (see [20, 40] for two examples of economic analysis on such data).

---

[6]World Wide Web Consortium (W3C): https://www.w3.org/.

[7]Federal National Mortgage Association (Fannie Mae): https://www.fanniemae.com.

[8]Federal Home Loan Mortgage Corporation (Freddie Mac): http://www.freddiemac.com.

[9]European Datawarehouse: https://www.eurodw.eu/.

2.2 Data Quantity and Ground Truth

Economic and financial data are growing at staggering rates that have not been seen in the past [33]. Organizations today are gathering large volume of data from both proprietary and public sources, such as social media and open data, and eventually use them for economic and financial analysis. The increasing data volume and velocity pose new technical challenges that researchers and analysts can face by leveraging on data science. A general data science scenario consists of a series of observations, often called instances, each of which is characterized by the realization of a group of variables, often referred to as attributes, which could take the form of, e.g., a string of text, an alphanumeric code, a date, a time, or a number. Data volume is exploding in various directions: there are more and more available data sets, each with an increasing number of instances; technological advances allow to collect information on a vast number of features, also in the form of images and videos.

Data scientists commonly distinguish between two types of data, unlabeled and labeled [15]. Given an attribute of interest (label), unlabeled data are not associated with an observed value of the label and they are used in unsupervised learning problems, where the goal is to extract the most information available from the data itself, like with clustering and association rules problems [15]. For the second type of data, there is instead a label associated with each data instance that can be used in a supervised learning task: one can use the information available in the data set to predict the value of the attribute of interest that have not been observed yet. If the attribute of interest is categorical, the task is called classification, while if it is numerical, the task is called regression [15]. Breakthrough technologies, such as deep learning, require large quantities of labelled data for training purposes, that is data need to come with annotations, often referred to as *ground truth* [15].

In finance, e.g., numerous works of unsupervised and supervised learning have been explored in the fraud detection literature [3, 11], whose goal is to identify whether a potential fraud has occurred in a certain financial transaction. Within this field, the well-known Credit Card Fraud Detection data set[10] is often used to compare the performance of different algorithms in identifying fraudulent behaviors (e.g., [17, 32]). It contains 284,807 transactions of European cardholders executed in 2 days of 2013, where only 492 of them have been marked as fraudulent, i.e., 0.17% of the total. This small number of positive cases need to be consistently divided into training and test sets via stratified sampling, such that both sets contain some fraudulent transactions to allow for a fair comparison of the out-of-sample forecasting performance. Due to the growing data volume, it is more and more common to work with such highly unbalanced data set, where the number of positive cases is just a small fraction of the full data set: in these cases, standard econometric analysis might bring poor results and it could be useful investigating rebalancing

---

[10]https://www.kaggle.com/mlg-ulb/creditcardfraud.

techniques like undersampling, oversampling or a combination of the both, which could be used to possibly improve the classification accuracy [15, 36].

2.3   Data Quality and Provenance

Data quality generally refers to whether the received data are fit for their intended use and analysis. The basis for assessing the quality of the provided data is to have an updated metadata section, where there is a proper description of each feature in the analysis. It must be stressed that a large part of the data scientist's job resides in checking whether the data records actually correspond to the metadata descriptions. Human errors and inconsistent or biased data could create discrepancies with respect to what the data receiver was originally expecting. Take, for instance, the European Datawarehouse presented in Sect. 2.1: loan-level data are reported by each financial institution, gathered in a centralized platform and published under a common data structure. Financial institutions are properly instructed on how to provide data; however, various error types may occur. For example, rates could be reported as fractions instead of percentages, and loans may be indicated as defaulted according to a definition that varies over time and/or country-specific legislation.

Going further than standard data quality checks, *data provenance* aims at collecting information on the whole data generating process, such as the software used, the experimental steps undertaken in gathering the data or any detail of the previous operations done on the raw input. Tracking such information allows the data receiver to understand the source of the data, i.e., how it was collected, under which conditions, but also how it was processed and transformed before being stored. Moreover, should the data provider adopt a change in any of the aspect considered by data provenance (e.g., a software update), the data receiver might be able to detect early a structural change in the quality of the data, thus preventing their potential misuse and analysis. This is important not only for the reproducibility of the analysis but also for understanding the reliability of the data that can affect outcomes in economic research. As the complexity of operations grows, with new methods being developed quite rapidly, it becomes key to record and understand the origin of data, which in turn can significantly influence the conclusion of the analysis. For a recent review on the future of data provenance, we refer, among others, to [10].

2.4   Data Integration and Sharing

Data science works with structured and unstructured data that are being generated by a variety of sources and in different formats, and aims at integrating them into big data repositories or Data Warehouses [43]. There exists a large number of standardized ETL (Extraction, Transformation, and Loading) operations that help to identify and reorganize structural, syntactic, and semantic heterogeneity across different data sources [31]. Structural heterogeneity refers to different data and schema models, which require integration on the schema level. Syntactic heterogeneity appears in the form of different data access interfaces, which need to be reconciled. Semantic heterogeneity consists of differences in the interpretation of data values and can be overcome by employing semantic technologies, like graph-based knowledge bases and domain ontologies [8], which map concepts and definitions to the data source, thus facilitating collaboration, sharing, modeling, and reuse across applications [7].

A process of integration ultimately results in consolidation of duplicated sources and data sets. Data integration and linking can be further enhanced by properly exploiting information extraction algorithm, machine learning methods, and Semantic Web technologies that enable context-based information interpretation [26]. For example, authors in [12] proposed a semantic approach to generate industry-specific lexicons from news documents collected within the Dow Jones DNA dataset,[11] with the goal of dynamically capturing, on a daily basis, the correlation between words used in these documents and stock price fluctuations of industries of the Standard & Poor's 500 index. Another example is represented by the work in [37], which has used information extracted from the *Wall Street Journal* to show that high levels of pessimism in the news are relevant predictors of convergence of stock prices towards their fundamental values.

In macroeconomics, [24] has looked at the informational content of the Federal Reserve statements and the guidance that these statements provide about the future evolution of monetary policy.

Given the importance of data-sharing among researchers and practitioners, many institutions have already started working toward this goal. The European Commission (EC) has launched numerous initiatives, such as the EU Open Data[12] and the European Data[13] portals directly aimed at facilitating data sharing and interoperability.

2.5 Data Management and Infrastructures

To manage and analyze the large data volume appearing nowadays, it is necessary to employ new infrastructures able to efficiently address the four big data dimensions of volume, variety, veracity, and velocity. Indeed, massive data sets require to be stored in specialized distributed computing environments that are essential for building the data pipes that slice and aggregate this large amount of information. Large unstructured data are stored in distributed file systems (DFS), which join

---

[11] Dow Jones DNA: https://www.dowjones.com/dna/.

[12] EU Open Data Portal: https://data.europa.eu/euodp/en/home/.

[13] European Data Portal: https://www.europeandataportal.eu/en/homepage.

together many computational machines (nodes) over a network [36]. Data are broken into blocks and stored on different nodes, such that the DFS allows to work with partitioned data, that otherwise would become too big to be stored and analyzed on a single computer. Frameworks that heavily use DFS include Apache Hadoop[14] and Amazon S3,[15] the backbone of storage on AWS. There are a variety of platforms for wrangling and analyzing distributed data, the most prominent of which perhaps is Apache Spark.[16] When working with big data, one should use specialized algorithms that avoid having all of the data in a computer's working memory at a single time [36]. For instance, the MapReduce[17] framework consists of a series of algorithms that can prepare and group data into relatively small chunks (Map) before performing an analysis on each chunk (Reduce). Other popular DFS platforms today are MongoDB,[18] Apache Cassandra,[19] and ElasticSearch,[20] just to name a few. As an example in economics, the authors of [38] presented a NO-SQL infrastructure based on ElasticSearch to store and interact with the huge amount of news data contained in the Global Database of Events, Language and Tone (GDELT),[21] consisting of more than 8 TB of textual information from around 500 million news articles worldwide since 2015. The authors showed an application exploiting GDELT to construct news-based financial sentiment measures capturing investor's opinions for three European countries: Italy, Spain, and France [38].

Even though many of these big data platforms offer proper solutions to businesses and institutions to deal with the increasing amount of data and information available, numerous relevant applications have not been designed to be dynamically scalable, to enable distributed computation, to work with nontraditional databases, or to interoperate with infrastructures. Existing cloud infrastructures will have to massively invest in solutions designed to offer dynamic scalability, infrastructures interoperability, and massive parallel computing in order to effectively enable reliable execution of, e.g., machine learning algorithms and AI techniques. Among other actions, the importance of cloud computing was recently highlighted by the EC through its European Cloud Initiative,[22] which led to the birth of the European Open Science Cloud,[23] a trusted open environment for the scientific community for

---

[14] Apache Hadoop: https://hadoop.apache.org/.

[15] Amazon AWS S3: https://aws.amazon.com/s3/.

[16] Apache Spark: https://spark.apache.org/.

[17] https://hadoop.apache.org/docs/r1.2.1/mapred_tutorial.html.

[18] MongoDB: https://www.mongodb.com/.

[19] Apache Cassandra: https://cassandra.apache.org/.

[20] ElasticSearch: https://www.elastic.co/.

[21] GDELT website: https://blog.gdeltproject.org/.

[22] European Cloud Initiative: https://ec.europa.eu/digital-single-market/en/%20european-cloud-initiative.

[23] European Open Science Cloud: https://ec.europa.eu/research/openscience/index.cfm?pg=open-science-cloud.

storing, sharing, and reusing scientific data and results, and of the European Data Infrastructure,[24], which targets the construction of an EU super-computing capacity.

3 Data Analytics Methods

Traditional nowcasting and forecasting economic models are not dynamically scalable to manage and maintain big data structures, including raw logs of user actions, natural text from communications, images, videos, and sensors data. This high volume of data is arriving in inherently complex high-dimensional formats, and their use for economic analysis requires new tool sets [36]. Traditional techniques, in fact, do not scale well when the data dimensions are big or growing fast. Relatively simple tasks such as data visualization, model fitting, and performance checks become hard. Classical hypothesis testing aimed to check the importance of a variable in a model (T-test), or to select one model across different alternatives (F-test), have to be used with caution in a big data environment [26, 30]. In this complicated setting, it is not possible to rely on precise guarantees upon standard low-dimensional strategies, visualization approaches, and model specification diagnostics [36, 26]. In these contexts, social scientists can benefit from using data science techniques and in recent years the efforts to make those applications accepted within the economic modeling space have increased exponentially. A focal point consists in opening up the black-box machine learning solutions and building interpretable models [22]. Indeed, data science algorithms are useless for policy-making when, although easily scalable and highly performing, they turn out to be hardly comprehensible. Good data science applied to economics and finance requires a balance across these dimensions and typically involves a mix of domain knowledge and analysis tools in order to reach the level of model performance, interpretability, and automation required by the stakeholders. Therefore, it is good practice for economists to figure out what can be modeled as a prediction task and reserving statistical and economic efforts for the tough structural questions. In the following, we provide an high-level overview of maybe the two most popular families of data science technologies used today in economics and finance.

3.1 Deep Machine Learning

Despite long-established machine learning technologies, like Support Vector Machines, Decision Trees, Random Forests, and Gradient Boosting have shown high potential to solve a number of data mining (e.g., classification, regression) problems around organizations, governments, and individuals. Nowadays the

---

[24]European Data Infrastructure: https://www.eudat.eu/.

technology that has obtained the largest success among both researchers and practitioners is *deep learning* [29]. Deep learning is a general-purpose machine learning technology, which typically refers to a set of machine learning algorithms based on learning data representations (capturing highly nonlinear relationships of low level unstructured input data to form high-level concepts). Deep learning approaches made a real breakthrough in the performance of several tasks in the various domains in which traditional machine learning methods were struggling, such as speech recognition, machine translation, and computer vision (object recognition). The advantage of deep learning algorithms is their capability to analyze very complex data, such as images, videos, text, and other unstructured data.

Deep hierarchical models are Artificial Neural Networks (ANNs) with deep structures and related approaches, such as Deep Restricted Boltzmann Machines, Deep Belief Networks, and Deep Convolutional Neural Networks. ANN are computational tools that may be viewed as being inspired by how the brain functions and applying this framework to construct mathematical models [30]. Neural networks estimate functions of arbitrary complexity using given data. Supervised Neural Networks are used to represent a mapping from an input vector onto an output vector. Unsupervised Neural Networks are used instead to classify the data without prior knowledge of the classes involved. In essence, Neural Networks can be viewed as generalized regression models that have the ability to model data of arbitrary complexities [30]. The most common ANN architectures are the multilayer perceptron (MLP) and the radial basis function (RBF). In practice, sequences of ANN layers in cascade form a deep learning framework. The current success of deep learning methods is enabled by advances in algorithms and high-performance computing technology, which allow analyzing the large data sets that have now become available. One example is represented by robot-advisor tools that currently make use of deep learning technologies to improve their accuracy [19]. They perform stock market forecasting by either solving a regression problem or by mapping it into a classification problem and forecast whether the market will go up or down.

There is also a vast literature on the use of deep learning in the context of time series forecasting [29, 6, 27, 5]. Although it is fairly straightforward to use classic MLP ANN on large data sets, its use on medium-sized time series is more difficult due to the high risk of overfitting. Classical MLPs can be adapted to address the sequential nature of the data by treating time as an explicit part of the input. However, such an approach has some inherent difficulties, namely, the inability to process sequences of varying lengths and to detect time-invariant patterns in the data. A more direct approach is to use recurrent connections that connect the neural networks' hidden units back to themselves with a time delay. This is the principle at the base of Recurrent Neural Networks (RNNs) [29] and, in particular, of Long Short-Term Memory Networks (LSTMs) [25], which are ANNs specifically designed to handle sequential data that arise in applications such as time series, natural language processing, and speech recognition [34].

In finance, deep learning has been already exploited, e.g., for stock market analysis and prediction (see e.g. [13] for a review). Another proven ANNs approach for financial time-series forecasting is the Dilated Convolutional Neural Network presented in [9], wherein the underlying architecture comes from DeepMind's WaveNet project [41]. The work in [5] exploits an ensemble of Convolutional Neural Networks, trained over Gramian Angular Fields images generated from time series related to the Standard & Poor's 500 Future index, where the aim is the prediction of the future trend of the US market.

Next to deep learning, *reinforcement learning* has gained popularity in recent years: it is based on a paradigm of learning by trial and error, solely from rewards or punishments. It was successfully applied in breakthrough innovations, such as the AlphaGo system[25] of Deep Mind that won the Go game against the best human player. It can also be applied in the economic domain, e.g., to dynamically optimize portfolios [23] or for financial assert trading [18]. All these advanced machine learning systems can be used to learn and relate information from multiple economic sources and identify hidden correlations not visible when considering only one source of data. For instance, combining features from images (e.g., satellites) and text (e.g., social media) can yield to improve economic forecasting.

Developing a complete deep learning or reinforcement learning pipeline, including tasks of great importance like processing of data, interpretation, framework design, and parameters tuning, is far more of an art (or a skill learnt from experience) than an exact science. However the job is facilitated by the programming languages used to develop such pipelines, e.g., R, Scala, and Python, that provide great work spaces for many data science applications, especially those involving unstructured data. These programming languages are progressing to higher levels, meaning that it is now possible with short and intuitive instructions to automatically solve some fastidious and complicated programming issues, e.g., memory allocation, data partitioning, and parameters optimization. For example, the currently popular Gluon library[26] wraps (i.e., provides higher-level functionality around) MXNet,[27] a deep learning framework that makes it easier and faster to build deep neural networks. MXNet itself wraps C++, the fast and memory-efficient code that is actually compiled for execution. Similarly, Keras,[28] another widely used library, is an extension of Python that wraps together a number of other deep learning frameworks, such as Google's TensorFlow.[29] These and future tools are creating a world of user friendly interfaces for faster and simplified (deep) machine learning [36].

---

[25]Deep Mind AlphaGo system: https://deepmind.com/research/case-studies/alphago-the-story-so-far.

[26]Gluon: https://gluon.mxnet.io/.

[27]Apache MXNet: https://mxnet.apache.org/.

[28]Keras: https://keras.io/.

[29]TensorFlow: https://www.tensorflow.org/.

3.2   Semantic Web Technologies

From the perspectives of data content processing and mining, textual data belongs to the so-called unstructured data. Learning from this type of complex data can yield more concise, semantically rich, descriptive patterns in the data, which better reflect their intrinsic properties. Technologies such as those from the Semantic Web, including Natural Language Processing (NLP) and Information Retrieval, have been created for facilitating easy access to a wealth of textual information. The Semantic Web, often referred to as "Web 3.0," is a system that enables machines to "understand" and respond to complex human requests based on their meaning. Such an "understanding" requires that the relevant information sources be semantically structured [7]. Linked Open Data (LOD) has gained significant momentum over the past years as a best practice of promoting the sharing and publication of structured data on the Semantic Web [8], by providing a formal description of concepts, terms, and relationships within a given knowledge domain, and by using Uniform Resource Identifiers (URIs), Resource Description Framework (RDF), and Web Ontology Language (OWL), whose standards are under the care of the W3C.

LOD offers the possibility of using data across different domains for purposes like statistics, analysis, maps, and publications. By linking this knowledge, interrelations and associations can be inferred and new conclusions drawn. RDF/OWL allows for the creation of triples about anything on the Semantic Web: the decentralized data space of all the triples is growing at an amazing rate since more and more data sources are being published as semantic data. But the size of the Semantic Web is not the only parameter of its increasing complexity. Its distributed and dynamic character, along with the coherence issues across data sources, and the interplay between the data sources by means of reasoning, contribute to turning the Semantic Web into a complex, big system [7, 8].

One of the most popular technology used to tackle different tasks within the Semantic Web is represented by NLP, often referred to with synonyms like text mining, text analytics, or knowledge discovery from text. NLP is a broad term referring to technologies and methods in computational linguistics for the automatic detection and analysis of relevant information in unstructured textual content (free text). There has been significant breakthrough in NLP with the introduction of advanced machine learning technologies (in particular deep learning) and statistical methods for major text analytics tasks like: linguistic analysis, named entity recognition, co-reference resolution, relations extraction, and opinion and sentiment analysis [15].

In economics, NLP tools have been adapted and further developed for extracting relevant concepts, sentiments, and emotions from social media and news (see, e.g., [37, 24, 14, 4], among others). These technologies applied in the economic context facilitate data integration from multiple heterogeneous sources, enable the development of information filtering systems, and support knowledge discovery tasks.

4   Conclusions

In this chapter we have introduced the topic of data science applied to economic and financial modeling. Challenges like economic data handling, quality, quantity, protection, and integration have been presented as well as the major big data management infrastructures and data analytics approaches for prediction, interpretation, mining, and knowledge discovery tasks. We summarized some common big data problems in economic modeling and relevant data science methods.

There is clear need and high potential to develop data science approaches that allow for humans and machines to cooperate more closely to get improved models in economics and finance. These technologies can handle, analyze, and exploit the set of very diverse, interlinked, and complex data that already exist in the economic universe to improve models and forecasting quality, in terms of guarantee on the trustworthiness of information, a focus on generating actionable advice, and improving the interactivity of data processing and analytics.

References

1. Aruoba, S. B., Diebold, F. X., & Scotti, C. (2009). Real-time measurement of business conditions. *Journal of Business & Economic Statistics, 27*(4), 417–427.
2. Babii, A., Chen, X., & Ghysels, E. (2019). Commercial and residential mortgage defaults: Spatial dependence with frailty. *Journal of Econometrics, 212*, 47–77.
3. Baesens, B., Van Vlasselaer, V., & Verbeke, W. (2015). *Fraud analytics using descriptive, predictive, and social network techniques: a guide to data science for fraud detection.* Chichester: John Wiley & Sons.
4. Barbaglia, L., Consoli, S., & Manzan, S. (2020). Monitoring the business cycle with fine-grained, aspect-based sentiment extraction from news. In V. Bitetta et al. (Eds.), *Mining Data for Financial Applications (MIDAS 2019), Lecture Notes in Computer Science* (Vol. 11985, pp. 101–106). Cham: Springer. https://doi.org/10.1007/978-3-030-37720-5_8
5. Barra, S., Carta, S., Corriga, A., Podda, A. S., & Reforgiato Recupero, D. (2020). Deep learning and time series-to-image encoding for financial forecasting. *IEEE Journal of Automatica Sinica, 7*, 683–692.
6. Benidis, K., Rangapuram, S. S., Flunkert, V., Wang, B., Maddix, D. C., Türkmen, C., Gasthaus, J., Bohlke-Schneider, M., Salinas, D., Stella, L., Callot, L., & Januschowski, T. (2020). Neural forecasting: Introduction and literature overview. CoRR, abs/2004.10240.
7. Berners-Lee, T., Chen, Y., Chilton, L., Connolly, D., Dhanaraj, R., Hollenbach, J., Lerer, A., & Sheets, D. (2006). Tabulator: Exploring and analyzing linked data on the semantic web. In *Proc. 3rd International Semantic Web User Interaction Workshop (SWUI 2006)*.
8. Bizer, C., Heath, T., & Berners-Lee, T. (2009). Linked Data - The story so far. *International Journal on Semantic Web and Information Systems, 5*, 1–22.
9. Borovykh, A., Bohte, S., & Oosterlee, C. W. (2017). Conditional time series forecasting with convolutional neural networks. *Lecture Notes in Computer Science, 10614*, 729–730.
10. Buneman, P., & Tan, W.-C. (2019). Data provenance: What next? *ACM SIGMOD Record, 47*(3), 5–16.
11. Carta, S., Fenu, G., Reforgiato Recupero, D., & Saia, R. (2019). Fraud detection for e-commerce transactions by employing a prudential multiple consensus model. *Journal of Information Security and Applications, 46*, 13–22.

12. Carta, S., Consoli, S., Piras, L., Podda, A. S., & Reforgiato Recupero, D. (2020). Dynamic industry specific lexicon generation for stock market forecast. In G. Nicosia et al. (Eds.), *Machine Learning, Optimization, and Data Science (LOD 2020), Lecture Notes in Computer Science* (Vol. 12565, pp. 162–176). Cham: Springer. https://doi.org/10.1007/978-3-030-64583-0_16

13. Chong, E., Han, C., & Park, F. C. (2017). Deep learning networks for stock market analysis and prediction: Methodology, data representations, and case studies. *Expert Systems with Applications, 83*, 187–205.

14. Consoli, S., Tiozzo Pezzoli, L., & Tosetti, E. (2020). Using the GDELT dataset to analyse the Italian bond market. In G. Nicosia et al. (Eds.), *Machine learning, optimization, and data science (LOD 2020), Lecture Notes in Computer Science* (Vol. 12565, pp. 190–202). Cham: Springer. https://doi.org/10.1007/978-3-030-64583-0_18.

15. Consoli, S., Reforgiato Recupero, D., & Petkovic, M. (2019). *Data science for healthcare - Methodologies and applications*. Berlin: Springer Nature.

16. Daily, J., & Peterson, J. (2017). Predictive maintenance: How big data analysis can improve maintenance. In *Supply chain integration challenges in commercial aerospace* (pp. 267–278). Cham: Springer.

17. Dal Pozzolo, A., Caelen, O., Johnson, R. A., & Bontempi, G. (2015). Calibrating probability with undersampling for unbalanced classification. In *2015 IEEE Symposium Series on Computational Intelligence* (pp. 159–166). Piscataway: IEEE.

18. Deng, Y., Bao, F., Kong, Y., Ren, Z., & Dai, Q. (2017). Deep direct reinforcement learning for financial signal representation and trading. *IEEE Transactions on Neural Networks and Learning Systems, 28*(3), 653–664.

19. Ding, X., Zhang, Y., Liu, T., & Duan, J. (2015). Deep learning for event-driven stock prediction. In *IJCAI International Joint Conference on Artificial Intelligence* (Vol. 2015, pp. 2327–2333).

20. Ertan, A., Loumioti, M., & Wittenberg-Moerman, R. (2017). Enhancing loan quality through transparency: Evidence from the European central bank loan level reporting initiative. *Journal of Accounting Research, 55*(4), 877–918.

21. Giannone, D., Reichlin, L., & Small, D. (2008). Nowcasting: The real-time informational content of macroeconomic data. *Journal of Monetary Economics, 55*(4), 665–676.

22. Gilpin, L. H., Bau, D., Yuan, B. Z., Bajwa, A., Specter, M., & Kagal, L. (2019). Explaining explanations: An overview of interpretability of machine learning. In *IEEE International Conference on Data Science and Advanced Analytics (DSAA 2018)* (pp. 80–89).

23. Goodfellow, I., Bengio, Y., & Courville, A. (2016). *Deep Learning*. Cambridge: MIT Press.

24. Hansen, S., & McMahon, M. (2016). Shocking language: Understanding the macroeconomic effects of central bank communication. *Journal of International Economics, 99*, S114–S133.

25. Hochreiter, S., & Schmidhuber, J. (1997). Long short-term memory. *Neural Computation, 9*, 1735–1780.

26. Jabbour, C. J .C., Jabbour, A. B. L. D. S., Sarkis, J., & Filho, M. G. (2019). Unlocking the circular economy through new business models based on large-scale data: An integrative framework and research agenda. *Technological Forecasting and Social Change, 144*, 546–552.

27. Januschowski, T., Gasthaus, J., Wang, Y., Salinas, D., Flunkert, V., Bohlke-Schneider, M., & Callot, L. (2020). Criteria for classifying forecasting methods. *International Journal of Forecasting, 36*(1), 167–177.

28. Kuzin, V., Marcellino, M., & Schumacher, C. (2011). MIDAS vs. mixed-frequency VAR: Nowcasting GDP in the euro area. *International Journal of Forecasting, 27*(2), 529–542.

29. LeCun, Y., Bengio, Y., & Hinton, G. (2015). Deep Learning. *Nature, 521*(7553), 436–444.

30. Marwala, T. (2013). *Economic modeling using Artificial Intelligence methods*. Heidelberg: Springer.

31. Marx, V. (2013). The big challenges of big data. *Nature, 498*, 255–260.

32. Oblé, F., & Bontempi, G. (2019). Deep-learning domain adaptation techniques for credit cards fraud detection. In *Recent Advances in Big Data and Deep Learning: Proceedings of the INNS Big Data and Deep Learning Conference* (Vol. 1, pp. 78–88). Cham: Springer.

33. OECD. (2015). Data-driven innovation: Big data for growth and well-being. *OECD Publishing, Paris*. https://doi.org/10.1787/9789264229358-en
34. Salinas, D., Flunkert, V., Gasthaus, J., & Januschowski, T. (2020). Deepar: Probabilistic forecasting with autoregressive recurrent networks. *International Journal of Forecasting, 36*(3), 1181–1191.
35. Sirignano, J., Sadhwani, A., & Giesecke, K. (2018). Deep learning for mortgage risk. Technical report, Working paper available at SSRN: https://doi.org/10.2139/ssrn.2799443
36. Taddy, M. (2019). *Business data science: Combining machine learning and economics to optimize, automate, and accelerate business decisions*. New York: McGraw-Hill, US.
37. Tetlock, P. C. (2007). Giving content to investor sentiment: The role of media in the stock market. *The Journal of Finance, 62*(3), 1139–1168.
38. Tiozzo Pezzoli, L., Consoli, S., & Tosetti, E. (2020). Big data financial sentiment analysis in the European bond markets. In V. Bitetta et al. (Eds.), *Mining Data for Financial Applications (MIDAS 2019), Lecture Notes in Computer Science* (Vol. 11985, pp. 122–126). Cham: Springer. https://doi.org/10.1007/978-3-030-37720-5_10
39. Tiwari, S., Wee, H. M., & Daryanto, Y. (2018). Big data analytics in supply chain management between 2010 and 2016: Insights to industries. *Computers & Industrial Engineering, 115*, 319–330.
40. Van Bekkum, S., Gabarro, M., & Irani, R. M. (2017). Does a larger menu increase appetite? Collateral eligibility and credit supply. *The Review of Financial Studies, 31*(3), 943–979.
41. van den Oord, A., Dieleman, S., Zen, H., Simonyan, K., Vinyals, O., Graves, A., et al. (2016). WaveNet: A generative model for raw audio. *CoRR, abs/1609.03499*.
42. Wilkinson, M., Dumontier, M., Aalbersberg, I., Appleton, G., Axton, M., Baak, A., et al. (2016). The FAIR guiding principles for scientific data management and stewardship. *Scientific Data, 3*, 1.
43. Wu, X., Zhu, X., Wu, G., & Ding, W. (2014). Data mining with Big Data. *IEEE Transactions on Knowledge and Data Engineering, 26*(1), 97–107.

Supervised Learning for the Prediction of Firm Dynamics

Falco J. Bargagli-Stoffi, Jan Niederreiter, and Massimo Riccaboni

Abstract Thanks to the increasing availability of granular, yet high-dimensional, firm level data, machine learning (ML) algorithms have been successfully applied to address multiple research questions related to firm dynamics. Especially supervised learning (SL), the branch of ML dealing with the prediction of labelled outcomes, has been used to better predict firms' performance. In this chapter, we will illustrate a series of SL approaches to be used for prediction tasks, relevant at different stages of the company life cycle. The stages we will focus on are (1) startup and innovation, (2) growth and performance of companies, and (3) firms' exit from the market. First, we review SL implementations to predict successful startups and R&D projects. Next, we describe how SL tools can be used to analyze company growth and performance. Finally, we review SL applications to better forecast financial distress and company failure. In the concluding section, we extend the discussion of SL methods in the light of targeted policies, result interpretability, and causality.

Keywords Machine learning · Firm dynamics · Innovation · Firm performance

1 Introduction

In recent years, the ability of machines to solve increasingly more complex tasks has grown exponentially [86]. The availability of learning algorithms that deal with tasks such as facial and voice recognition, automatic driving, and fraud detection makes the various applications of machine learning a hot topic not just in the specialized literature but also in media outlets. Since many decades, computer scientists have been using algorithms that automatically update their course of

F. J. Bargagli-Stoffi
Harvard University, Boston, MA, USA
e-mail: fbargaglistoffi@hsph.harvard.edu

J. Niederreiter · M. Riccaboni (✉)
IMT School for Advanced Studies Lucca, Lucca, Italy
e-mail: jan.niederreiter@alumni.imtlucca.it; massimo.riccaboni@imtlucca.it

© The Author(s) 2021   19
S. Consoli et al. (eds.), *Data Science for Economics and Finance*,
https://doi.org/10.1007/978-3-030-66891-4_2

action to better their performance. Already in the 1950s, Arthur Samuel developed a program to play checkers that improved its performance by learning from its previous moves. The term "machine learning" (ML) is often said to have originated in that context. Since then, major technological advances in data storage, data transfer, and data processing have paved the way for learning algorithms to start playing a crucial role in our everyday life.

Nowadays, the usage of ML has become a valuable tool for enterprises' management to predict key performance indicators and thus to support corporate decision-making across the value chain, including the appointment of directors [33], the prediction of product sales [7], and employees' turnover [1, 85]. Using data which emerges as a by-product of economic activity has a positive impact on firms' growth [37], and strong data analytic capabilities leverage corporate performance [75]. Simultaneously, publicly accessible data sources that cover information across firms, industries, and countries open the door for analysts and policy-makers to study firm dynamics on a broader scale such as the fate of start-ups [43], product success [79], firm growth [100], and bankruptcy [12].

Most ML methods can be divided into two main branches: (1) *unsupervised learning* (UL) and (2) *supervised learning* (SL) models. UL refers to those techniques used to draw inferences from data sets consisting of input data without labelled responses. These algorithms are used to perform tasks such as clustering and pattern mining. SL refers to the class of algorithms employed to make predictions on labelled response values (i.e., discrete and continuous outcomes). In particular, SL methods use a known data set with input data and response values, referred to as training data set, to learn how to successfully perform predictions on labelled outcomes. The learned decision rules can then be used to predict unknown outcomes of new observations. For example, an SL algorithm could be trained on a data set that contains firm-level financial accounts and information on enterprises' solvency status in order to develop decision rules that predict the solvency of companies.

SL algorithms provide great added value in predictive tasks since they are specifically designed for such purposes [56]. Moreover, the nonparametric nature of SL algorithms makes them suited to uncover hidden relationships between the predictors and the response variable in large data sets that would be missed out by traditional econometric approaches. Indeed, the latter models, e.g., ordinary least squares and logistic regression, are built assuming a set of restrictions on the functional form of the model to guarantee statistical properties such as estimator unbiasedness and consistency. SL algorithms often relax those assumptions and the functional form is dictated by the data at hand (data-driven models). This characteristic makes SL algorithms more "adaptive" and inductive, therefore enabling more accurate predictions for future outcome realizations.

In this chapter, we focus on the traditional usage of SL for predictive tasks, excluding from our perspective the growing literature that regards the usage of SL for causal inference. As argued by Kleinberg et al. [56], researchers need to answer to both causal and predictive questions in order to inform policy-makers. An example that helps us to draw the distinction between the two is provided by a policy-maker facing a pandemic. On the one side, if the policy-maker wants to assess whether a quarantine will prevent a pandemic to spread, he needs to answer a purely causal question (i.e., "what is the effect of quarantine on the chance that the pandemic will spread?"). On the other side, if the policy-maker wants to know if he should start a vaccination campaign, he needs to answer a purely predictive question (i.e., "Is the pandemic going to spread within the country?"). SL tools can help policy-makers navigate both these sorts of policy-relevant questions [78]. We refer to [6] and [5] for a critical review of the causal machine learning literature.

Before getting into the nuts and bolts of this chapter, we want to highlight that our goal is not to provide a comprehensive review of all the applications of SL for prediction of firm dynamics, but to describe the alternative methods used so far in this field. Namely, we selected papers based on the following inclusion criteria: (1) the usage of SL algorithm to perform a predictive task in one of our fields of interest (i.e., enterprises success, growth, or exit), (2) a clear definition of the outcome of the model and the predictors used, (3) an assessment of the quality of the prediction. The purpose of this chapter is twofold. First, we outline a general SL framework to ready the readers' mindset to think about prediction problems from an SL-perspective (Sect. 2). Second, equipped with the general concepts of SL, we turn to real-world applications of the SL predictive power in the field of firms' dynamics. Due to the broad range of SL applications, we organize Sect. 3 into three parts according to different stages of the firm life cycle. The prediction tasks we will focus on are about the success of new enterprises and innovation (Sect. 3.1), firm performance and growth (Sect. 3.2), and the exit of established firms (Sect. 3.3). The last section of the chapter discusses the state of the art, future trends, and relevant policy implications (Sect. 4).

2 Supervised Machine Learning

In a famous paper on the difference between model-based and data-driven statistical methodologies, Berkeley professor Leo Breiman, referring to the statistical community, stated that "there are two cultures in the use of statistical modeling to reach conclusions from data. One assumes that the data are generated by a given stochastic data model. The other uses algorithmic models and treats the data mechanism as unknown. [. . .] If our goal as a field is to use data to solve problems, then we need to move away from exclusive dependence on data models and adopt a diverse set of tools" [20, p. 199]. In this quote, Breiman catches the essence of SL algorithms: their ability to capture hidden patterns in the data by directly learning from them, without the restrictions and assumptions of model-based statistical methods.

SL algorithms employ a set of data with input data and response values, referred as training sample, to learn and make predictions (in-sample predictions), while another set of data, referred as test sample, is kept separate to validate the predictions (out-of-sample predictions). Training and testing sets are usually built by randomly sampling observations from the initial data set. In the case of panel data, the testing sample should contain only observations that occurred later in time than the observations used to train the algorithm to avoid the so-called *look-ahead bias*. This ensures that future observations are predicted from past information, not vice versa.

When the dependent variable is categorical (e.g., yes/no or category 1–5) the task of the SL algorithm is referred as a "classification" problem, whereas in "regression" problems the dependent variable is continuous.

The common denominator of SL algorithms is that they take an information set $\mathbf{X}_{N \times P}$, i.e., a matrix of features (also referred to as attributes or predictors), and map it to an N-dimensional vector of outputs y (also referred to as actual values or dependent variable), where N is the number of observations $i = 1, \ldots, N$ and P is the number of features. The functional form of this relationship is very flexible and gets updated by evaluating a loss function. The functional form is usually modelled in two steps [78]:

1. pick the best in-sample loss-minimizing function $f(\cdot)$:

$$argmin \sum_{i=1}^{N} L\big(f(x_i), y_i\big) \quad over \quad f(\cdot) \in F \quad s.t. \quad R\big(f(\cdot)\big) \leq c \tag{1}$$

   where $\sum_{i=1}^{N} L\big(f(x_i), y_i\big)$ is the in-sample loss functional to be minimized (i.e., the mean squared error of prediction), $f(x_i)$ are the predicted (or fitted) values, y_i are the actual values, $f(\cdot) \in F$ is the function class of the SL algorithm, and $R\big(f(\cdot)\big)$ is the complexity functional that is constrained to be less than a certain value $c \in \mathbb{R}$ (e.g., one can think of this parameter as a budget constraint);
2. estimate the optimal level of complexity using empirical tuning through cross-validation.

Cross-validation refers to the technique that is used to evaluate predictive models by training them on the training sample, and evaluating their performance on the test sample.[1] Then, on the test sample the algorithm's performance is evaluated on how well it has learned to predict the dependent variable y. By construction, many SL algorithms tend to perform extremely well on the training data. This phenomenon is commonly referred as "overfitting the training data" because it combines very high predictive power on the training data with poor fit on the test data. This lack of generalizability of the model's prediction from one sample to another can be addressed by penalizing the model's complexity. The choice of a good penalization algorithm is crucial for every SL technique to avoid this class of problems.

In order to optimize the complexity of the model, the performance of the SL algorithm can be assessed by employing various performance measures on the test sample. It is important for practitioners to choose the performance measure that

---

[1] This technique (hold-out) can be extended from two to k folds. In k-folds cross-validation, the original data set is randomly partitioned into k different subsets. The model is constructed on $k - 1$ folds and evaluated on onefold, repeating the procedure until all the k folds are used to evaluate the predictions.

		Observed Outcome		
		Positive	**Negative**	
Predicted Outcome	**Positive**	80 (True positive)	17 (False positive)	Positive predicted value (PPV, Precision): 80/97= 82.5%
	Negative	2 (False negative)	1 (True negative)	Negative predicted value (NPV): 1/3= 33.3%
		True positive rate (TPR, Recall, Sensitivity) 80/82= 97.6%	True negative rate (TNR, Specificity): 1/18= 5.6%	Accuracy (ACC): 81/100= 81% Balanced Accuracy (BACC): (TPR+TNR)/2= 51.6%

Fig. 1 Exemplary confusion matrix for assessment of classification performance

best fits the prediction task at hand and the structure of the response variable. In regression tasks, different performance measures can be employed. The most common ones are the mean squared error (MSE), the mean absolute error (MAE), and the R^2. In classification tasks the most straightforward method is to compare true outcomes with predicted ones via confusion matrices from where common evaluation metrics, such as true positive rate (TPR), true negative rate (TNR), and accuracy (ACC), can be easily calculated (see Fig. 1). Another popular measure of prediction quality for binary classification tasks (i.e., positive vs. negative response), is the Area Under the receiver operating Curve (AUC) that relates how well the trade-off between the models TPR and TNR is solved. TPR refers to the proportion of positive cases that are predicted correctly by the model, while TNR refers to the proportion of negative cases that are predicted correctly. Values of AUC range between 0 and 1 (perfect prediction), where 0.5 indicates that the model has the same prediction power as a random assignment. The choice of the appropriate performance measure is key to communicate the fit of an SL model in an informative way.

Consider the example in Fig. 1 in which the testing data contains 82 positive outcomes (e.g., firm survival) and 18 negative outcomes, such as firm exit, and the algorithm predicts 80 of the positive outcomes correctly but only one of the negative ones. The simple accuracy measure would indicate 81% correct classifications, but the results suggest that the algorithm has not successfully learned how to detect negative outcomes. In such a case, a measure that considers the unbalance of outcomes in the testing set, such as balanced accuracy (BACC, defined as $((TPR + TNR/2) = 51.6\%)$, or the F1-score would be more suited. Once the algorithm has been successfully trained and its out-of-sample performance has been properly tested, its decision rules can be applied to predict the outcome of new observations, for which outcome information is not (yet) known.

Choosing a specific SL algorithm is crucial since performance, complexity, computational scalability, and interpretability differ widely across available implementations. In this context, easily interpretable algorithms are those that provide comprehensive decision rules from which a user can retrace results [62]. Usually, highly complex algorithms require the discretionary fine-tuning of some model hyperparameters, more computational resources, and their decision criteria are less straightforward. Yet, the most complex algorithms do not necessarily deliver the best predictions across applications [58]. Therefore, practitioners usually run a *horse race* on multiple algorithms and choose the one that provides the best balance between interpretability and performance on the task at hand. In some learning applications for which prediction is the sole purpose, different algorithms are combined and the contribution of each chosen so that the overall predictive performance gets maximized. Learning algorithms that are formed by multiple self-contained methods are called ensemble learners (e.g., the super-learner algorithm by Van der Laan et al. [97]).

Moreover, SL algorithms are used by scholars and practitioners to perform predictors selection in high-dimensional settings (e.g., scenarios where the number of predictors is larger than the number of observations: small N large P settings), text analytics, and natural language processing (NLP). The most widely used algorithms to perform the former task are the least absolute shrinkage and selection operator (Lasso) algorithm [93] and its related versions, such as stability selection [74] and C-Lasso [90]. The most popular supervised NLP and text analytics SL algorithms are support vector machines [89], Naive Bayes [80], and Artificial Neural Networks (ANN) [45].

Reviewing SL algorithms and their properties in detail would go beyond the scope of this chapter; however, in Table 1 we provide a basic intuition of the most widely used SL methodologies employed in the field of firm dynamics. A more detailed discussion of the selected techniques, together with a code example to implement each one of them in the statistical software R, and a toy application on real firm-level data, is provided in the following web page: http://github.com/fbargaglistoffi/machine-learning-firm-dynamics.

3 SL Prediction of Firm Dynamics

Here, we review SL applications that have leveraged inter firm data to predict various company dynamics. Due to the increasing volume of scientific contributions that employ SL for company-related prediction tasks, we split the section in three parts according to the life cycle of a firm. In Sect. 3.1 we review SL applications that deal with early-stage firm success and innovation, in Sect. 3.2 we discuss growth and firm-performance-related work, and lastly, in Sect. 3.3, we turn to firm exit prediction problems.

Table 1 SL algorithms commonly applied in predicting firm dynamics

Method	Description	Interpretability
Decision Tree (DT)	Decision trees (DT) consist of a sequence of binary decision rules (nodes) on which the tree splits into branches (edges). At each final branch (leaf node) a decision regarding the outcome is estimated. The sequence and definition of nodes is based on minimizing a measure of node purity (e.g., Gini index, or entropy for classification tasks and MSE for regression tasks). Decision trees are easy to interpret but sensitive to changes in the features that frequently lower their predictive performance (see also [21]).	High
Random Forest (RF)	Instead of estimating just one DT, random forest (RF) re-samples the training set observations to estimate multiple trees. For each tree at each node a set of m (with $m < P$) predictors is chosen randomly from the features space. To obtain the final prediction, the outcomes of all trees are averaged or, in the case of classification tasks, chosen by majority vote (see also [19]).	Medium
Support vector machines (SVM)	Support vector machine (SVM) algorithms estimate a hyperplane over the feature space to classify observations. The vectors that span the hyperplane are called support vectors. They are chosen such that the overall distance (referred to as margin) between the data points and the hyperplane as well as the prediction accuracy is maximized (see also [89]).	Medium
(Deep) Artificial Neural Networks (ANN)	Inspired by biological networks, every artificial neural network (ANN) consists of, at least, three layers (deep ANNs are ANNs with more than three layers): an input layer with feature information, one or more hidden layers, and an output layer returning the predicted values. Each layer consists of nodes (neurons) that are connected via edges across layers. During the learning process, edges that are more important are reinforced. Neurons may then only send a signal if the signal received is strong enough (see also [45]).	Low

3.1 Entrepreneurship and Innovation

The success of young firms (referred to as startups) plays a crucial role in our economy since these firms often act as net creators of new jobs [46] and push, through their product and process innovations, the societal frontier of technology. Success stories of Schumpeterian entrepreneurs that reshaped entire industries are very salient, yet from a probabilistic point of view it is estimated that only 10% of startups stay in business long term [42, 59].

Not only is startup success highly uncertain, but it also escapes our ability to identify the factors to predict successful ventures. Numerous contributions have used traditional regression-based approaches to identify factors associated with the success of small businesses (e.g., [69, 68, 44]), yet do not test the predictive quality of their methods out of sample and rely on data specifically collected for the research purpose. Fortunately, open access platforms such as *Chrunchbase.com* and *Kickstarter.com* provide company- and project-specific data whose high dimensionality can be exploited using predictive models [29]. SL algorithms, trained on a large amount of data, are generally suited to predict startup success, especially because success factors are commonly unknown and their interactions complex. Similarly to the prediction of success at the firm level, SL algorithms can be used to predict success for singular projects. Moreover, unstructured data, e.g., business plans, can be combined with structured data to better predict the odds of success.

Table 2 summarizes the characteristics of recent contributions in various disciplines that use SL algorithms to predict startup success (upper half of the table) and success on the project level (lower half of the table). The definition of success varies across these contributions. Some authors define successful startups as firms that receive a significant source of external funding (this can be additional financing via venture capitalists, an initial public offering, or a buyout) that would allow to scale operations [4, 15, 87, 101, 104]. Other authors define successful startups as companies that simply survive [16, 59, 72] or coin success in terms of innovative capabilities [55, 43]. As data on the project level is usually not publicly available [51, 31], research has mainly focused on two areas for which it is, namely, the project funding success of crowdfunding campaigns [34, 41, 52] and the success of pharmaceutical projects to pass clinical trials [32, 38, 67, 79].[2]

To successfully distinguish how to classify successes from failures, algorithms are usually fed with company-, founder-, and investor-specific inputs that can range from a handful of attributes to a couple of hundred. Most authors find the information that relate to the source of funds predictive for startup success (e.g., [15, 59, 87]), but also entrepreneurial characteristics [72] and engagement in social networks [104] seem to matter. At the project level, funding success depends on the number of investors [41] as well as on the audio/visual content provided by the owner to pitch the project [52], whereas success in R&D projects depends on an interplay between company-, market-, and product-driven factors [79].

Yet, it remains challenging to generalize early-stage success factors, as these accomplishments are often context dependent and achieved differently across heterogeneous firms. To address this heterogeneity, one approach would be to first categorize firms and then train SL algorithms for the different categories. One can manually define these categories (i.e., country, size cluster) or adopt a data-driven approach (e.g., [90]).

---

[2]Since 2007 the US Food and Drug Administration (FDA) requires that the outcome of clinical trials that passed "Phase I" be publicly disclosed [103]. Information on these clinical trials, and pharmaceutical companies in general, has since then been used to train SL methods to classify the outcome of R&D projects.

Table 2 SL literature on firms' early success and innovation

References	Domain	Output	Country, time	Data set size	Primary SL-method	Attributes	GOF
Arroyo et al. [4]	CS	Startup funding	INT (2011–2018)	120,507	GTB	105	82% (ACC)
Bento [15]	BI	Startup funding	USA (1985–2014)	143,348	RF	158	93% (AUC)
Böhm et al. [16]	BI	Startup survival, -growth	USA, GER (1999–2015)	181	SVM	69	67–84% (ACC)
Guerzoni et al. [43]	ECON	Startup innovativeness	ITA (2013)	45,576	bagging, ANN	262	56% (TPR), 95% (TNR)
Kinne and Lenz [55]	ECON	Firm innovativeness	GER (2012–2016)	4481	ANN	N/A	80% (F-score)
Krishna et al. [59]	CS	Startup survival	INT (1999–2014)	13,000	RF, LR	70	73–96% (ACC)
McKenzie and Sansone [72]	ECON	Startup survival	NIG (2014–2015)	2506	SVM	393	64% (ACC)
Sharchilev et al. [87]	CS	Startup funding	INT	21,947	GTB	49	85% (AUC)
Xiang et al. [101]	BI	Startup M&A	INT (1970–2007)	59,631	BN	27	68–89% (AUC)
Yankov et al. [102]	ECON	Startup survival	BUL	142	DT	15	67% (ACC)
Zhang et al. [104]	CS	Startup funding	INT (2015–2016)	4001	SVM	14	84% (AM)
DiMasi et al. [32]	PHARM	Project success (oncology drugs)	INT (1999–2007)	98	RF	4	92% (AUC)
Etter et al. [34]	CS	Project funding	INT (2012–2013)	16,042	Ensemble SVM	12	> 76% (ACC)
Feijoo et al. [38]	PHARM	Project success (clinical trials)	INT (1993–2018)	6417	RF	17	80% (ACC)
Greenberg et al. [41]	CS	Project funding	INT (2012)	13,000	RF	12	67% (ACC)
Kaminski and Hopp [52]	ECON	Project funding	INT (2009–2017)	20,188	LR	200	65–71% (ACC)
Kyebambe et al. [60]	BMA	Emerging Technologies	USA (1979–2010)	11,000	SVM	7	71% (ACC)
Lo et al. [67]	CS	Project success (drugs)	INT (2003–2015)	27,800	KNN,RF	140	74–81% (AUC)
Munos et al. [79]	PHARM	Project success (drugs)	USA (2008–2018)	8.800	BART	37	91–96% (AUC)
Rouhani and Ravasan [84]	ENG	Project success (IT system)	ME (2011)	171	ANN	24	69% (ACC)

Abbreviations used—Domain: ECON: Economics, CS: Computer Science, BI: Business Informatics, ENG: Engineering, BMA: Business, Management and Accounting, PHARM: Pharmacology. Country: ITA: Italy, GER: Germany, INT: International, BUL: Bulgaria, USA: United states of America, NIG: Nigeria, ME: Middle East. Primary SL-method: ANN: (deep) neural network, SL: supervised learner, GTB: gradient tree boosting, DT: Decision Tree, SVM: support vector machine, BN: Bayesian Network, IXL: induction on eXtremely Large databases, RF: random forest, KNN: k-nearest neighbor, BART: Bayesian additive regression tree, LR: Logistic regression, TPR: true positive rate, TNR: true negative rate, ACC: Accuracy, AUC: Area under the receiver operating curve, BACC: Balanced Accuracy (average between TPR and TNR). The year was not reported when it was not possible to recover this information from the papers

The SL methods that best predict startup and project success vary vastly across reviewed applications, with random forest (RF) and support vector machine (SVM) being the most commonly used approaches. Both methods are easily implemented (see our web appendix), and despite their complexity still deliver interpretable results, including insights on the importance of singular attributes. In some applications, easily interpretable logistic regressions (LR) perform at par or better than more complex methods [36, 52, 59]. This might first seem surprising, yet it largely depends on whether complex interdependencies in the explanatory attributes are present in the data at hand. As discussed in Sect. 2 it is therefore recommendable to run a horse race to explore the prediction power of multiple algorithms that vary in terms of their interpretability.

Lastly, even if most contributions report their goodness of fit (GOF) using standard measures such as ACC and AUC, one needs to be cautions when cross-comparing results because these measures depend on the underlying data set characteristics, which may vary. Some applications use data samples, in which successes are less frequently observed than failures. Algorithms that perform well when identifying failures but have limited power when it comes to classifying successes would then be better ranked in terms of ACC and AUC than algorithms for which the opposite holds (see Sect. 2). The GOF across applications simply reflects that SL methods, on average, are useful for predicting startup and project outcomes. However, there is still considerable room for improvement that could potentially come from the quality of the used features as we do not find a meaningful correlation between data set size and GOF in the reviewed sample.

3.2   Firm Performance and Growth

Despite recent progress [22] firm growth is still an elusive problem. Table 3 schematizes the main supervised learning works in the literature on firms' growth and performance. Since the seminal contribution of Gibrat [40] firm growth is still considered, at least partially, as a random walk [28], there has been little progress in identifying the main drivers of firm growth [26], and recent empirical models have a small predictive power [98]. Moreover, firms have been found to be persistently heterogeneous, with results varying depending on their life stage and marked differences across industries and countries. Although a set of stylized facts are well established, such as the negative dependency of growth on firm age and size, it is difficult to predict the growth and performance from previous information such as balance sheet data—i.e., it remains unclear what are good predictors for what type of firm.

SL excels at using high-dimensional inputs, including nonconventional unstructured information such as textual data, and using them all as predictive inputs. Recent examples from the literature reveal a tendency in using multiple SL tools to make better predictions out of publicly available data sources, such as financial reports [82] and company web pages [57]. The main goal is to identify the key

Table 3 SL literature on firms' growth and performance

References	Domain	Output	Country, time	Data set size	Primary SL-method	Attributes	GOF
Weinblat [100]	BMA	High growth firms	INT (2004–2014)	179,970	RF	30	52%-81% (AUC)
Megaravalli and Sampagnaro [73]	BMA	High growth firms	ITA (2010–2014)	22,333	PR*	5	71% (AUC)
Coad and Srhoj [27]	BMA	High growth firms	HRV (2003–2016)	79,109	Lasso	172	76% (ACC)
Miyakawa et al. [76]	ECON	Firm exit, sales growth, profit growth	JPN (2006–2014)	1,700,000	weighted RF	50	70%,68%,61% (AUC)
Lam [61]	BI	ROE	USA (1985–1995)	364 firms per set	ANN	27	Portfolio return comparison
Kolkman and van Witteloostuijn [57]	ECON	Asset growth	NL	8163 firms	RF	113	16% (R^2)
Qiu et al. [82]	CS	Groups of SAR, EPS growth	USA (1997–2003)	1276 firms	SVM	From annual reports	50% (ACC)
Bakar and Tahir [8]	BMA	ROA	MYS (2001–2006)	91	ANN	7	66.9% (R^2)
Baumann et al. [13]	CS	Customer Churn	INT	5000–93,893	Ensemble	20–359	1.5–6.8 (L_1)
Ravi et al. [83]	CS	Profit of banks	INT (1991–1993)	1000	Ensemble	54	80–93% (ACC)

Abbreviations used—Domain: ECON: Economics, CS: Computer Science, BI: Business Informatics, BMA: Business, Management and Accounting. Country: ITA: Italy, INT: International, HRV: Croatia, USA: United states of America, JPN: Japan, NL: Netherlands, MYS: Malaysia. Primary SL-method: ANN: (deep) neural network, SVM: support vector machine, RF: random forest, PR: Probit regression (simplest form of SL if out of sample performance analysis used), Lasso: Least absolute shrinkage and selection operator, Ensemble: Ensemble Learner. GOF: Accuracy, AUC: Area under the receiver operating curve, L_1: Top decile lift. R^2 R-squared. The year was not reported when it was not possible to recover this information from the papers drivers of superior firm performance in terms of profits, growth rates, and return on investments. This is particularly relevant for stakeholders, including investors and policy-makers, to devise better strategies for sustainable competitive advantage. For example, one of the objectives of the European commission is to incentivize high growth firms (HGFs) [35], which could get facilitated by classifying such companies adequately.

A prototypical example of application of SL methods to predict HGFs is Weinblat [100], who uses an RF algorithm trained on firm characteristics for different EU countries. He finds that HGFs have usually experienced prior accelerated growth and should not be confused with startups that are generally younger and smaller. Predictive performance varies substantially across country samples, suggesting that the applicability of SL approaches cannot be generalized. Similarly, Miyakawa et al. [76] show that RF can outperform traditional credit score methods to predict firm exit, growth in sales, and profits of a large sample of Japanese firms. Even if the reviewed SL literature on firms' growth and performance has introduced approaches that increment predictive performance compared to traditional forecasting methods, it should be noted that this performance stays relatively low across applications in the firms' life cycle and does not seem to correlate significantly with the size of the data sets. A firm's growth seems to depend on many interrelated factors whose quantification might still be a challenge for researchers who are interested in performing predictive analysis.

Besides identifying HGFs, other contributions attempt to maximize predictive power of future performance measures using sophisticated methods such as ANN or ensemble learners (e.g., [83, 61]). Even though these approaches achieve better results than traditional benchmarks, such as financial returns of market portfolios, a lot of variation of the performance measure is left unexplained. More importantly, the use of such "black-box" tools makes it difficult to derive useful recommendations on what options exist to better individual firm performance. The fact that data sets and algorithm implementation are usually not made publicly available adds to our impotence at using such results as a base for future investigations.

Yet, SL algorithms may help individual firms improve their performance from different perspectives. A good example in this respect is Erel et al. [33], who showed how algorithms can contribute to appoint better directors.

3.3 Financial Distress and Firm Bankruptcy

The estimation of default probabilities, financial distress, and the predictions of firms' bankruptcies based on balance sheet data and other sources of information on firms viability is a highly relevant topic for regulatory authorities, financial institutions, and banks. In fact, regulatory agencies often evaluate the ability of banks to assess enterprises viability, as this affects their capacity of best allocating financial resources and, in turn, their financial stability. Hence, the higher predictive power of SL algorithms can boost targeted financing policies that lead to safer allocation of credit either on the extensive margin, reducing the number of borrowers by lending money just to the less risky ones, or on the intensive margin (i.e., credit granted) by setting a threshold to the amount of credit risk that banks are willing to accept.

In their seminal works in this field, Altman [3] and Ohlson [81] apply standard econometric techniques, such as multiple discriminant analysis (MDA) and logistic regression, to assess the probability of firms' default. Moreover, since the Basel II Accord in 2004, default forecasting has been based on standard reduced-form regression approaches. However, these approaches may fail, as for MDA the assumptions of linear separability and multivariate normality of the predictors may be unrealistic, and for regression models there may be pitfalls in (1) their ability to capture sudden changes in the state of the economy, (2) their limited model complexity that rules out nonlinear interactions between the predictors, and (3) their narrow capacity for the inclusion of large sets of predictors due to possible multicollinearity issues.

SL algorithms adjust for these shortcomings by providing flexible models that allow for nonlinear interactions in the predictors space and the inclusion of a large number of predictors without the need to invert the covariance matrix of predictors, thus circumventing multicollinearity [66]. Furthermore, as we saw in Sect. 2, SL models are directly optimized to perform predictive task and this leads, in many situations, to a superior predictive performance. In particular, Moscatelli et al. [77] argue that SL models outperform standard econometric models when the predictions of firms' distress is (1) based solely on financial accounts data as predictors and (2) relies on a large amount of data. In fact, as these algorithms are "model free," they need large data sets ("data-hungry algorithms") in order to extract the amount of information needed to build precise predictive models. Table 4 depicts a number of papers in the field of economics, computer science, statistics, business, and decision sciences that deal with the issue of predicting firms' bankruptcy or financial distress through SL algorithms. The former stream of literature (bankruptcy prediction)— which has its foundations in the seminal works of Udo [96], Lee et al. [63], Shin et al. [88], and Chandra et al. [23]—compares the binary predictions obtained with SL algorithms with the actual realized failure outcomes and uses this information to calibrate the predictive models. The latter stream of literature (financial distress prediction)—pioneered by Fantazzini and Figini [36]—deals with the problem of predicting default probabilities (DPs) [77, 12] or financial constraint scores [66]. Even if these streams of literature approach the issue of firms' viability from slightly different perspectives, they train their models on dependent variables that range from firms' bankruptcy (see all the "bankruptcy" papers in Table 4) to firms' insolvency [12], default [36, 14, 77], liquidation [17], dissolvency [12] and financial constraint [71, 92].

In order to perform these predictive tasks, models are built using a set of *structured* and *unstructured* predictors. With structured predictors we refer to balance sheet data and financial indicators, while unstructured predictors are, for instance, auditors' reports, management statements, and credit behavior indicators. Hansen et al. [71] show that the usage of unstructured data, in particular, auditors

Table 4 SL literature on firms' failure and financial distress

References	Domain	Output	Country, time	Data set size	Primary SL-method	Attributes	GOF
Alaka et al. [2]	CS	Bankruptcy	UK (2001–2015)	30,000	NN	5	88% (AUC)
Barboza et al. [9]	CS	Bankruptcy	USA (1985–2014)	10,000	SVM, RF, BO, BA	11	93% (AUC)
Bargagli-Stoffi et al. [12]	ECON	Fin. distress	ITA (2008–2017)	304,000	BART	46	97% (AUC)
Behr and Weinblat [14]	ECON	Bankruptcy	INT (2010–2011)	945,062	DT, RF	20	85% (AUC)
Bonello et al. [17]	ECON	Fin. distress	USA (1996–2016)	1848	NB, DT, NN	96	78% (ACC)
Brédart [18]	BMA	Bankruptcy	BEL (2002–2012)	3728	NN	3	81%(ACC)
Chandra et al. [23]	CS	Bankruptcy	USA (2000)	240	DT	24	75%(ACC)
Cleofas-Sánchez et al. [25]	CS	Fin. distress	INT (2007)	240–8200	SVM, NN, LR	12–30	78% (ACC)
Danenas and Garsva [30]	CS	Fin. distress	USA (1999–2007)	21,487	SVM, NN, LR	51	93% (ACC)
Fantazzini and Figini [36]	STAT	Fin. distress	DEU (1996–2004)	1003	SRF	16	93% (ACC)
Hansen et al. [71]	ECON	Fin. distress	DNK (2013–2016)	278,047	CNN, RNN	50	84% (AUC)
Heo and Yang [47]	CS	Bankruptcy	KOR (2008–2012)	30,000	ADA	12	94% (AUC)
Hosaka [48]	CS	Bankruptcy	JPN (2002–2016)	2703	CNN	14	18% (F-score)
Kim and Upneja [54]	CS	Bankruptcy	KOR (1988–2010)	10,000	ADA, DT	30	95% (ACC)
Lee et al. [63]	BMA	Bankruptcy	KOR (1979–1992)	166	NN	57	82% (ACC)
Liang et al. [65]	ECON	Bankruptcy	TWN (1999–2009)	480	SVM, KNN, DT, NB	190	82% (ACC)
Linn and Weagley [66]	ECON	Fin. distress	INT (1997–2015)	48,512	DRF	16	15% (R^2)
Moscatelli et al. [77]	ECON	Fin. distress	ITA (2011–2017)	250,000	RF	24	84%(AUC)
Shin et al. [88]	CS	Bankruptcy	KOR (1996–1999)	1160	SVM	52	77%(ACC)
Sun and Li [91]	CS	Bankruptcy	CHN	270	CBR, KNN	5	79% (ACC)
Sun et al. [92]	BMA	Fin. distress	CHN (2005–2012)	932	ADA, SVM	13	87%(ACC)
Tsai and Wu [94]	CS	Bankruptcy	INT	690–1000	NN	14–20	79–97%(ACC)
Tsai et al. [95]	CS	Bankruptcy	TWN	440	ANN, SVM, BO, BA	95	86% (ACC)
Wang et al. [99]	CS	Bankruptcy	POL (1997–2001)	240	DT, NN, NB, SVM	30	82% (ACC)
Udo [96]	CS	Bankruptcy	KOR (1996–2016)	300	NN	16	91% (ACC)
Zikeba et al. [105]	CS	Bankruptcy	POL (2000–2013)	10,700	BO	64	95% (ACC)

Abbreviations used—Domain: ECON: Economics, CS: Computer Science, BMA: Business, Management, Accounting, STAT: Statistics. Country: BEL: Belgium, ITA: Italy, DEU: Germany, INT: International, KOR: Korea, USA: United states of America, TWN: Taiwan, CHN: China, UK: United Kingdom, POL: Poland. Primary SL-method: ADA: AdaBoost, ANN: Artificial neural network, CNN: Convolutional neural network, NN: Neural network, GTB: gradient tree boosting, RF: Random forest, DRF: Decision random forest, SRF: Survival random forest, DT: Decision Tree, SVM: support vector machine, NB: Naïve Bayes, BO: Boosting, BA: Bagging, KNN: k-nearest neighbor, BART: Bayesian additive regression tree, DT: decision tree, LR: Logistic regression. Rate: ACC: Accuracy, AUC: Area under the receiver operating curve. The year was not reported when it was not possible to recover this information from the papers reports, can improve the performance of SL algorithms in predicting financial distress. As SL algorithms do not suffer from multicollinearity issues, researchers can keep the set of predictors as large as possible. However, when researcher wish to incorporate just a set of "meaningful" predictors, Behr and Weinblat [14] suggest to include indicators that (1) were found to be useful to predict bankruptcies in previous studies, (2) are expected to have a predictive power based on firms' dynamics theory, and (3) were found to be important in practical applications. As, on the one side, informed choices of the predictors can boost the performance of the SL model, on the other side, economic intuition can guide researchers in the choice of the best SL algorithm to be used with the disposable data sources. Bargagli-Stoffi et al. [12] show that an SL methodology that incorporates the information on missing data into its predictive model—i.e., the BART-mia algorithm by Kapelner and Bleich [53]—can lead to staggering increases in predictive performances when the predictors are missing not at random (MNAR) and their missingness patterns are correlated with the outcome.[3]

As different attributes can have different predictive powers with respect to the chosen output variable, it may be the case that researchers are interested in providing to policy-makers interpretable results in terms of which are the most important variables or the marginal effects of a certain variable on the predictions. Decision-tree-based algorithms, such as random forest [19], survival random forests [50], gradient boosted trees [39], and Bayesian additive regression trees [24], provide useful tools to investigate the aforementioned dimensions (i.e., variables importance, partial dependency plots, etc.). Hence, most of the economics papers dealing with bankruptcy or financial distress predictions implement such techniques [14, 66, 77, 12] in service of policy-relevant implications. On the other side, papers in the fields of computer science and business, which are mostly interested in the quality of predictions, de-emphasizing the interpretability of the methods, are built on black box methodologies such as artificial neural networks [2, 18, 48, 91, 94, 95, 99, 63, 96]. We want to highlight that, from the analyses of selected papers, we find no evidence of a positive correlation between the number of observations and predictors included in the model and the performance of the model. Indicating that "more" is not always better in SL applications to firms' failures and bankruptcies.

4 Final Discussion

SL algorithms have advanced to become effective tools for prediction tasks relevant at different stages of the company life cycle. In this chapter, we provided a general introduction into the basics of SL methodologies and highlighted how they can be

---

[3]Bargagli-Stoffi et al. [12] argue that oftentimes the decision not to release financial account information is driven by firms' financial distress.

applied to improve predictions on future firm dynamics. In particular, SL methods improve over standard econometric tools in predicting firm success at an early stage, superior performance, and failure. High-dimensional, publicly available data sets have contributed in recent years to the applicability of SL methods in predicting early success on the firm level and, even more granular, success at the level of single products and projects. While the dimension and content of data sets varies across applications, SVM and RF algorithms are oftentimes found to maximize predictive accuracy. Even though the application of SL to predict superior firm performance in terms of returns and sales growth is still in its infancy, there is preliminary evidence that RF can outperform traditional regression-based models while preserving interpretability. Moreover, shrinkage methods, such as Lasso or stability selection, can help in identifying the most important drivers of firm success. Coming to SL applications in the field of bankruptcy and distress prediction, decision-tree-based algorithms and deep learning methodologies dominate the landscape, with the former widely used in economics due to their higher interpretability, and the latter more frequent in computer science where usually interpretability is de-emphasized in favor of higher predictive performance.

In general, the predictive ability of SL algorithms can play a fundamental role in boosting targeted policies at every stage of the lifespan of a firm—i.e., (1) identifying projects and companies with a high success propensity can aid the allocation of investment resources; (2) potential high growth companies can be directly targeted with supportive measures; (3) the higher ability to disentangle valuable and non-valuable firms can act as a screening device for potential lenders.

As granular data on the firm level becomes increasingly available, it will open many doors for future research directions focusing on SL applications for prediction tasks. To simplify future research in this matter, we briefly illustrated the principal SL algorithms employed in the literature of firm dynamics, namely, decision trees, random forests, support vector machines, and artificial neural networks. For a more detailed overview of these methods and their implementation in R we refer to our GitHub page (http://github.com/fbargaglistoffi/machine-learning-firm-dynamics), where we provide a simple tutorial to predict firms' bankruptcies.

Besides reaching a high-predictive power, it is important, especially for policy-makers, that SL methods deliver retractable and interpretable results. For instance, the US banking regulator has introduced the obligation for lenders to inform borrowers about the underlying factors that influenced their decision to not provide access to credit.[4] Hence, we argue that different SL techniques should be evaluated, and researchers should opt for the most interpretable method when the predictive performance of competing algorithms is not too different. This is central, as the understanding of which are the most important predictors, or which is the marginal effect of a predictor on the output (e.g., via partial dependency plots), can provide useful insights for scholars and policy-makers. Indeed, researchers and practitioners

---

[4]These obligations were introduced by recent modification in the Equal Credit Opportunity Act (ECOA) and the Fair Credit Reporting Act (FCRA).

can enhance models' interpretability using a set of ready-to-use models and tools that are designed to provide useful insights on the SL black box. These tools can be grouped into three different categories: tools and models for (1) complexity and dimensionality reduction (i.e., variables selection and regularization via Lasso, ridge, or elastic net regressions, see [70]); (2) model-agnostic variables' importance techniques (i.e., permutation feature importance based on how much the accuracy decreases when the variable is excluded, Shapley values, SHAP [SHapley Additive exPlanations], decrease in Gini impurity when a variable is chosen to split a node in tree-based methodologies); and (3) model-agnostic marginal effects estimation methodologies (average marginal effects, partial dependency plots, individual conditional expectations, accumulated local effects).[5]

In order to form a solid knowledge base derived from SL applications, scholars should put an effort in making their research as replicable as possible in the spirit of Open Science. Indeed, in the majority of papers that we analyzed, we did not find possible to replicate the reported analyses. Higher standards of replicability should be reached by releasing details about the choice of the model hyperparameters, the codes, and software used for the analyses as well as by releasing the training/testing data (to the extent that this is possible), anonymizing them in the case that the data are proprietary. Moreover, most of the datasets used for the SL analyses that we covered in this chapter were not disclosed by the authors as they are linked to proprietary data sources collected by banks, financial institutions, and business analytics firms (i.e., Bureau Van Dijk).

Here, we want to stress once more time that SL learning per se is not informative about the causal relationships between the predictors and the outcome; therefore researchers who wish to draw causal inference should carefully check the standard identification assumptions [49] and inspect whether or not they hold in the scenario at hand [6]. Besides not directly providing causal estimands, most of the reviewed SL applications focus on pointwise predictions where inference is de-emphasized. Providing a measure of uncertainty about the predictions, e.g., via confidence intervals, and assessing how sensitive predictions appear to unobserved points, are important directions to explore further [11].

In this chapter, we focus on the analysis of how SL algorithms predict various firm dynamics on "intercompany data" that cover information across firms. Yet, nowadays companies themselves apply ML algorithms for various clustering and predictive tasks [62], which will presumably become more prominent for small and medium-sized companies (SMEs) in the upcoming years. This is due to the fact that (1) SMEs start to construct proprietary data bases, (2) develop the skills to perform in-house ML analysis on this data, and (3) powerful methods are easily implemented using common statistical software.

Against this background, we want to stress that applying SL algorithms and economic intuition regarding the research question at hand should ideally com-

---

[5]For a more extensive discussion on interpretability, models' simplicity, and complexity, we refer the reader to [10] and [64].

plement each other. Economic intuition can aid the choice of the algorithm and the selection of relevant attributes, thus leading to better predictive performance [12]. Furthermore, it requires a deep knowledge of the studied research question to properly interpret SL results and to direct their purpose so that *intelligent machines are driven by expert human beings.*

References

1. Ajit, P. (2016). Prediction of employee turnover in organizations using machine learning algorithms. *International Journal of Advanced Research in Artificial Intelligence, 5*(9), 22–26.
2. Alaka, H. A., Oyedele, L. O., Owolabi, H. A., Kumar, V., Ajayi, S. O., Akinade, O. O., et al. (2018). Systematic review of bankruptcy prediction models: Towards a framework for tool selection. *Expert Systems with Applications, 94*, 164–184.
3. Altman, E. I. (1968). Financial ratios, discriminant analysis and the prediction of corporate bankruptcy. *The Journal of Finance, 23*(4), 589–609.
4. Arroyo, J., Corea, F., Jimenez-Diaz, G., & Recio-Garcia, J. A. (2019). Assessment of machine learning performance for decision support in venture capital investments. *IEEE Access, 7*, 124233–124243.
5. Athey, S. (2018). The impact of machine learning on economics. In *The economics of artificial intelligence: An agenda* (pp. 507–547). Chicago: University of Chicago Press.
6. Athey, S. & Imbens, G. (2019). *Machine learning methods economists should know about*, arXiv, CoRR abs/1903.10075.
7. Bajari, P., Chernozhukov, V., Hortaçsu, A., & Suzuki, J. (2019). The impact of big data on firm performance: An empirical investigation. *AEA Papers and Proceedings, 109*, 33–37.
8. Bakar, N. M. A., & Tahir, I. M. (2009). Applying multiple linear regression and neural network to predict bank performance. *International Business Research, 2*(4), 176–183.
9. Barboza, F., Kimura, H., & Altman, E. (2017). Machine learning models and bankruptcy prediction. *Expert Systems with Applications, 83*, 405–417.
10. Bargagli-Stoffi, F. J., Cevolani, G., & Gnecco, G. (2020). Should simplicity be always preferred to complexity in supervised machine learning? In *6th International Conference on machine Learning, Optimization Data Science (LOD2020), Lecture Notes in Computer Science.* (Vol. 12565, pp. 55–59). Cham: Springer.
11. Bargagli-Stoffi, F. J., De Beckker, K., De Witte, K., & Maldonado, J. E. (2021). Assessing sensitivity of predictions. A novel toolbox for machine learning with an application on financial literacy. arXiv, CoRR abs/2102.04382
12. Bargagli-Stoffi, F. J., Riccaboni, M., & Rungi, A. (2020). Machine learning for zombie hunting. firms' failures and financial constraints. *FEB Research Report Department of Economics DPS20. 06.*
13. Baumann, A., Lessmann, S., Coussement, K., & De Bock, K. W. (2015). Maximize what matters: Predicting customer churn with decision-centric ensemble selection. In *ECIS 2015 Completed Research Papers*, Paper number 15. Available at: https://aisel.aisnet.org/ecis2015_cr/15
14. Behr, A., & Weinblat, J. (2017). Default patterns in seven EU countries: A random forest approach. *International Journal of the Economics of Business, 24*(2), 181–222.
15. Bento, F. R. d. S. R. (2018). *Predicting start-up success with machine learning.* B.S. thesis, Universidade NOVA de Lisboa. Available at: https://run.unl.pt/bitstream/10362/33785/1/TGI0132.pdf

16. Böhm, M., Weking, J., Fortunat, F., Müller, S., Welpe, I., & Krcmar, H. (2017). The business model DNA: Towards an approach for predicting business model success. In *Int. En Tagung Wirtschafts Informatik* (pp. 1006–1020).
17. Bonello, J., Brédart, X., & Vella, V. (2018). Machine learning models for predicting financial distress. *Journal of Research in Economics, 2*(2), 174–185.
18. Brédart, X. (2014). Bankruptcy prediction model using neural networks. *Accounting and Finance Research, 3*(2), 124–128.
19. Breiman, L. (2001). Random forests. *Machine Learning, 45*(1), 5–32.
20. Breiman, L. (2001). Statistical modeling: The two cultures (with comments and a rejoinder by the author). *Statistical Science, 16*(3), 199–231.
21. Breiman, L. (2017). *Classification and regression trees*. New York: Routledge.
22. Buldyrev, S., Pammolli, F., Riccaboni, M., & Stanley, H. (2020). *The rise and fall of business firms: A stochastic framework on innovation, creative destruction and growth*. Cambridge: Cambridge University Press.
23. Chandra, D. K., Ravi, V., & Bose, I. (2009). Failure prediction of dotcom companies using hybrid intelligent techniques. *Expert Systems with Applications, 36*(3), 4830–4837.
24. Chipman, H. A., George, E. I., McCulloch, R. E. (2010). Bart: Bayesian additive regression trees. *The Annals of Applied Statistics, 4*(1), 266–298.
25. Cleofas-Sánchez, L., García, V., Marqués, A., & Sánchez, J. S. (2016). Financial distress prediction using the hybrid associative memory with translation. *Applied Soft Computing, 44*, 144–152.
26. Coad, A. (2009). *The growth of firms: A survey of theories and empirical evidence*. Northampton: Edward Elgar Publishing.
27. Coad, A., & Srhoj, S. (2020). Catching gazelles with a lasso: Big data techniques for the prediction of high-growth firms. *Small Business Economics, 55*, 541–565. https://doi.org/10.1007/s11187-019-00203-3
28. Coad, A., Frankish, J., Roberts, R. G., & Storey, D. J. (2013). Growth paths and survival chances: An application of gambler's ruin theory. *Journal of Business Venturing, 28*(5), 615–632.
29. Dalle, J.-M., Den Besten, M., & Menon, C. (2017). Using crunchbase for economic and managerial research. In *OECD SCience, Technology and Industry Working Papers*, 2017/08. https://doi.org/10.1787/6c418d60-en
30. Danenas, P., & Garsva, G. (2015). Selection of support vector machines based classifiers for credit risk domain. *Expert Systems with Applications, 42*(6), 3194–3204.
31. Dellermann, D., Lipusch, N., Ebel, P., Popp, K. M., & Leimeister, J. M. (2017). Finding the unicorn: Predicting early stage startup success through a hybrid intelligence method. In *International Conference on Information Systems (ICIS), Seoul*. Available at: https://doi.org/10.2139/ssrn.3159123
32. DiMasi, J., Hermann, J., Twyman, K., Kondru, R., Stergiopoulos, S., Getz, K., et al. (2015). A tool for predicting regulatory approval after phase ii testing of new oncology compounds. *Clinical Pharmacology & Therapeutics, 98*(5), 506–513.
33. Erel, I., Stern, L. H., Tan, C., & Weisbach, M. S. (2018). *Selecting directors using machine learning*. Technical report, National Bureau of Economic Research. Working paper 24435. https://doi.org/10.3386/w24435
34. Etter, V., Grossglauser, M., & Thiran, P. (2013). Launch hard or go home! predicting the success of kickstarter campaigns. In *Proceedings of the First ACM Conference on Online Social Networks* (pp. 177–182).
35. European Commission. (2010). *Communication from the commission: Europe 2020: A strategy for smart, sustainable and inclusive growth*. Publications Office of the European Union, 52010DC2020. Available at: https://eur-lex.europa.eu/legal-content/en/ALL/?uri=CELEX%3A52010DC2020
36. Fantazzini, D., & Figini, S. (2009). Random survival forests models for SME credit risk measurement. *Methodology and Computing in Applied Probability, 11*(1), 29–45.

37. Farboodi, M., Mihet, R., Philippon, T., & Veldkamp, L. (2019). Big data and firm dynamics. In *AEA Papers and Proceedings* (Vol. 109, pp. 38–42).
38. Feijoo, F., Palopoli, M., Bernstein, J., Siddiqui, S., & Albright, T. E. (2020). Key indicators of phase transition for clinical trials through machine learning. *Drug Discovery Today, 25*(2), 414–421.
39. Friedman, J. H. (2001). Greedy function approximation: a gradient boosting machine. *Annals of Statistics, 29*(5), 1189–1232.
40. Gibrat, R. (1931). *Les inégalités économiques: applications aux inégalités des richesses, à la concentration des entreprises... d'une loi nouvelle, la loi de l'effet proportionnel.* Paris: Librairie du Recueil Sirey.
41. Greenberg, M. D., Pardo, B., Hariharan, K., & Gerber, E. (2013). Crowdfunding support tools: predicting success & failure. In *CHI'13 Extended Abstracts on Human Factors in Computing Systems* (pp. 1815–1820). New York: ACM.
42. Griffith, E. (2014). Why startups fail, according to their founders. *Fortune Magazine*, Last accessed on 12 March, 2021. Available at: https://fortune.com/2014/09/25/why-startups-fail-according-to-their-founders/
43. Guerzoni, M., Nava, C. R., & Nuccio, M. (2019). *The survival of start-ups in time of crisis. a machine learning approach to measure innovation.* Preprint. arXiv:1911.01073.
44. Halabi, C. E., & Lussier, R. N. (2014). A model for predicting small firm performance. *Journal of Small Business and Enterprise Development, 21*(1), 4–25.
45. Hassoun, M. H. (1995). *Fundamentals of artificial neural networks.* Cambridge: MIT Press.
46. Henrekson, M., & Johansson, D. (2010). Gazelles as job creators: a survey and interpretation of the evidence. *Small Business Economics, 35*(2), 227–244.
47. Heo, J., & Yang, J. Y. (2014). Adaboost based bankruptcy forecasting of Korean construction companies. *Applied Soft Computing, 24*, 494–499.
48. Hosaka, T. (2019). Bankruptcy prediction using imaged financial ratios and convolutional neural networks. *Expert Systems with Applications, 117*, 287–299.
49. Imbens, G. W., & Rubin, D. B. (2015). *Causal inference for statistics, social, and biomedical sciences: An introduction.* New York: Cambridge University Press.
50. Ishwaran, H., Kogalur, U. B., Blackstone, E. H., & Lauer, M. S. (2008). Random survival forests. *The Annals of Applied Statistics, 2*(3), 841–860.
51. Janssen, N. E. (2019). *A machine learning proposal for predicting the success rate of IT-projects based on project metrics before initiation.* B.Sc. thesis, University of Twente. Available at: https://essay.utwente.nl/78526/
52. Kaminski, J. C., & Hopp, C. (2020). Predicting outcomes in crowdfunding campaigns with textual, visual, and linguistic signals. *Small Business Economics, 55*, 627–649.
53. Kapelner, A., & Bleich, J. (2015). Prediction with missing data via Bayesian additive regression trees. *Canadian Journal of Statistics, 43*(2), 224–239.
54. Kim, S. Y., & Upneja, A. (2014). Predicting restaurant financial distress using decision tree and adaboosted decision tree models. *Economic Modelling, 36*, 354–362.
55. Kinne, J., & Lenz, D. (2019). Predicting innovative firms using web mining and deep learning. In *ZEW-Centre for European Economic Research Discussion Paper*, (19-01).
56. Kleinberg, J., Ludwig, J., Mullainathan, S., & Obermeyer, Z. (2015). Prediction policy problems. *American Economic Review, 105*(5), 491–495.
57. Kolkman, D., & van Witteloostuijn, A. (2019). Data science in strategy: Machine learning and text analysis in the study of firm growth. In *Tinbergen Institute Discussion Paper 2019-066/VI*. Available at: https://doi.org/10.2139/ssrn.3457271
58. Kotthoff, L. (2016). Algorithm selection for combinatorial search problems: A survey. In *Data Mining and Constraint Programming*, LNCS (Vol. 10101, pp. 149–190). Cham: Springer.
59. Krishna, A., Agrawal, A., & Choudhary, A. (2016). Predicting the outcome of startups: less failure, more success. In *2016 IEEE 16th International Conference on Data Mining Workshops (ICDMW)* (pp. 798–805). Piscataway: IEEE.

60. Kyebambe, M. N., Cheng, G., Huang, Y., He, C., & Zhang, Z. (2017). Forecasting emerging technologies: A supervised learning approach through patent analysis. *Technological Forecasting and Social Change, 125*, 236–244.
61. Lam, M. (2004). Neural network techniques for financial performance prediction: integrating fundamental and technical analysis. *Decision support systems, 37*(4), 567–581.
62. Lee, I., & Shin, Y. J. (2020). Machine learning for enterprises: Applications, algorithm selection, and challenges. *Business Horizons, 63*(2), 157–170.
63. Lee, K. C., Han, I., & Kwon, Y. (1996). Hybrid neural network models for bankruptcy predictions. *Decision Support Systems, 18*(1), 63–72.
64. Lee, K., Bargagli-Stoffi, F. J., & Dominici, F. (2020). *Causal rule ensemble: Interpretable inference of heterogeneous treatment effects*, arXiv, CoRR abs/2009.09036
65. Liang, D., Lu, C.-C., Tsai, C.-F., & Shih, G.-A. (2016). Financial ratios and corporate governance indicators in bankruptcy prediction: A comprehensive study. *European Journal of Operational Research, 252*(2), 561–572.
66. Linn, M., & Weagley, D. (2019). Estimating financial constraints with machine learning. In *SSRN*, paper number 3375048. https://doi.org/10.2139/ssrn.3375048
67. Lo, A. W., Siah, K. W., & Wong, C. H. (2019). Machine learning with statistical imputation for predicting drug approvals. *Harvard Data Science Review, 1*(1). https://doi.org/10.1162/99608f92.5c5f0525
68. Lussier, R. N., & Halabi, C. E. (2010). A three-country comparison of the business success versus failure prediction model. *Journal of Small Business Management, 48*(3), 360–377.
69. Lussier, R. N., & Pfeifer, S. (2001). A cross-national prediction model for business success. *Journal of Small Business Management, 39*(3), 228–239.
70. Martínez, J. M., Escandell-Montero, P., Soria-Olivas, E., MartíN-Guerrero, J. D., Magdalena-Benedito, R., & GóMez-Sanchis, J. (2011). Regularized extreme learning machine for regression problems. *Neurocomputing, 74*(17), 3716–3721.
71. Matin, R., Hansen, C., Hansen, C., & Molgaard, P. (2019). Predicting distresses using deep learning of text segments in annual reports. *Expert Systems with Applications, 132*(15), 199–208.
72. McKenzie, D., & Sansone, D. (2017). Man vs. machine in predicting successful entrepreneurs: evidence from a business plan competition in Nigeria. In *World Bank Policy Research Working Paper No. 8271*. Available at: https://ssrn.com/abstract=3086928
73. Megaravalli, A. V., & Sampagnaro, G. (2019). Predicting the growth of high-growth SMEs: evidence from family business firms. *Journal of Family Business Management, 9*(1), 98–109. https://doi.org/10.1108/JFBM-09-2017-0029
74. Meinshausen, N., & Bühlmann, P. (2010). Stability selection. *Journal of the Royal Statistical Society: Series B (Statistical Methodology), 72*(4), 417–473.
75. Mikalef, P., Boura, M., Lekakos, G., & Krogstie, J. (2019). Big data analytics and firm performance: Findings from a mixed-method approach. *Journal of Business Research, 98*, 261–276.
76. Miyakawa, D., Miyauchi, Y., & Perez, C. (2017). *Forecasting firm performance with machine learning: Evidence from Japanese firm-level data*. Technical report, Research Institute of Economy, Trade and Industry (RIETI). Discussion Paper Series 17-E-068. Available at: https://www.rieti.go.jp/jp/publications/dp/17e068.pdf
77. Moscatelli, M., Parlapiano, F., Narizzano, S., & Viggiano, G. (2020). Corporate default forecasting with machine learning. *Expert Systems with Applications, 161*(15), art. num. 113567
78. Mullainathan, S., & Spiess, J. (2017). Machine learning: an applied econometric approach. *Journal of Economic Perspectives, 31*(2), 87–106.
79. Munos, B., Niederreiter, J., & Riccaboni, M. (2020). Improving the prediction of clinical success using machine learning. In *EIC Working Paper Series*, number 3/2020. Available at: http://eprints.imtlucca.it/id/eprint/4079
80. Ng, A. Y., & Jordan, M. I. (2002). On discriminative vs. generative classifiers: A comparison of logistic regression and naive bayes. In *Advances in neural information processing systems*,

NIPS 2001 (Vol. 14, pp. 841–848), art code 104686. Available at: https://papers.nips.cc/paper/ 2001/file/7b7a53e239400a13bd6be6c91c4f6c4e-Paper.pdf

81. Ohlson, J. A. (1980). Financial ratios and the probabilistic prediction of bankruptcy. *Journal of Accounting Research, 18*(1), 109–131.
82. Qiu, X. Y., Srinivasan, P., & Hu, Y. (2014). Supervised learning models to predict firm performance with annual reports: An empirical study. *Journal of the Association for Information Science and Technology, 65*(2), 400–413.
83. Ravi, V., Kurniawan, H., Thai, P. N. K., & Kumar, P. R. (2008). Soft computing system for bank performance prediction. *Applied Soft Computing, 8*(1), 305–315.
84. Rouhani, S., & Ravasan, A. Z. (2013). ERP success prediction: An artificial neural network approach. *Scientia Iranica, 20*(3), 992–1001.
85. Saradhi, V. V., & Palshikar, G. K. (2011). Employee churn prediction. *Expert Systems with Applications, 38*(3), 1999–2006.
86. Sejnowski, T. J. (2018). *The deep learning revolution.* Cambridge: MIT Press.
87. Sharchilev, B., Roizner, M., Rumyantsev, A., Ozornin, D., Serdyukov, P., & de Rijke, M. (2018). Web-based startup success prediction. In *Proceedings of the 27th ACM International Conference on Information and Knowledge Management* (pp. 2283–2291).
88. Shin, K.-S., Lee, T. S., & Kim, H.-j. (2005). An application of support vector machines in bankruptcy prediction model. *Expert Systems with Applications, 28*(1), 127–135.
89. Steinwart, I., & Christmann, A. (2008). *Support vector machines.* New York: Springer Science & Business Media.
90. Su, L., Shi, Z., & Phillips, P. C. (2016). Identifying latent structures in panel data. *Econometrica, 84*(6), 2215–2264.
91. Sun, J., & Li, H. (2011). Dynamic financial distress prediction using instance selection for the disposal of concept drift. *Expert Systems with Applications, 38*(3), 2566–2576.
92. Sun, J., Fujita, H., Chen, P., & Li, H. (2017). Dynamic financial distress prediction with concept drift based on time weighting combined with Adaboost support vector machine ensemble. *Knowledge-Based Systems, 120*, 4–14.
93. Tibshirani, R. (1996). Regression shrinkage and selection via the lasso. *Journal of the Royal Statistical Society: Series B (Methodological), 58*(1), 267–288.
94. Tsai, C.-F., & Wu, J.-W. (2008). Using neural network ensembles for bankruptcy prediction and credit scoring. *Expert Systems with Applications, 34*(4), 2639–2649.
95. Tsai, C.-F., Hsu, Y.-F., & Yen, D. C. (2014). A comparative study of classifier ensembles for bankruptcy prediction. *Applied Soft Computing, 24*, 977–984.
96. Udo, G. (1993). Neural network performance on the bankruptcy classification problem. *Computers & Industrial Engineering, 25*(1–4), 377–380.
97. Van der Laan, M. J., Polley, E. C., & Hubbard, A. E. (2007). Super learner. *Statistical Applications in Genetics and Molecular Biology, 6*(1), Article No. 25. https://doi.org/10.2202/ 1544-6115.1309
98. van Witteloostuijn, A., & Kolkman, D. (2019). Is firm growth random? A machine learning perspective. *Journal of Business Venturing Insights, 11*, e00107.
99. Wang, G., Ma, J., & Yang, S. (2014). An improved boosting based on feature selection for corporate bankruptcy prediction. *Expert Systems with Applications, 41*(5), 2353–2361.
100. Weinblat, J. (2018). Forecasting European high-growth firms-a random forest approach. *Journal of Industry, Competition and Trade, 18*(3), 253–294.
101. Xiang, G., Zheng, Z., Wen, M., Hong, J., Rose, C., & Liu, C. (2012). A supervised approach to predict company acquisition with factual and topic features using profiles and news articles on techcrunch. In *Sixth International AAAI Conference on Weblogs and Social Media (ICWSM 2012).* Menlo Park: The AAAI Press. Available at: http://dblp.uni-trier.de/db/conf/icwsm/ icwsm2012.html#XiangZWHRL12
102. Yankov, B., Ruskov, P., & Haralampiev, K. (2014). Models and tools for technology start-up companies success analysis. *Economic Alternatives, 3*, 15–24.
103. Zarin, D. A., Tse, T., Williams, R. J., & Carr, S. (2016). Trial Reporting in ClinicalTrials.gov – The Final Rule. *New England Journal of Medicine, 375*(20), 1998–2004.

104. Zhang, Q., Ye, T., Essaidi, M., Agarwal, S., Liu, V., & Loo, B. T. (2017). Predicting startup crowdfunding success through longitudinal social engagement analysis. In *Proceedings of the 2017 ACM on Conference on Information and Knowledge Management* (pp. 1937–1946).
105. Zikeba, M., Tomczak, S. K., & Tomczak, J. M. (2016). Ensemble boosted trees with synthetic features generation in application to bankruptcy prediction. *Expert Systems with Applications, 58*, 93–101.

Opening the Black Box: Machine Learning Interpretability and Inference Tools with an Application to Economic Forecasting

Marcus Buckmann, Andreas Joseph, and Helena Robertson

Abstract We present a comprehensive comparative case study for the use of machine learning models for macroeconomics forecasting. We find that machine learning models mostly outperform conventional econometric approaches in forecasting changes in US unemployment on a 1-year horizon. To address the black box critique of machine learning models, we apply and compare two variables attribution methods: permutation importance and Shapley values. While the aggregate information derived from both approaches is broadly in line, Shapley values offer several advantages, such as the discovery of unknown functional forms in the data generating process and the ability to perform statistical inference. The latter is achieved by the Shapley regression framework, which allows for the evaluation and communication of machine learning models akin to that of linear models.

1 Introduction

Machine learning provides a toolbox of powerful methods that excel in static prediction problems such as face recognition [37], language translation [12], and playing board games [41]. The recent literature suggests that machine learning methods can also outperform conventional models in forecasting problems; see, e.g., [4] for bond risk premia, [15] for recessions, and [5] for financial crises. Predicting macroeconomic dynamics is challenging. Relationships between variables may not hold over time, and shocks such as recessions or financial crises might lead to a breakdown of previously observed relationships. Nevertheless, several studies have shown that machine learning methods outperform econometric baselines in predicting unemployment, inflation, and output [38, 9].

M. Buckmann · A. Joseph (✉)
Bank of England, London, UK
e-mail: marcus.buckmann@bankofengland.co.uk; andreas.joseph@bankofengland.co.uk

H. Robertson
Financial Conduct Authority, London, UK
e-mail: helena.robertson2@fca.org.uk

S. Consoli et al. (eds.), *Data Science for Economics and Finance*,
https://doi.org/10.1007/978-3-030-66891-4_3

While they learn meaningful relationships between variables from the data, these are not directly observable, leading to the criticism that machine learning models such as random forests and neural networks are opaque black boxes. However, as we demonstrate, there exist approaches that can make machine learning predictions transparent and even allow for statistical inference.

We have organized this chapter as a guiding example for how to combine improved performance and statistical inference for machine learning models in the context of macroeconomic forecasting.

We start by comparing the forecasting performance and inference on various machine learning models to more commonly used econometric models. We find that machine learning models outperform econometric benchmarks in predicting 1-year changes in US unemployment. Next, we address the black box critique by using Shapley values [44, 28] to depict the nonlinear relationships learned by the machine learning models and then test their statistical significance [24]. Our method closes the gap between two distinct data modelling objectives, using black box machine learning methods to maximize predictive performance and statistical techniques to infer the data-generating process [8].

While several studies have shown that multivariate machine learning models can be useful for macroeconomic forecasting [38, 9, 31], only a little research has tried to explain the machine learning predictions. Coulombe et al. [13] shows generally that the success of machine learning models in macro-forecasting can be attributed to their ability to exploit nonlinearities in the data, particularly at longer time horizons. However, we are not aware of any macroeconomic forecasting study that attempted to identify the functional form learned by the machine learning models.[1] However, addressing the explainability of models is important when model outputs inform decisions, given the intertwined ethical, safety, privacy, and legal concerns about the application of opaque models [14, 17, 20]. There exists a debate about the level of model explainability that is necessary. Lipton [27] argues that a complex machine learning model does not need to be less interpretable than a simpler linear model if the latter operates on a more complex space, while Miller [32] suggests that humans prefer simple explanations, i.e., those providing fewer causes and explaining more general events—even though these may be biased.

Therefore, with our focus on explainability, we consider a small but diverse set of variables to learn a forecasting model, while the forecasting literature often relies on many variables [21] or latent factors that summarize individual variables [43]. In the machine learning literature, approaches to interpreting machine learning models usually focus on measuring how important input variables are for prediction. These *variable attributions* can be either global, assessing variable importance across the whole data set [23, 25] or local, by measuring the importance of the variables at the level of individual observations. Popular global methods are permutation importance or Gini importance for tree-based models [7]. Popular local methods are

---

[1] See Bracke et al. [6], Bluwstein et al. [5] for examples that explain machine learning predictions in economic prediction problems.

LIME[2] [34], DeepLIFT[3] [40] and Shapley values [44]. Local methods decompose *individual* predictions into variable contributions [36, 45, 44, 34, 40, 28, 35]. The main advantage of local methods is that they uncover the functional form of the association between a feature and the outcome as learned by the model. Global methods cannot reveal the direction of association between a variable and the outcome of interest. Instead, they only identify variables that are relevant on average across all predictions, which can also be achieved via local methods and averaging attributions across all observations.

For model explainability in the context of macroeconomic forecasting, we suggest that local methods that uncover the functional form of the data generating process are most appropriate. Lundberg and Lee [28] demonstrate that local method Shapley values offer a unified framework of LIME and DeepLIFT with appealing properties. We chose to use Shapely values in this chapter because of their important property of *consistency*. Here, consistency is when on increasing the impact of a feature in a model, the feature's estimated attribution for a prediction does not decrease, independent of all other features. Originally, Shapley values were introduced in game theory [39] as a way to determine the contribution of individual players in a cooperative game. Shapely values estimate the increase in the collective pay-off when a player joins all possible coalitions with other players. Štrumbelj and Kononenko [44] used this approach to estimate the contribution of variables to a model prediction, where the variables and the predicted value are analogous to the players and payoff in a game.

The global and local attribution methods mentioned here are descriptive—they explain the drivers of a model's prediction but they do not assess a model's goodness-of-fit or the predictors' statistical significance. These concepts relate to statistical inference and require two steps: (1) measuring or estimating some quantity, such as a regression coefficient, and (2) inferring how certain one is in this estimate, e.g., how likely is it that the true coefficient in the population is different from zero.

The econometric approach of statistical inference for machine learning is mostly focused on measuring low-dimensional parameters of interest [10, 11], such as treatment effects in randomized experiments [2, 47]. However, in many situations we are interested in estimating the effects for *all* variables included in a model. To the best of our knowledge, there exists only one general framework that performs statistical inference jointly on all variables used in a machine learning prediction model to test for their statistical significance [24]. The framework is called *Shapley regressions*, where an auxiliary regression of the outcome variable on the Shapley values of individual data points is used to identify those variables that significantly improve the predictions of a nonlinear machine learning model. We will discuss this framework in detail in Sect. 4. Before that, we will describe the data and the

---

[2]Local Interpretable Model-agnostic Explanations.
[3]Deep Learning Important FeaTures for NN.

forecasting methodology (Sect. 2) and present the forecasting results (Sect. 3). We conclude in Sect. 5.

2 Data and Experimental Setup

We first introduce the necessary notation. Let y and $\hat{y} \in \mathbb{R}^m$ be the observed and predicted continuous outcome, respectively, where m is the number of observations in the time series.[4] The feature matrix is denoted by $x \in \mathbb{R}^{m \times n}$, where n is the number of features in the dataset. The feature vector of observation i is denoted by x_i. Generally, we use i to index the point in time of the observation and k to index features. While our empirical analysis is limited to numerical features, the forecasting methods as well as the techniques to interpret their predictions also work when the data contains categorical features. These just need to be transformed into binary variables, each indicating membership of a category.

2.1 Data

We use the *FRED-MD* macroeconomic database [30]. The data contains monthly series of 127 macroeconomic indicators of the USA between 1959 and 2019. Our outcome variable is unemployment and we choose nine variables as predictors, each capturing a different macroeconomic channel. We add the slope of the yield curve as a variable by computing the difference of the interest rates of the 10-year treasury note and the 3-month treasury bill. The authors of the database suggest specific transformations to make each series stationary. We use these transformations, which are (for a variable a:) (1) changes ($a_i - a_{i-l}$), (2) log changes ($\log_e a_i - \log_e a_{i-l}$), and (3) second-order log changes (($\log_e a_i - \log_e a_{i-l}) - (\log_e a_{i-l} - \log_e a_{i-2l})$). As we want to predict the year-on-year change in unemployment, we set l to 12 for the outcome and the lagged outcome when used as a predictor. For the remaining predictors, we set $l = 3$ in our baseline setup. This generally leads to the best performance (see Table 3 for other choices of l). Table 1 shows the variables, with the respective transformations and the series names in the original database. The augmented Dickey-Fuller test confirms that all transformed series are stationary ($p < 0.01$).

---

[4]That is, we are in the setting of a regression problem in machine learning speak, while classification problems operate on categorical targets. All approaches presented here can be applied to both situations.

Table 1 Series used in the forecasting experiment. The middle column shows the transformations suggested by the authors of the FRED-MD database and the right column shows the names in that database

Variable	Transformation	Name in the FRED-MD database
Unemployment	Changes	UNRATE
3-month treasury bill	Changes	TB3MS
Slope of the yield curve	Changes	–
Real personal income	Log changes	RPI
Industrial production	Log changes	INDPRO
Consumption	Log changes	DPCERA3M086SBEA
S&P 500	Log changes	S&P 500
Business loans	Second-order log changes	BUSLOANS
CPI	Second-order log changes	CPIAUCSL
Oil price	Second-order log changes	OILPRICEx
M2 Money	Second-order log changes	M2SL

2.2 Models

We test three families of models that can be formalized in the following way assuming that all variables have been transformed according to Table 1.

- The **simple linear lag model** only uses the 1-year lag of the outcome variable as a predictor: $\hat{y}_i = \alpha + \theta_0 y_{i-12}$.
- The **autoregressive model (AR)** uses several lags of the response as predictors: $\hat{y}_i = \alpha + \sum_{l=1}^{h} \theta_i y_{i-l}$. We test AR models with a horizon $1 \leq h \leq 12$, chosen by the Akaike Information Criterion [1].
- The **full information models** use the 1-year lag of the outcome and 1-year lags of the other features as independent variables: $\hat{y}_t = f(y_{i-12}; x_{i-12})$, where f can be any prediction model. For example, if f is a linear regression, $f(y_i, x_i) = \alpha + \theta_0 y_{i-12} + \sum_{k=1}^{n} \theta_k x_{i-12,k}$. To simplify this notation we imply that the lagged outcome is included in the feature matrix x in the following. We test five full information models: Ordinary least squares regression and Lasso regularized regression [46], and three machine learning regressors—random forest [7], support vector regression [16], and artificial neural networks [22].[5]

---

[5]In machine learning, classification is arguably the most relevant and most researched prediction problem, and while models such as random forests and support vector machines are best known as classification, their variants being used in regression problems are also known to perform well.

2.3 Experimental Procedure

We evaluate how all models predict changes in unemployment 1 year ahead. After transforming the variables (see Table 1) and removing missing values, the first observation in the training set is February 1962. All methods are evaluated on the 359 data points of the forecasts between January 1990 and November 2019 using an expanding window approach. We recalibrate the full information and simple linear lag models every 12 months such that each model makes 12 predictions before it is updated. The autoregressive model is updated every month. Due to the lead-lag structure of the full information and simple linear lag models, we have to create an initial gap between training and test set when making predictions to avoid a look-ahead bias. For a model trained on observations $1 \ldots i$, the earliest observation in the test set that provides a true 12-month forecast is $i + 12$. For observations $i + 1, \ldots, i + 11$, the time difference to the last observed outcome in the training set is smaller than a year.

All machine learning models that we tested have hyperparameters. We optimize their values in the training sets using fivefold cross-validation.[6] As this is computationally expensive, we conduct the hyperparameter search every 36 months with the exception of the computationally less costly Lasso regression, whose hyperparameters are updated every 12 months.

To increase the stability of the full information models, we use bootstrap aggregation, also referred to as bagging. We train 100 models on different bootstrapped samples (of the same size as the training set) and average their predictions. We do not use bagging for the random forest as, by design, each individual tree is already calibrated on a different bootstrapped sample of the training set.

3 Forecasting Performance

3.1 Baseline Setting

Table 2 shows three measures of forecasting performance: the correlation of the observed and predicted response, the mean absolute error (MAE), and the root mean squared error (RMSE). The latter is the main metric considered, as most models minimize RMSE during training. The models are ordered by decreasing RMSE on the whole test period between 1990 and 2019. The random forest performs best and we divide the MAE and RMSE of all models by that of the random forest for ease of comparison.

---

[6]For the hyperparameter search, we also consider partitionings of the training set that take the temporal dependency of our data into account [3]. We use block cross-validation [42] and hv-block cross-validation [33]. However, both methods do not improve the forecasting accuracy.

Table 2 Forecasting performance for the different prediction models. The models are ordered by decreasing RMSE on the whole sample with the errors of the random forest set to unity. The forest's MAE and RMSE (full period) are 0.574 and 0.763, respectively. The asterisks indicate the statistical significance of the Diebold-Mariano test, comparing the performance of the random forest with the other models, with significance levels $*p < 0.1$; $**p < 0.05$; $***p < 0.01$

	Corr. MAE	RMSE (normalized by first row)			
	01/1990– 11/2019	01/1990– 11/2019	01/1990– 12/1999	01/2000– 08/2008	09/2008– 11/2019
Random forest	0.609 1.000	1.000	1.000	1.000	1.000
Neural network	0.555 1.009	1.049	0.969	0.941	1.114**
Linear regression	0.521 1.094***	1.082**	1.011	0.959	1.149***
Lasso regression	0.519 1.094***	1.083***	1.007	0.949	1.156***
Ridge regression	0.514 1.099***	1.087***	1.019	0.952	1.157***
SVR	0.475 1.052	1.105**	1.000	1.033	1.169**
AR	0.383 1.082(*)	1.160(***)	1.003	1.010	1.265(***)
Linear regression (lagged response)	0.242 1.163***	1.226***	1.027	1.057	1.352***

Table 2 also breaks down the performance in three periods: the 1990s and the period before and after the onset of the global financial crisis in September 2008. We statistically compare the RMSE and MAE of the best model, the random forest, against all other models using a Diebold-Mariano test. The asterisks indicate the p-value of the tests.[7]

Apart from support vector regression (SVR), all machine learning models outperform the linear models on the whole sample. The inferior performance of SVR is not surprising as it does not minimize a squared error metric such as RMSE but a metric similar to MAE which is lower for SVR than for the linear models. In the 1990s and the periods before the global financial crisis, there are only small differences in performance between the models, with the neural network being the most accurate model. Only after the onset of the crisis does the random forest outperform the other models by a large and statistically significant margin.

Figure 1 shows the observed response variable and the predictions of the random forest, the linear regression, and the AR. The vertical dashed lines indicate the different time periods distinguished in Table 2. The predictions of the random forest are more volatile than that of the regression and the AR.[8] All models underestimate unemployment during the global financial crisis and overestimate it during the recovery. However, the random forest is least biased in those periods and forecasts high unemployment earliest during the crisis. This shows that its relatively high

---

[7]The horizon of the Diebold-Mariano test is set to 1 for all tests. Note, however, that the horizon of the AR model is 12 so that the p-values for this comparison are biased and thus reported in parentheses. Setting the horizon of the Diebold-Mariano test to 12, we do not observe significant differences between the RMSE of the random forest and AR.

[8]The mean absolute deviance from the models' mean prediction are 0.439, 0.356, and 0.207 for the random forest, regression, and AR, respectively.

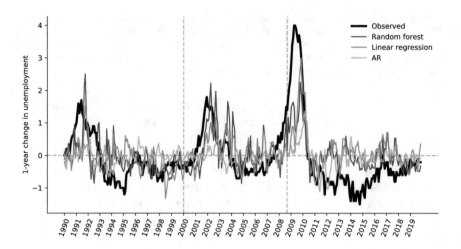

Fig. 1 Observed and predicted 1-year change in unemployment for the whole forecasting period comparing different models

forecast volatility can be useful in registering negative turning points. A similar observation can be made after the burst of the dotcom bubble in 2000. This points to an advantage of machine learning models associated with their greater flexibility incorporating new information as it arrives. This can be intuitively understood as adjusting model predictions locally, e.g., in regions (periods) of high unemployment, while a linear model needs to realign the full (global) model hyperplane.

3.2 Robustness Checks

We altered several parameters in our baseline setup to investigate their effects on the forecasting performance. The results are shown in Table 3. The RMSE of alternative specifications is again divided by the RMSE of the random forest in the baseline setup for a clearer comparison.

- **Window size.** In the baseline setup, the training set grows over time (expanding window). This can potentially improve the performance over time as more observations may facilitate a better approximation of the true data generating process. On the other hand, it may also make the model sluggish and prevent quick adaptation to structural changes. We test sliding windows of 60, 120, and 240 months. Only the simplest model, linear regression with only a lagged response, profits from a short horizon; the remaining models perform best with the biggest possible training set. This is not surprising for machine learning models, as they can "memorize" different sets of information through the incorporation of multiple specification in the same model. For instance, different

Table 3 Performance for different parameter specifications. The shown metric is RMSE divided by the RMSE of the random forest in the baseline setup

	Random forest	Neural network	Linear regression	SVR	AR	Linear regression (lagged response)
Training set size (in months)						
Max (baseline)	1.000	1.049	1.082	1.105	1.160	1.226
60	1.487	1.497	1.708	1.589	2.935	1.751
120	1.183	1.163	1.184	1.248	1.568	1.257
240	1.070	1.051	1.087	1.106	1.304	1.198
Change horizon (in months)						
3 (baseline)	1.000	1.049	1.082	1.105	1.160	1.226
1	1.077	1.083	1.128	1.148	–	–
6	1.043	1.111	1.142	1.162	–	–
9	1.216	1.321	1.251	1.344	–	–
12	1.345	1.278	1.336	1.365	–	–
Bootstrap aggregation						
No	1.000	1.179	1.089	1.117	1.160	1.226
100 models	–	1.049	1.082	1.105	–	–

paths down a tree model, or different trees in a forest, are all different submodels, e.g., characterizing different time periods in our setting. By contrast, a simple linear model cannot adjust in this way and needs to fit the best hyperplane to the current situation, explaining its improved performance for some fixed window sizes.

- **Change horizon.** In the baseline setup, we use a horizon of 3 months, when calculating changes, log changes, and second-order log changes of the predictors (see Table 1). Testing the horizons of 1, 6, 9, and 12 months, we find that 3 months generally leads to the best performance of all full information models. This is useful from a practical point of view, as quarterly changes are one of the main horizons considered for short-term economic projections.

- **Bootstrap aggregation (bagging).** The linear regression, neural network, and SVR all benefit from averaging the prediction of 100 bootstrapped models. The intuition is that our relatively small dataset likely leads to models with high variance, i.e., overfitting. The bootstrap aggregation of models reduces the models' variance and the degree of overfitting. Note that we do not expect much improvement for bagged linear models, as different draws from the training set are likely to lead to similar slope parameters resulting in almost identical models. This is confirmed by the almost identical performance of the single and bagged model.

4   Model Interpretability

4.1   Methodology

We saw in the last section that machine learning models outperform conventional linear approaches in a comprehensive economic forecasting exercise. Improved model accuracy is often the principal reason for applying machine learning models to a problem. However, especially in situations where model results are used to inform decisions, it is crucial to both understand and clearly communicate modelling results. This brings us to a second step when using machine learning models— explaining them.

Here, we introduce and compare two different methods for interpreting machine learning forecasting models *permutation importance* [7, 18] and *Shapley values and regressions* [44, 28, 24]. Both approaches are *model-agnostic*, meaning that they can be applied to *any* model, unlike other approaches, such as Gini impurity [25, 19], which are only compatible with specific machine learning methods. Both methods allow us to understand the relative importance of model features. For permutation importance, variable attribution is at the global level while Shapley values are constructed locally, i.e., for each single prediction. We note that both importance measures require column-wise independence of the features, i.e., contemporaneous independence in our forecasting experiments, an assumption that will not hold under all contexts.[9]

4.1.1   Permutation Importance

The permutation importance of a variable measures the change of model perfor- mance when the values of that variable are randomly scrambled. Scrambling or permuting a variable's values can either be done within a particular sample or by swapping values between samples. If a model has learnt a strong dependency between the model outcome and a given variable, scrambling the value of the variable leads to very different model predictions and thus affects performance. A variable k is said to be important in a model, if the test error e after scrambling feature k is substantially higher than the test error when using the original value for k, i.e., $e_k^{perm} >> e$. Clearly, the value of the permutation error e_k^{perm} depends on the realization of the permutation, and variation in its value can be large, particularly in small datasets. Therefore, it is recommended to average e_k^{perm} over several random draws for more accurate estimates and to assess sampling variability.[10]

---

[9]Lundberg et al. [29] proposed TREESHAP, which correctly estimates the Shapley values when features are dependent for tree models only.

[10]Considering a test set of size m with each observation having a unique value, there are $m!$ permutations to consider for an exhaustive evaluation, which is intractable to compute for larger m.

The following procedure estimates the permutation importance.

1. For each feature x_k:

   (a) Generate a permutation sample x_k^{perm} with the values of x_k permuted across observations (or swapped between samples).
   (b) Reevaluate the test score for x_k^{perm}, resulting in e_k^{perm}.
   (c) The permutation importance of x_k is given by $I(x_k) = e_k^{perm}/e$.[11]
   (d) Repeat and average over Q iterations and average $I_k = 1/Q \sum_q I_q(x_k)$.

2. If I_q is given by the ratio of errors, consider the normalized quantity $\bar{I}_k = (I_k - 1)\sum_k (I_k - 1) \in (0, 1)$.[12]
3. Sort features by I_k (or, $\bar{I}_k$).

Permutation importance is an intuitive measure that is relatively cheap to compute, requiring only new predictions generated on the permuted data and not model retraining. However, this ease of use comes at some cost. First, and foremost, permutation importance is *inconsistent*. For example, if two features contain similar information, permuting either of them will not reflect the actual importance of this feature relative to all other features in the model. Only permuting both or excluding one would do so. This situation is accounted for by Shapley values because they identify the individual marginal effect of a feature, accounting for its interaction with all other features. Additionally, the computation of permutation importance necessitates access to true outcome values and in many situations, e.g., when working with models trained on sensitive or confidential data, these may not be available. As a global measure, permutation importance only explains *which* variables are important but not *how* they contribute to the model, i.e., we cannot uncover the functional form or even the direction of the association between features and outcome that was learned by the model.

4.1.2 Shapley Values and Regressions

Shapley values originate from game theory [39] as a general solution to the problem of attributing a payoff obtained in a cooperative game to the individual players based on their contribution to the game. Štrumbelj and Kononenko [44] introduced the analogy between players in a cooperative game and variables in a general supervised model, where variables jointly generate a prediction, the payoff. The calculation is analogous in both cases (see also [24]),

$$\Phi^S\left[f(x_i)\right] \equiv \phi_0^S + \sum_{k=1}^n \phi_k^S(x_i) = f(x_i), \qquad (1)$$

---

[11] Alternatively, the difference $e_j^{perm} - e$ can be considered.
[12] Note, $I_k \geq 1$ in general. If not, there may be problems with model optimization.

$$\phi_k^S(x_i; f) = \sum_{x' \subseteq \mathcal{C}(x)\setminus\{k\}} \frac{|x'|!(n - |x'| - 1)!}{n!} \left[f(x_i|x' \cup \{k\}) - f(x_i|x') \right], \quad (2)$$

$$= \sum_{x' \subseteq \mathcal{C}(x)\setminus\{k\}} \omega_{x'} \left[\mathbb{E}_b[f(x_i)|x' \cup \{k\}] - \mathbb{E}_b[f(x_i)|x'] \right], \quad (3)$$

$$\text{with} \quad \mathbb{E}_b[f(x_i)|x'] \equiv \int f(x_i)\, db(\bar{x}') = \frac{1}{|b|} \sum_b f(x_i|\bar{x}').$$

Equation 1 states that the Shapley decomposition $\Phi^S[f(x_i)]$ of model f is local at x_i and exact, i.e., it precisely adds up to the actually predicted value $f(x_i)$. In Eq. 2, $\mathcal{C}(x) \setminus \{k\}$ is the set of all possible variable combinations (coalitions) of $n - 1$ variables when excluding the kth variable. $|x'|$ denotes the number of variables included in that coalition, $\omega_{x'} \equiv |x'|!(n - |x'| - 1)!/n!$ is a combinatorial weighting factor summing to one over all possible coalition, b is a background dataset, and $\bar{x}'$ stands for the set of variables not included in x'.

Equation 2 is the weighted sum of marginal contributions of variable k accounting for the number of possible variable coalitions.[13] In a general model, it is usually not possible to put an arbitrary feature to missing, i.e., exclude it. Instead, the contributions from features not included in x' are integrated out over a suitable background dataset, where $\{x_i|\bar{x}'\}$ is the set of points with variables not in x' being replaced by values in b. The background provides an informative reference point by determining the intercept ϕ_0^S. A reasonable choice is the training dataset incorporating all information the model has learned from.

An obvious disadvantage of Shapley values compared to permutation importance is the considerably higher complexity of their calculation. Given the factorial in Eq. 2, an exhaustive calculation is generally not feasible with larger feature sets. This can be addressed by either sampling from the space of coalitions or by setting all "not important" variables to "others," i.e., treating them as single variables. This substantially reduces the number of elements in $\mathcal{C}(x)$.

Nevertheless, these computational costs come with significant advantages. Shapley values are the only feature attribution method which is model independent, local, accurate, linear, and consistent [28]. This means that it delivers a granular high-fidelity approach for assessing the contribution and importance of variables. By comparing the local attributions of a variable across all observations we can visualize the functional form learned by the model! For instance, we might see that observations with a high (low) value on the variable have a disproportionally high (low) Shapley value on that variable, indicating a positive nonlinear functional form.

---

[13]For example, assuming we have three players (variables) $\{A, B, C\}$, the Shapley value of player C would be $\phi_C^S(f) = 1/3[f(\{A, B, C\}) - f(\{A, B\})] + 1/6[f(\{A, C\}) - f(\{A\})] + 1/6[f(\{B, C\}) - f(\{B\})] + 1/3[f(\{C\}) - f(\{\emptyset\})]$.

Based on these properties, which are directly inherited from the game theoretic origins of Shapley values, we can formulate an inference framework using Eq. 1. Namely, the *Shapley regression* [24],

$$y_i = \sum_{k=0}^{n} \phi_k^S(f, x_i) \beta_k^S + \hat{\epsilon}_i \equiv \Phi_i^S \beta^S + \hat{\epsilon}_i, \qquad (4)$$

where $k = 0$ corresponds to the intercept and $\hat{\epsilon}_i \sim \mathcal{N}(0, \sigma_\epsilon^2)$. The surrogate coefficients β_k^S are tested against the null hypothesis

$$\mathcal{H}_0^k(\Omega) : \{\beta_k^S \leq 0 \,|\, \Omega\}, \qquad (5)$$

with $\Omega \in \mathbb{R}^n$ (a region of) the model input space. The intuition behind this approach is to test the alignment of Shapley components with the target variable. This is analogous to a linear model where we use "raw" feature values rather than their associated Shapley attributions. A key difference to the linear case is the regional dependence on Ω. We only make *local* statements about the significance of variable contributions, i.e., on those regions where it is tested against $\mathcal{H}_0$. This is appropriate in the context of potential nonlinearity, where the model plane in the original input-target space may be curved, unlike that of a linear model. Note that the Shapley value decomposition (Eqs. 1–3) absorbs the signs of variable attributions, such that only positive coefficient values indicate significance. When negative values occur, it indicates that a model has poorly learned from a variable and $\mathcal{H}_0$ cannot be rejected.

The coefficients β^S are only informative about variable alignment (the strength of association between the output variable and feature of interest), not the magnitude of importance of a variable. Both together can be summarized by *Shapley share coefficients*,

$$\Gamma_k^S(f, \Omega) \equiv \left[sign(\beta_k^{lin}) \left\langle \frac{|\phi_k^S(f)|}{\sum_{l=1}^{n} |\phi_l^S(f)|} \right\rangle_{\Omega} \right]^{(*)} \in [-1, 1], \qquad (6)$$

$$\stackrel{f(x)=x\beta}{=} \beta_k^{(*)} \left\langle \frac{|(x_k - \langle x_k \rangle)|}{\sum_{l=1}^{n} |\beta_k(x_l - \langle x_l \rangle)|} \right\rangle_{\Omega}, \qquad (7)$$

where $\langle \cdot \rangle_{\Omega}$ stands for the average over x_k in $\Omega_k \in \mathbb{R}$. The Shapley share coefficient $\Gamma_k^S(f, \Omega)$ is a summary statistic for the contribution of x_k to the model over a region $\Omega \subset \mathbb{R}^n$ for modelling y.

It consists of three parts. The first is the sign, which is the sign of the corresponding linear model. The motivation for this is to indicate the direction of alignment of a variable with the target y. The second part is coefficient size. It is defined as the fraction of absolute variable attribution allotted to x_k across Ω. The sum of the absolute value of Shapley share coefficients is one by construction.[14] It measures how much of the model output is explained by x_k. The last component is the significance level, indicated by the star notation ($*$), and refers to the standard notation used in regression analysis to indicate the certainty with which we can reject the null hypothesis (Eq. 5). This indicates the confidence one can have in information derived from variable x_k measured by the strength of alignment of the corresponding Shapley components and the target, which is the same as its interpretation in a conventional regression analysis.

Equation 7 provides the explicit form for the linear model, where an analytical form exists. The only difference to the conventional regression case is the normalizing factor.

4.2 Results

We explain the predictions of the machine learning models and the linear regression as calibrated in the baseline setup of our forecasting. Our focus is largely on explaining forecast predictions in a pseudo-real-world setting where the model is trained on earlier observations that predate the predictions. However, in some cases it can be instructive to explain the predictions of a model that was trained on observations across the whole time period. For that, we use fivefold block cross-validation [3, 42].[15] This cross-validation analysis is subject to look-ahead bias, as we use future data to predict the past, but it allows us to evaluate a model for the whole time series.

4.2.1 Feature Importance

Figure 2 shows the global variable importance based on the analysis of the forecasting predictions. It compares Shapley shares $|\Gamma^S|$ (left panel) with permutation importance $\bar{I}$ (middle panel). The variables are sorted by the Shapley shares of the best-performing model, the random forest. Vertical lines connect the lowest and highest share across models for each feature as a measure for disagreement between models.

The two importance measures only roughly agree in their ranking of feature importance. For instance, using a random forest model, past unemployment seems to be a key indicator according to permutation importance but relatively less crucial

---

[14]The normalization is not needed in binary classification problems where the model output is a probability. Here, the a Shapley contribution relative to a base rate can be interpreted as the expected change in probability due to that variable.

[15]The time series is partitioned in five blocks of consecutive points in time and each block is once used as the test set.

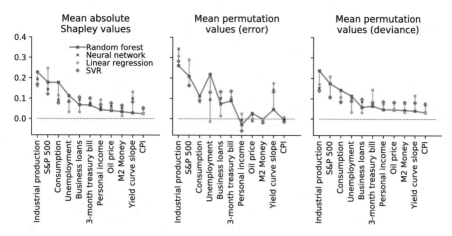

Fig. 2 Variable importance according to different measures. The left panel shows the importance according to the Shapley shares and the middle panel shows the variable importance according to permutation importance. The right panel shows an altered metric of permutation importance that measures the effect of permutation on the predicted value

according to Shapley calculations. Permutation importance is based on model forecasting error and so is a measure of a feature's predictive power (how much does its inclusion in a model improve predictive accuracy) and it is influenced by how the relationship between outcome and features may change over time. In contrast, Shapley values indicate which variables influence a predicted value, independent of predictive accuracy. The right panel of Fig. 2 shows an altered measure of permutation importance. Instead of measuring the change in the error due to permutations, we measure the change in the predicted value.[16] We see that this importance measure is more closely aligned with Shapley values. Furthermore, when we evaluate permutation importance using predictions based on block cross-validation, we find a strong alignment with Shapley values as the relationship between variables is not affected by the change between the training and test set (not shown).

Figure 3 plots Shapley values attributed to the S&P500 (vertical axis) against its input values (horizontal axis) for the random forest (left panel) and the linear regression (right panel) based on the block cross-validation analysis.[17] Each point reflects one of the observations between 1990 and 2019 and their respective value

---

[16]This metric computes the mean absolute difference between the observed predicted values and the predicted values after permuting feature $k : \frac{1}{m} \sum_{i=1}^{m} |\hat{y}_i - \hat{y}_{i(k)}^{perm}|$. The higher this difference, the higher the importance of the feature k (see [26, 36] for similar approaches to measure variable importance).

[17]Showing the Shapley values based on the forecasting predictions makes it difficult to disentangle whether nonlinear patterns are due to a nonlinear functional form or to (slow) changes of the functional form over time.

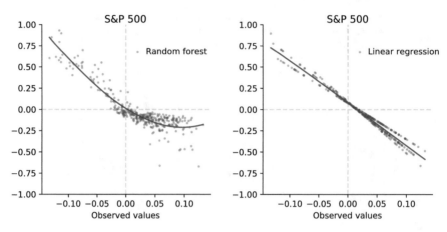

Fig. 3 Functional form learned by the random forest (left panel) and linear regression. The gray line shows a 3-degree polynomial fitted to the data. The Shapley values shown here are computed based on fivefold block cross-validation and are therefore subject to look-ahead bias

on the S&P500 variable. The approximate functional forms learned by both models are traced out by best-fit degree-3 polynomials. The linear regression learns a steep negative slope, i.e., higher stock market values are associated with lower unemployment 1 year down the road. This makes economic sense. However, we can make more nuanced observations for the random forest. There is satiation for high market valuations, i.e., changes beyond a certain point do not provide greater information for changes in unemployment.[18] A linear model is not able to reflect those nuances, while machine learning models provide a more detailed signal from the stock market and other variables.

4.2.2 Shapley Regressions

Shapley value-based inference allows to communicate machine learning models analogously to a linear regression analysis. The difference between the coefficients of a linear model and Shapley share coefficients is primarily the normalization of the latter. The reason for this is that nonlinear models do not have a "natural scale," for instance, to measure variation. We summarize the Shapley regression on the forecasting predictions (1990–2019) of the random forest and linear regression in Table 4.

The coefficients β^S measure the alignment of a variable with the target. Values close to one indicate perfect alignment and convergence of the learning process. Values larger than one indicate that a model underestimates the effect of a variable on the outcome. And the opposite is the case for values smaller than one. This

---

[18]Similar nonlinearities are learned by the SVR and the neural network.

Table 4 Shapley regression of random forest (left) and linear regression (right) for forecasting predictions between 1990–2019. Significance levels: $^*p < 0.1$; $^{**}p < 0.05$; $^{***}p < 0.01$

	Random forest			Linear regression		
	β^S	p-value	Γ^S	β^S	p-value	Γ^S
Industrial production	0.626	0.000	-0.228^{***}	0.782	0.000	-0.163^{***}
S&P 500	0.671	0.000	-0.177^{***}	0.622	0.000	-0.251^{***}
Consumption	1.314	0.000	-0.177^{***}	2.004	0.000	-0.115^{***}
Unemployment	1.394	0.000	$+0.112^{***}$	2.600	0.010	$+0.033^{***}$
Business loans	2.195	0.000	-0.068^{***}	2.371	0.024	-0.031^{**}
3-month treasury bill	1.451	0.008	-0.066^{***}	-1.579	1.000	-0.102
Personal income	-0.320	0.749	$+0.044$	-0.244	0.730	$+0.089$
Oil price	1.589	0.018	-0.040^{**}	-0.246	0.624	-0.052
M2 Money	0.168	0.363	-0.034	-4.961	0.951	-0.011
Yield curve slope	1.952	0.055	$+0.029^*$	0.255	0.171	$+0.132$
CPI	0.245	0.419	-0.024	-0.790	0.673	-0.022

can intuitively be understood from the model hyperplane of the Shapley regression either tilting more towards a Shapley component from a variable (underestimation, $\beta_k^S > 1$) or away from it (overestimation, $\beta_k^S < 1$). Significance decreases as the β_k^S approaches zero.[19]

Variables with lower p-values usually have higher Shapley shares $|\Gamma^S|$, which are equivalent to those shown in Fig. 2. This is intuitive as the model learns to rely more on features which are important for predicting the target. However this does not hold by construction. Especially in the forecasting setting where the relationships of variables change over time, the statistical significance may disappear in the test set, even for features with high shares.

In the Shapley regression, more variables are statistically significant for the random forest than for the linear regression model. This is expected, because the forest, like other machine learning models, can exploit nonlinear relationships that the regression cannot account for (as in Fig. 3), i.e., it is a more flexible model. These are then reflected in localized Shapley values providing a stronger, i.e., more significant, signal in the regression stage.

5 Conclusion

This chapter provided a comparative study of how machine learning models can be used for macroeconomic forecasting relative to standard econometric approaches. We find significantly better performance of machine learning models for forecasting

---

[19]The underlying technical details for this interpretation are provided in [24].

changes in US unemployment at a 1-year horizon, particularly in the period after the global financial crisis of 2008.

Apart from model performance, we provide an extensive explanation of model predictions, where we present two approaches that allow for greater machine learning interpretability—permutation feature importance and Shapley values. Both methods demonstrate that a range of machine learning models learn comparable signals from the data. By decomposing individual predictions into Shapley value attributions, we extract learned functional forms that allow us to visually demonstrate how the superior performance of machine learning models is explained by their enhanced ability to adapt to individual variable-specific nonlinearities. Our example allows for a more nuanced economic interpretation of learned dependencies compared to the interpretation offered by a linear model. The Shapley regression framework, which enables conventional parametric inference on machine learning models, allows us to communicate the results of machine learning models analogously to traditional presentations of regression results.

Nevertheless, as with conventional linear models, the interpretation of our results is not fixed. We observe some variation under different models, different model specifications, and the interpretability method chosen. This is in part due to small sample limitations; this modelling issue is common, but likely more aggravated when using machine learning models due to their nonparametric structure.

However, we believe that the methodology and results presented justify the use of machine learning models and such explainability methods to inform decisions in a policy-making context. The inherent advantages of their nonlinearity over conventional models are most evident in a situation where the underlying data-generating process is unknown and expected to change over time, such as in a forecasting environment as presented in the case study here. Overall, the use of machine learning in conjunction with Shapley value-based inference as presented in this chapter may offer a better trade-off between maximizing predictive performance and statistical inference thereby narrowing the gap between Breiman's two cultures.

References

1. Akaike, H. (1974). A new look at the statistical model identification. *IEEE Transactions on Automatic Control, 19*(6), 716–723.
2. Athey, S., & Imbens, G. (2016). Recursive partitioning for heterogeneous causal effects. *Proceedings of the National Academy of Sciences, 113*(27), 7353–7360.
3. Bergmeir, C., & Benítez, J. M. (2012). On the use of cross-validation for time series predictor evaluation. *Information Sciences, 191*, 192–213.
4. Bianchi, D., Büchner, M., & Tamoni, A. (2019). Bond risk premia with machine learning. In *USC-INET Research Paper*, No. 19–11.
5. Bluwstein, K., Buckmann, M., Joseph, A., Kang, M., Kapadia, S., & Simsek, Ö. (2020). Credit growth, the yield curve and financial crisis prediction: evidence from a machine learning approach. In *Bank of England Staff Working Paper, No. 848*.
6. Bracke, P., Datta, A., Jung, C., & Sen, S. (2019). Machine learning explainability in finance: an application to default risk analysis. In *Bank of England Staff Working Paper, No. 816*.
7. Breiman, L. (2001). Random forests. *Machine Learning, 45*(1), 5–32.

8. Breiman, L. (2001). Statistical modeling: The two cultures (with comments and a rejoinder by the author). *Statistical Science, 16*(3), 199–231.

9. Chen, J. C., Dunn, A., Hood, K. K., Driessen, A., & Batch, A. (2019). Off to the races: A comparison of machine learning and alternative data for predicting economic indicators. In *Big Data for 21st Century Economic Statistics*. Chicago: National Bureau of Economic Research, University of Chicago Press. Available at: http://www.nber.org/chapters/c14268.pdf

10. Chernozhukov, V., Chetverikov, D., Demirer, M., Duflo, E., Hansen, C., Newey, W., et al. (2018). Double/debiased machine learning for treatment and structural parameters. *The Econometrics Journal, 21*(1), C1–C68.

11. Chernozhukov, V., Demirer, M., Duflo, E., & Fernandez-Val, I. (2018). Generic machine learning inference on heterogenous treatment effects in randomized experiments. In *NBER Working Paper Series, No. 24678*.

12. Conneau, A., & Lample, G. (2019). Cross-lingual language model pretraining. In *Advances in Neural Information Processing Systems, NIPS 2019* (Vol. 32, pp. 7059–7069). Available at: https://proceedings.neurips.cc/paper/2019/file/c04c19c2c2474dbf5f7ac4372c5b9af1-Paper.pdf

13. Coulombe, P. G., Leroux, M., Stevanovic, D., & Surprenant, S. (2019). How is machine learning useful for macroeconomic forecasting. In *CIRANO Working Papers 2019s-22*. Available at: https://ideas.repec.org/p/cir/cirwor/2019s-22.html

14. Crawford, K. (2013). The hidden biases of big data. *Harvard Business Review*, art number H00ADR-PDF-ENG. Available at: https://hbr.org/2013/04/the-hidden-biases-in-big-data

15. Döpke, J., Fritsche, U., & Pierdzioch, C. (2017). Predicting recessions with boosted regression trees. *International Journal of Forecasting, 33*(4), 745–759.

16. Drucker, H., Burges, C. J. C., Kaufman, L., Smola, A. J., & Vapnik, V. (1997). Support vector regression machines. In *Advances in Neural Information Processing Systems, NIPS 2016* (Vol. 9, pp. 155–161). Available at: https://papers.nips.cc/paper/1996/file/d38901788c533e8286cb6400b40b386d-Paper.pdf

17. European Union. (2016). Regulation (EU) 2016/679 of the European Parliament, Directive 95/46/EC (General Data Protection Regulation). *Official Journal of the European Union, L119*, 1–88.

18. Fisher, A., Rudin, C., & Dominici, F. (2019). All models are wrong, but many are useful: Learning a variable's importance by studying an entire class of prediction models simultaneously. *Journal of Machine Learning Research, 20*(177), 1–81.

19. Friedman, J., Hastie, T., & Tibshirani, R. (2009). *The Elements of Statistical Learning. Springer Series in Statistics*. Berlin: Springer.

20. Fuster, A., Goldsmith-Pinkham, P., Ramadorai, T., & Walther, A. (2017). Predictably unequal? the effects of machine learning on credit markets. In *CEPR Discussion Papers* (No. 12448).

21. Giannone, D., Lenza, M., & Primiceri, G. E. (2017). Economic predictions with big data: The illusion of sparsity. In *CEPR Discussion Paper* (No. 12256).

22. Goodfellow, I., Bengio, Y., & Courville, A. (2016). *Deep learning*. Cambridge: MIT Press.

23. Henelius, A., Puolamäki, K., Boström, H., Asker, L., & Papapetrou, P. (2014). A peek into the black box: exploring classifiers by randomization. *Data Mining and Knowledge Discovery, 28*(5–6), 1503–1529.

24. Joseph, A. (2020). *Parametric inference with universal function approximators*, arXiv, CoRR abs/1903.04209

25. Kazemitabar, J., Amini, A., Bloniarz, A., & Talwalkar, A. S. (2017). Variable importance using decision trees. In *Advances in Neural Information Processing Systems, NIPS 2017* (Vol. 30, pp. 426–435). Available at: https://papers.nips.cc/paper/2017/file/5737c6ec2e0716f3d8a7a5c4e0de0d9a-Paper.pdf

26. Lemaire, V., Féraud, R., & Voisine, N. (2008). Contact personalization using a score understanding method. In *2008 IEEE International Joint Conference on Neural Networks (IEEE World Congress on Computational Intelligence)* (pp. 649–654).

27. Lipton, Z. C. (2016). *The mythos of model interpretability*, ArXiv, CoRR abs/1606.03490

28. Lundberg, S., & Lee, S.-I. (2017). A unified approach to interpreting model predictions. In *Advances in Neural Information Processing Systems, NIPS 2017* (Vol. 30, pp. 4765–4774). Available: https://papers.nips.cc/paper/2017/file/8a20a8621978632d76c43dfd28b67767-Paper.pdf

29. Lundberg, S., Erion, G., & Lee, S.-I. (2018). *Consistent individualized feature attribution for tree ensembles.* ArXiv, CoRR abs/1802.03888

30. McCracken, M. W., & Ng, S. (2016). FRED-MD: A monthly database for macroeconomic research. *Journal of Business & Economic Statistics, 34*(4), 574–589.

31. Medeiros, M. C., Vasconcelos, G. F. R., Veiga, Á., & Zilberman, E. (2019). Forecasting inflation in a data-rich environment: the benefits of machine learning methods. *Journal of Business & Economic Statistics, 39*(1), 98–119.

32. Miller, T. (2017). *Explanation in Artificial Intelligence: Insights from the Social Sciences.* ArXiv, CoRR abs/1706.07269

33. Racine, J. (2000). Consistent cross-validatory model-selection for dependent data: hv-block cross-validation. *Journal of Econometrics, 99*(1), 39–61.

34. Ribeiro, M., Singh, S., & Guestrin, C. (2016). "Why should I trust you?": Explaining the predictions of any classifier. In *Proceedings of the 22nd ACM SIGKDD* (pp. 1135–1134).

35. Ribeiro, M. T., Singh, S., & Guestrin, C. (2018). Anchors: High-precision model-agnostic explanations. In *Thirty-Second AAAI Conference on Artificial Intelligence, AAAI 2018* (pp. 1527–1535), art number 16982. Available at: https://www.aaai.org/ocs/index.php/AAAI/AAAI18/paper/view/16982

36. Robnik-Šikonja, M., & Kononenko, I. (2008). Explaining classifications for individual instances. *IEEE Transactions on Knowledge and Data Engineering, 20*(5), 589–600.

37. Schroff, F., Kalenichenko, D., & Philbin. J. (2015). Facenet: A unified embedding for face recognition and clustering. In *Proceedings of the IEEE Conference on Computer Vision and Pattern Recognition* (pp. 815–823).

38. Sermpinis, G., Stasinakis, C., Theofilatos, K., & Karathanasopoulos, A. (2014). Inflation and unemployment forecasting with genetic support vector regression. *Journal of Forecasting, 33*(6), 471–487.

39. Shapley, L. (1953). A value for n-person games. *Contributions to the Theory of Games, 2,* 307–317.

40. Shrikumar, A., Greenside, P., & Anshul, K. (2017). *Learning important features through propagating activation differences.* ArXiv, CoRR abs/1704.02685.

41. Silver, D., Hubert, T., Schrittwieser, J., Antonoglou, I., Lai, M., Guez, A., et al. (2018). A general reinforcement learning algorithm that masters chess, shogi, and go through self-play. *Science, 362*(6419), 1140–1144.

42. Snijders, T. A. B. (1988). On cross-validation for predictor evaluation in time series. In T. K. Dijkstra (Ed.), *On model uncertainty and its statistical implications, LNE* (Vol. 307, pp. 56–69). Berlin: Springer.

43. Stock, J. H., & Watson, M. W. (2002). Forecasting using principal components from a large number of predictors. *Journal of the American Statistical Association, 97*(460), 1167–1179.

44. Štrumbelj, E., & Kononenko, I. (2010). An efficient explanation of individual classifications using game theory. *Journal of Machine Learning Research, 11,* 1–18.

45. Štrumbelj, E., Kononenko, I., Robnik-Šikonja, M. (2009). Explaining instance classifications with interactions of subsets of feature values. *Data & Knowledge Engineering, 68*(10), 886–904.

46. Tibshirani, R. (1996). Regression shrinkage and selection via the lasso. *Journal of the Royal Statistical Society: Series B (Methodological), 58*(1), 267–288.

47. Wager, S., & Athey, S. (2018). Estimation and inference of heterogeneous treatment effects using random forests. *Journal of the American Statistical Association, 113*(523), 1228–1242.

Machine Learning for Financial Stability

Lucia Alessi and Roberto Savona

Abstract What we learned from the global financial crisis is that to get information about the underlying financial risk dynamics, we need to fully understand the complex, nonlinear, time-varying, and multidimensional nature of the data. A strand of literature has shown that machine learning approaches can make more accurate data-driven predictions than standard empirical models, thus providing more and more timely information about the building up of financial risks. Advanced machine learning techniques provide several advantages over empirical models traditionally used to monitor and predict financial developments. First, they are able to deal with high-dimensional datasets. Second, machine learning algorithms allow to deal with unbalanced datasets and retain all of the information available. Third, these methods are purely data driven. All of these characteristics contribute to their often better predictive performance. However, as "black box" models, they are still much underutilized in financial stability, a field where interpretability and accountability are crucial.

1 Introduction

What we learned from the global financial crisis is that to get information about the underlying financial risk dynamics, we need to fully understand the complex, nonlinear, time-varying, and multidimensional nature of the data. A strand of literature has shown that machine learning approaches can make more accurate data-driven predictions than standard empirical models, thus providing more and more timely information about the building up of financial risks.

L. Alessi (✉)
European Commission - Joint Research Centre, Ispra (VA), Italy
e-mail: lucia.alessi@ec.europa.eu

R. Savona
University of Brescia, Brescia, Italy
e-mail: roberto.savona@unibs.it

© The Author(s) 2021
S. Consoli et al. (eds.), *Data Science for Economics and Finance*,
https://doi.org/10.1007/978-3-030-66891-4_4

Advanced machine learning techniques provide several advantages over empirical models traditionally used to monitor and predict financial developments. First, they are able to deal with high-dimensional datasets, which is often the case in economics and finance. In fact, the information set of economic agents, be it central banks or financial market participants, comprises hundreds of indicators, which should ideally all be taken into account. Looking at the financial sphere more closely, as also mentioned by [25] and [9], banks should use, and are in fact using, advanced data technologies to ensure they are able to identify and address new sources of risks by processing large volumes of data. Financial supervisors should also use machine learning and advanced data analytics (so-called suptech) to increase their efficiency and effectiveness in dealing with large amounts of information. Second, and contrary to standard econometric models, machine learning algorithms allow to deal with unbalanced datasets, hence retaining all of the information available. In the era of big data, one might think that losing observations, i.e., information, is not anymore a capital sin as it used to be decades ago. Hence, one could afford cleaning the dataset from problematic observations to obtain, e.g., a balanced panel, given the large amount of observations one starts with. On the contrary, large datasets require even more flexible models, as they almost invariably feature large amounts of missing values or unpopulated fields, "ragged" edges, mixed frequencies or irregular periodic patterns, and all sorts of issues that standard techniques are not able to handle. Third, these methods are purely data-driven, as they do not require making ex ante crucial modelling choices. For example, standard econometric techniques require selecting a restricted number of variables, as the models cannot handle too many predictors. Factor models, which allow handling large datasets, still require the econometrician to set the number of the underlying driving forces. Another crucial assumption, often not emphasized, relates to the linearity of the relevant relations. While standard econometric models require the econometrician to explicitly control for nonlinearities and interactions, whose existence she should know or hypothesize a priori, machine learning methods are designed to address these types of dynamics directly. All of these characteristics contribute to their often better predictive performance.

Thanks to these characteristics, machine learning techniques are also more robust in handling the fitting versus forecasting trade-off, which is reminiscent of the so-called "forecasting versus policy dilemma" [21], to indicate the separation between models used for forecasting and models used for policymaking. Presumably, having a model that overfits in-sample when past data could be noisy leads to the retention of variables that are spuriously significant, which produces severe deficiencies in forecasting. The noise could also affect the dependent variable when the definition of "crisis event" is unclear or when, notwithstanding a clear and accepted definition of crisis, the event itself is misclassified due to a sort of noisy transmission of the informational set used to classify that event. Machine learning gives an opportunity to overcome this problem.

While offering several advantages, however, machine learning techniques also suffer from some shortcomings. The most important one, and probably the main reason why these models are still far from dominating in the economic and financial literature, is that they are "black box" models. Indeed, while the modeler can surely control inputs, and obtain generally accurate outputs, she is not really able to explain the reasons behind the specific result yielded by the algorithm. In this context, it becomes very difficult, if not impossible, to build a story that would help users make sense of the results. In economics and finance, however, this aspect is at least as important as the ability to make accurate predictions.

Machine learning approaches are used in several very diverse disciplines, from chemometrics to geology. With some years delay, the potential of data mining and machine learning is also becoming apparent in the economics and finance profession. Focusing on the financial stability literature, some papers have appeared in relatively recent years, which use machine learning techniques for an improved predictive performance. Indeed, one of the areas where machine learning techniques have been more successful in finance is the construction of early warning models and the prediction of financial crises. This chapter focuses on the two supervised machine learning approaches becoming increasingly popular in the finance profession, i.e., decision trees and sparse models, including regularization-based approaches. After explaining how these algorithms work, this chapter offers an overview of the literature using these models to predict financial crises.

The chapter is structured as follows. The next section presents an overview of the main machine learning approaches. Section 3 explains how decision tree ensembles work, describing the most popular approaches. Section 4 deals with sparse models, in particular the LASSO, as well as related alternatives, and the Bayesian approach. Section 5 discusses the use of machine learning as a tool for financial stability policy. Section 6 provides an overview of papers that have used these methods to assess the probability of financial crises. Section 7 concludes and offers suggestions for further research.

2   Overview of Machine Learning Approaches

Machine learning pertains to the algorithmic modeling culture [17], for which data predictions are assumed to be the output of a partly unknowable system, in which a set of variables act as inputs. The objective is to find a rule (algorithm) that operates on inputs in order to predict or classify units more effectively without any a priori belief about the relationships between variables. The common feature of machine learning approaches is that algorithms are realized to learn from data with minimal human intervention. The typical taxonomy used to categorize machine learning algorithms is based on their learning approach, and clusters them in supervised and unsupervised learning methods.[1]

---

[1] See [7] for details on this classification and a comprehensive discussion on the relevance of the recent machine learning literature for economics and econometrics.

Supervised machine learning focuses on the problem of predicting a response variable, y, given a set of predictors, x. The goal of such algorithms is to make good out-of-sample predictions, rather than estimating the structural relationship between y and x. Technically, these algorithms are based on the *cross-validation* procedure. This latter involves the repeated rotation of subsamples of the entire dataset, whereby the analysis is performed on one subsample (the *training set*), then the output is tested on the other subset(s) (the *test set*). Such a rotational estimation procedure is conceived with the aim of improving the out-of-sample predictability (accuracy), while avoiding problems of overfitting and selection bias, this one induced by the distortion resulting from collecting nonrandomized samples.

Supervised machine learning methods include the following classes of algorithms:

- Decision tree algorithms. Decision trees are decision methods based on the actual attributes observed in the explored dataset. The objective is to derive a series of rules of thumb, visualized in a tree structure, which drive from observations to conclusions expressed as predicted values/attributes. When the response variable is continuous, decision trees are named regression trees. When instead the response variable is categorical, we have classification trees. The most popular algorithm in this class is CART (Classification and Regression Trees) introduced in [18].[2] CART partitions the space of predictors x in a series of homogeneous and disjoint regions with respect to the response variable y, whose nature defines the tree as classification (when y is a categorical variable) or regression (when y is a continuous variable) tree.
- Ensemble algorithms. Tree ensembles are extensions of single trees based on the concept of prediction averaging. They aim at providing more accurate predictions than those obtained with a single tree. The most popular ensemble methods are the following: Boosting, Bootstrapped Aggregation (Bagging), AdaBoost, Gradient Boosting Machines (GBM), Gradient Boosted Regression Trees (GBRT). and Random Forest.
- Instance-based algorithms. These methods generate classification predictions using only specific instances and finding the best match through a similarity measure between subsamples. A list of the most used algorithms includes the following: k-Nearest Neighbor (kNN), Learning Vector Quantization (LVQ), Locally Weighted Learning (LWL), and Support Vector Machines (SVM). SVM, in particular, are particularly flexible as they are used both for classification and regression analysis. They are an extension of the vector classifier, which provides clusters of observations identified through partitioning the space via linear boundaries. In addition, SVM provide nonlinear classifications by mapping their inputs into high-dimensional feature spaces through nonlinear boundaries. In more technical terms, SVM are based on a hyperplane (or a set of hyperplanes— which in a two-dimensional space are just lines) in a high/infinite-dimensional

---

[2]Less popular decision trees algorithms are: Chi-squared Automatic Interaction Detection (CHAID), Iterative Dichotomiser 3 (ID3); C4.5 and C5.0.

space. Hyperplane(s) are used to partition the space into classes and are optimally defined by assessing distances between pairs of data points in different classes. These distances are based on a kernel, i.e., a similarity function over pairs of data points.

- Regularization algorithms. Regularization-based models offer alternative fitting procedures to the least square method, leading to better prediction ability. The standard linear model is commonly used to describe the relationship between the y and a set of $x_1, x_2, \ldots, x_p$ variables. Ridge regression, Least Absolute Shrinkage and Selection Operator (LASSO), and Elastic Net are all based on detecting the optimal constraint on parameter estimations in order to discard redundant covariates and select those variables that most contribute to better predict the dependent variable out-of-sample.
- Bayesian algorithms. These methods apply Bayes Theorem for both classification and regression problems. The most popular Bayesian algorithms are: Naive Bayes, Gaussian Naive Bayes, Multinomial Naive Bayes, Averaged One-Dependence Estimators (AODE), Bayesian Belief Network (BBN), and Bayesian Network (BN).
- Supervised Artificial Neural Networks. Artificial neural networks (ANN) are models conceived to mimic the learning mechanisms of the human brain—specifically, supervised ANN run by receiving inputs, which activate "neurons" and ultimately lead to an output. The error between the estimation output and the target is used to adjust the weights used to connect the neurons, hence minimizing the estimation error.

Unsupervised machine learning applies in contexts where we explore only x without having a response variable. The goal of this type of algorithm is to understand the inner structure of x, in terms of relationships between variables, homogeneous clustering, and dimensional reduction. The approach involves pattern recognition using all available variables, with the aim of identifying intrinsic groupings, and subsequently assigning a label to each data point. Unsupervised machine learning includes clusters and networks.

The first class of algorithms pertains to clustering, in which the goal is, given a set of observations on features, to partition the feature space into homogeneous/natural subspaces. Cluster detection is useful when we wish to estimate parsimonious models conditional to homogeneous subspaces, or simply when the goal is to detect natural clusters based on the joint distribution of the covariates.

Networks are the second major class of unsupervised approaches, where the goal is to estimate the joint distribution of the x variables. Network approaches can be split in two subcategories: traditional networks and Unsupervised Artificial Neural Networks (U-ANN). Networks are a flexible approach that gained popularity in complex settings, where extremely large number of features have to be disentangled and connected in order to understand inner links and time/spatial dynamics. Finally, Unsupervised Artificial Neural Networks (U-ANN) are used when dealing with unlabeled data sets. Different from Supervised Artificial Neural Networks, here the objective is to find patterns in the data and build a new model based on a smaller set of relevant features, which can represent well enough the information in the data.[3] Self-Organizing Maps (SOM), e.g., are a popular U-ANN-based approach which provides a topographic organization of the data, with nearby locations in the map representing inputs with similar properties.

3 Tree Ensembles

This section provides a brief overview of the main tree ensemble techniques, starting from the basics, i.e., the construction of an individual decision tree. We start from CART, originally proposed by [18]. This seminal paper has spurred a literature reaching increasingly high levels of complexity and accuracy: among the most used ensemble approaches, one can cite as examples bootstrap aggregation (Bagging, [15]), boosting methods such as Adaptive Boosting (AdaBoost, [29]), Gradient Boosting [30], and [31], Multiple Additive Regression Trees (MART, [32]), as well as Random Forest [16].[4] Note, however, that some of the ensemble methods we describe below are not limited to CART and can be used in a general classification and regression context.

We only present the most well-known algorithms, as the aim of this section is not to provide a comprehensive overview of the relevant statistical literature. Indeed, many other statistical techniques have been proposed in the literature, that are similar to the ones we describe, and improve over the original proposed models in some respects. The objective of this section is to explain the main ideas at the root of the methods, in nontechnical terms.

Tree ensemble algorithms are generally characterized by a very good predictive accuracy, often better than that of the most widely used regression models in economics and finance, and contrary to the latter, are very flexible in handling problematic datasets. However, the main issue with tree ensemble learning models is that they are perceived as black boxes. As a matter of fact, it is ultimately not possible to explain what a particular result is due to. To make a comparison with a popular model in economics and finance, while in regression analysis one knows the contribution of each regressor to the predicted value, in tree ensembles one is not able to map a particular predicted value to one or more key determinants. In policymaking, this is often seen as a serious drawback.

---

[3]On supervised and unsupervised neural networks see [57].
[4]See [56] for a review of the relevant literature.

Fig. 1 Example of binary
tree structure

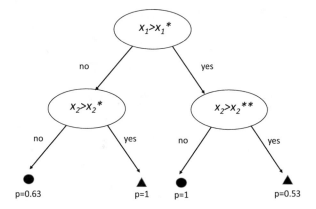

3.1 Decision Trees

Decision trees are nonparametric models constructed by recursively partitioning a
dataset through its predictor variables, with the objective of optimally predicting a
response variable. The response variable can be continuous (for regression trees)
or categorical (for classification trees). The output of the predictive model is a tree
structure like the one shown in Fig. 1. CART are binary trees, with one *root node*,
only two branches departing from each *parent node*, each entering into a *child node*,
and multiple *terminal nodes* (or "leaves"). There can also be nonbinary decision
trees, where more than two branches can attach to a node, as, e.g., those based on
Chi-square automatic interaction detection (CHAID, [43]). The tree in Fig. 1 has
been developed to classify observations, which can be circles, triangles, or squares.
The classification is based on two features, or predictors, x_1 and x_2. In order to
classify an observation, starting from the root node, one needs to check whether the
value of feature x_1 for this observation is higher or lower than a particular threshold
x^*. Next, the value of feature x_2 becomes relevant.[5] Based on this, the tree will
eventually classify the observation as either a circle or a triangle. In the case of the
tree in Fig. 1, for some terminal nodes the probability attached to the outcome is
100%, while for some other terminal nodes, it is lower. Notice that this simple tree
is not able to correctly classify squares, as a much deeper tree would be needed for
that. In other words, more splits would be needed to identify a partition of the space
where observations are more likely to be squares than anything else. The reason will
become clearer looking at the way the tree is grown.

Figure 2 explains how the tree has been constructed, starting from a sample
of circles, triangles, and squares. For each predictor, the procedure starts by
considering all the possible binary splits obtained from the sample as a whole.
In our example, where we only have two predictors, this implies considering all

---

[5]Notice that this is not necessarily the case, as the same variable can be relevant in the tree at
consecutive nodes.

Fig. 2 Recursive partitioning

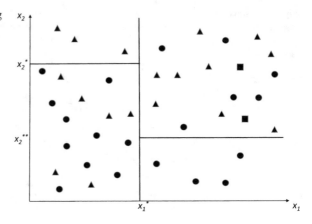

possible values for x_1 and x_2. For each possible split, the relevant impurity measure of the child nodes is calculated. The impurity of a node can be measured by the Mean Squared Error (MSE), in the case of regression trees, or the Gini index, for classification trees, or information entropy. In our case, the impurity measure will be based on the number of circles and triangles in each subspace associated with each split. The best split is the value for a specific predictor, which attains the maximum reduction in node impurity. In other words, the algorithm selects the predictor and the associated threshold value which split the sample into the two purest subsamples. In the case of classification trees, e.g., the aim is to obtain child nodes which ideally only contain observations belonging to one class, in which case the Gini index corresponds to zero. Looking at Fig. 2, the first best split corresponds to the threshold value x_1^*. Looking at the two subspaces identified by this split, the best split for $x_1 < x_1^*$ is x_2^*, which identifies a pure node for $x_2 > x_2^*$. The best split for $x_1 > x_1^*$ is x_2^{**}, which identifies a pure node for $x_2 < x_2^{**}$. The procedure is run for each predictor at each split and could theoretically continue until each terminal node is pure. However, to avoid overfitting, normally a stopping rule is imposed, which, e.g., requires a minimum size for terminal nodes. Alternatively, one can ex post "prune" large trees, by iteratively merging two adjoining terminal nodes.[6]

Decision trees are powerful algorithms that present many advantages. For example, in terms of data preparation, one does not need to clean the dataset from missing values or outliers, as they are both handled by the algorithm, nor does one need to normalize the data. Moreover, once the tree structure is built, the model output can be operationalized also by the nontechnical user, who will simply need to assess her observation of interest against the tree. However, they also suffer from one major shortcoming, i.e., the tree structure is often not robust to small variations in the data. This is due to the fact that the tree algorithm is recursive, hence a different split at any level of the structure is likely to yield different splits at any lower level. In

---

[6]See [38] for technical details, including specific model choice rules.

extreme cases, even a small change in the value of one predictor for one observation could generate a different split.

3.2 Random Forest

Tree ensembles have been proposed to improve the robustness of predictions realized through single models. They are collections of single trees, each one grown on a subsample of observations. In particular, tree ensembles involve drawing subsets of are collections of regression trees, where each tree is grown on a subsample of observations. In particular, tree ensembles involve drawing subsets of observations with replacement, i.e., Bootstrapping and Aggregating (also referred to as "BAGGING") the predictions from a multitude of trees. The Random Forest [16] is one of the most popular ensemble learning inference procedures. The Random Forest algorithm involves the following steps:

1. Selecting a random subsample of the observations[7]
2. Selecting a random small subset of the predictors[8]
3. Growing a single tree based on this restricted sample
4. Repeating the first two steps thousands of times[9]

Predictions for out-of-sample observations are based on the predictions from all the trees in the Forest.

On top of yielding a good predictive performance, the Random Forest allows to identify the key predictors. To do so, the predictive performance of each tree in the ensemble needs to be measured. This is done based on how good each tree is at correctly classifying or estimating the data that are not used to grow it, namely the so-called out-of-bag (OOB) observations. In practice, it implies computing the MSE or the Gini impurity index for each tree. To assess the importance of a predictor variable, one has to look at how it impacts on predictions in terms of MSE or Gini index reduction. To do so, one needs to check whether the predictive performance worsens by randomly permuting the values of a specific predictor in the OOB data. If the predictor does not bring a significant contribution in predicting the outcome variable, it should not make a difference if its values are randomly permuted before the predictions are generated. Hence, one can derive the importance of each predictor by checking to what extent the impurity measure worsens.

---

[7]It is common practice to use 60% of the total observations (see [38]).

[8]The number of the selected predictors is generally around to the square root of the total number of predictors, while [16] tests with one variable at a time and with a number of features equal to the first integer less than $log2M + 1$, where M is the total number of features.

[9]The accuracy of the Random Forest algorithm is heuristically proven to converge with around 3000 trees (see [38]).

3.3 Tree Boosting

Another approach to the construction of Tree ensembles is Tree Boosting. Boosting means creating a strong prediction model based on a multitude of weak prediction models, which could be, e.g., CARTs. Adaptive Boosting (AdaBoost, [29]) is one of the first and the most popular boosting methods, used for classification problems. It is called *adaptive* because the trees are built iteratively, with each consecutive tree increasing the predictive accuracy over the previous one. The simplest AdaBoost algorithm works as follows:

1. Start with growing a tree with just one split.[10]
2. Consider the misclassified observations and assign them a higher weight compared to the correctly classified ones.
3. Grow another tree on all the (weighted) observations.
4. Update the weights.
5. Repeat 3–4 until overfitting starts to occur. This will be reflected in a worsening of the predictive performance of the tree ensemble on out-of-sample data.

Later, the Gradient Boosting algorithm was proposed as a generalization of AdaBoost [30]. The simplest example would involve the following steps:

1. Grow a regression tree with a few attributes.[11]
2. Compute the prediction residuals from this tree and the resulting mean squared error.[12]
3. On the residuals, grow another tree which optimizes the mean squared error.
4. Repeat 1–3 until overfitting starts to occur.

To avoid overfitting, it has been proposed to include an element of randomness. In particular, in Stochastic Gradient Boosting [31], each consecutive simple tree is grown on the residuals from the previous trees, but based only on a subset of the full data set. In practice, each tree is built on a different subsample, similarly to the Random Forest.

3.4 CRAGGING

The approaches described above are designed for independent and identically distributed (i.i.d.) observations. However, often this is not the case in economics and finance. Often, the data has a panel structure, e.g., owing to a set of variables

---

[10]Freund and Schapire [29] do not use CART and also propose two more complex algorithms, where the trees are grown by using more than one attribute.

[11]Typically between 4 and 8, see [38].

[12]More generally, one can use other loss functions than the mean squared error, such as the mean absolute error.

being collected for several countries. In this case, observations are not independent; hence there is information in the data that can be exploited to improve the predictive performance of the algorithm. To this aim, the CRAGGING (CRoss-validation AGGregatING) algorithm has been developed as a generalization of regression trees [66]. In the case of a panel comprising a set of variables for a number of countries observed through time, the CRAGGING algorithm works as follows:

1. Randomly partition the whole sample into subsets of equal size. The number of subsets needs to be smaller than the number of countries.
2. One of the subsets is reserved for testing, while the others are used to train the algorithm. From the training set, one country is removed and a regression tree is grown and pruned.
3. The test set is used to compute predictions based on the tree.
4. The country is reinserted in the training set and steps 2–3 are repeated for all the countries.
5. A cross-validation procedure is run over the test set to obtain a tree which minimizes prediction errors. Hence, CRAGGING combines two types of cross-validation, namely, the leave-one-unit-out cross-validation, in which the units are removed one at a time from the training set and then perturbed, and the usual cross-validation on the test sets, run to minimize the prediction error out-of-sample. (see [66] for details).
6. Steps 1–5 are repeated thousands of times and predictions from the thousands of trees are aggregated by computing the arithmetic average of those predictions.
7. As a final step, a regression tree is estimated on the predictions' average (computed at step 6) using the same set of the original covariates.

This algorithm eventually yields one single tree, thereby retaining the interpretability of the model. At the same time, its predictions are based on an ensemble of trees, which increases its predictive accuracy and stability.

4   Regularization, Shrinkage, and Sparsity

In the era of Big Data, standard regression models increasingly face the "curse of dimensionality." This relates to the fact that they can only include a relatively small number of regressors. Too many regressors would lead to overfitting and unstable estimates. However, often we have a large number of predictors, or candidate predictors. For example, this is the case for policymakers in economics and finance, who base their decisions on a wide information set, including hundreds of macroeconomic and macrofinancial data through time. Still, they can ultimately only consider a limited amount of information; hence variable selection becomes crucial.

Sparse models offer a solution to deal with a large number of predictor variables. In these models, regressors are many but relevant coefficients are few. The Least Absolute Shrinkage and Selection Operator (LASSO), introduced by [58] and popularized by [64], is one of the most used models in this literature. Also in this case, from this seminal work an immense statistical literature has developed with increasingly sophisticated LASSO-based models. Bayesian shrinkage is another way to achieve sparsity, very much used, e.g., in empirical macroeconomics, when variables are often highly collinear. Instead of yielding a point estimate for the model parameters, it yields a probability distribution, hence incorporating the uncertainty surrounding the estimates. In the same spirit, Bayesian Model Averaging is becoming popular also in finance to account for model uncertainty.

4.1 Regularization

Regularization is a supervised learning strategy that overcomes this problem. It reduces the complexity of the model by *shrinking* the parameters toward some value. In practice, it penalizes more complex models in favor of more parsimonious ones. The Least Absolute Shrinkage and Selection Operator (LASSO, [58] and [64]), increasingly popular in economics and finance, uses L1 regularization. In practice, it limits the size of the regression coefficients by imposing a penalty equal to the absolute value of the magnitude of the coefficients. This implies shrinking the smallest coefficients to zero, which is removing some regressors altogether. For this reason, the LASSO is used as a variable selection method, allowing to identify key predictors from a pool of several candidate ones. A tuning parameter λ in the penalty function controls the level of shrinkage: for $\lambda = 0$ we have the OLS solution, while for increasing values of λ more and more coefficients are set to zero, thus yielding a sparse model.

Ridge regression involves L2 regularization, as it uses the squared magnitude of the coefficients as penalty term in the loss function. This type of regularization does not shrink parameters to zero. Also in this case, a crucial modeling choice relates to the value of the tuning parameter λ in the penalty function.

The Elastic Net has been proposed as an improvement over the LASSO [38], and combines the penalty from LASSO with that of the Ridge regression. The Elastic Net seems to be more efficient than the LASSO, while maintaining a similar sparsity of representation, in two cases. The first one is when the number of predictor variables is larger than the number of observations: in this case, the LASSO tends to select at most all the variables before it saturates. The second case is when there is a set of regressors whose pairwise correlations are high: in this case, the LASSO tends to select only one predictor at random from the group.[13]

The Adaptive LASSO [68] is an alternative model also proposed to improve over the LASSO, by allowing for different penalization factors of the regression coefficients. By doing so, the Adaptive LASSO addresses potential weaknesses of

---

[13] See [69].

the classical LASSO under some conditions, such as the tendency to select inactive predictors, or over-shrinking the coefficients associated with correct predictors.

4.2 Bayesian Learning

In Bayesian learning, shrinkage is defined in terms of a parameter's prior probability distribution, which reflects the modeler's beliefs.[14] In the case of Bayesian linear regression, in particular, the prior probability distribution for the coefficients may reflect how certain one is about some coefficients being zero, i.e., about the associated regressors being unimportant. The posterior probability of a given parameter is derived based on both the prior and the information that is contained in the data. In practice, estimating a linear regression using a Bayesian approach involves the following steps:

1. Assume a prior probability distribution for the dependent variable, the coefficients, and the variance of the error term.
2. Specify the likelihood function, which is defined as the probability of the data given the parameters.
3. Derive the posterior distribution, which is proportional to the likelihood times the prior.
4. If the likelihood is such that the posterior cannot be derived analytically, use sampling techniques such as Markov Chain Monte Carlo (MCMC) method to generate a large sample (typically based on thousands of draws) from the posterior distribution.
5. The predicted value for the dependent variable, as well as the associated highest posterior density interval (e.g., at the 95% level), are derived based on the coefficients' posterior distribution.

By yielding probability distributions for the coefficients instead of point estimates, Bayesian linear regression accounts for the uncertainty around model estimates. In the same spirit, Bayesian Model Averaging (BMA, [46]) adds one layer by considering the uncertainty around the model specification. In practice, it assumes a prior distribution over the set of all considered models, reflecting the modeler's beliefs about each model's accuracy in describing the data. In the context of linear regression, model selection amounts to selecting subsets of regressors from the set of all candidate variables. Based on the posterior probability associated with each model, which takes observed data into account, one is able to select and combine the best models for prediction purposes. Stochastic search algorithms help

---

[14]The book by [34] covers Bayesian inference from first principles to advanced approaches, including regression models.

reduce the dimension of the model space when the number of candidate regressors is not small.[15]

Finally, some approaches have more recently been proposed which link the LASSO-based literature with the Bayesian stream. This avenue was pioneered by the Bayesian LASSO [53], which connects the Bayesian and LASSO approaches by interpreting the LASSO estimates as Bayesian estimates, based on a particular prior distribution for the regression coefficients. As a Bayesian method, the Bayesian LASSO yields interval estimates for the LASSO coefficients. The Bayesian adaptive LASSO (BaLASSO, [47]) generalizes this approach by allowing for different parameters in the prior distributions of the regression coefficients. The Elastic Net has also been generalized in a Bayesian setting [40], providing an efficient algorithm to handle correlated variables in high-dimensional sparse models.

5 Critical Discussion on Machine Learning as a Tool for Financial Stability Policy

As discussed in [5], standard approaches are usually unable to fully understand the risk dynamics within financial systems in which structural relationships interact in nonlinear and state-contingent ways. And indeed, traditional models assume that risk dynamics, e.g., those eventually leading to banking or sovereign crises, can be reduced to common data models in which data are generated by independent draws from predictor variables, parameters, and random noise. Under these circumstances, the conclusions we can draw from these models are "about the models mechanism, and not about natures mechanism" [17]. To put the point into perspective, let us consider the goal of realizing a risk stratification for financial crisis prediction using regression trees. Here the objective should be based on identifying a series of "red flags" for potential observable predictors that help to detect an impending financial crisis through a collection of binary rules of thumb such as the value of a given predictor being larger or lower than a given threshold for a given observation. In doing this, we can realize a pragmatic rating system that can capture situations of different risk magnitudes, from low to extreme risk, whenever the values of the selected variables lead to risky terminal nodes. And the way in which such a risk stratification is carried out is, by itself, a guarantee to get the best risk mapping in terms of most important variables, optimal number of risk clusters (final nodes), and corresponding risk predictions (final nodes' predictions). In fact, since the estimation process of the regression tree, as all machine learning algorithms, is

---

[15]With M candidate regressors, the number of possible models is equal to 2^M.

based on cross-validation,[16] the rating system is validated by construction, as the risk partitions are realized in terms of maximum predictability.

However, machine learning algorithms also have limitations. The major caveat sits on the same connotation of data-driven algorithms. In a sense, machine learning approaches limit their knowledge to the data they process, no matter how and why those data lead to specific models. In statistical language, the point relates to the question about the underlying data-generation process. In more depth, machine learning is expected to say little about the causal effect between y and x, but rather it is conceived with the end to predict y using and selecting x. The issue is extremely relevant when exploring the underlying structure of the relationship and trying to make inference about the inner nature of the specific economic process under study. A clear example of how this problem materializes is in [52]. These authors make a repeated house-value prediction exercise on subsets of a sample from the American Housing Survey, by randomly partitioning the sample, next re-estimating the LASSO predictor. In doing so, they document that a variable used in one partition may be unused in another while maintaining a good prediction quality (the R2 remains roughly constant from partition to partition). A similar instability is also present in regression trees. Indeed, since these models are sequential in nature and locally optimal at each node split, the final tree may not guarantee a global optimal solution, with small changes in the input data translating into large changes in the estimation results (the final tree).

Because of these issues, machine learning tools should then be used carefully.[17] To overcome the limitations of machine learning techniques, one promising avenue is to use them in combination with existing model-based and theory-driven approaches.[18] For example, [60] focus on sovereign debt crises prediction and explanation proposing a procedure that mixes a pure algorithmic perspective, without making assumptions about the data generating process, with a parametric approach (see Sect. 6.1). This mixed approach allows to bypass the problem of reliability of the predictive model, thanks to the use of an advanced machine learning technique. At the same time, it allows to estimate a distance-to-default, via a standard probit regression. By doing so, the empirical analysis is contextualized within a theoretical-based process similar to the Merton-based distance-to-default.

---

[16]The data are partitioned into subsets such that the analysis is initially performed on a single subset (the training sets), while the other subset(s) are retained for subsequent use in confirming and validating the initial analysis (the validation or testing sets).

[17]See [6] on the use of big data for policy.

[18]See [7] for an overview of recently proposed methods at the intersection of machine learning and econometrics.

6 Literature Overview

This section provides an overview of a growing literature, which applies the models described in the previous section—or more sophisticated versions—for financial stability purposes. This literature has developed in the last decade, with more advanced techniques being applied in finance only in recent years. This is the so-called second generation of Early Warning Models (EWM), developed after the global financial crisis. While the first generation of EWM, popular in the 1990s, was based on rather simple approaches such as the signaling approach, the second generation of EWM implement machine learning techniques, including tree-based approaches and parametric multiple-regime models. In Sect. 6.1 we will review papers using decision trees, while Sect. 6.2 deals with financial stability applications of sparse models.

6.1 Decision Trees for Financial Stability

There are several success stories on the use of decision trees to address financial stability issues. Several papers propose EWM for banking crises. One of the first papers applying classification trees in this field is [22], where the authors use a binary classification tree to analyze banking crises in 50 emerging markets and developed economies. The tree they grow identifies the conditions under which a banking crisis becomes likely, which include high inflation, low bank profitability, and highly dollarized bank deposits together with nominal depreciation or low bank liquidity. The beauty of this tool stands in the ease of use of the model, which also provides specific threshold values for the key variables. Based on the proposed tree, policymakers only need to monitor whether the relevant variables exceed the warning thresholds in a particular country. [50] also aim at detecting vulnerabilities that could lead to banking crises, focusing on emerging markets. They apply the CRAGGING approach to test 540 candidate predictors and identify two banking crisis' "danger zones": the first occurs when high interest rates on bank deposits interact with credit booms and capital flights; the second occurs when an investment boom is financed by a large rise in banks' net foreign exposure. In a recent working paper by [33], the author uses the same CRAGGING algorithm to identify vulnerabilities to systemic banking crises, based on a sample of 15 European Union countries. He finds that high credit aggregates and a low market risk perception are amongst the key predictors. [1] also develop an early warning system for systemic banking crises, which focuses on the identification of unsustainable credit developments. They consider 30 predictor variables for all EU countries and apply the Random Forest approach, showing that it outperforms competing logit models out-of-sample. [63] also apply the Random Forest to assess vulnerabilities in the banking sector, including bank-level financial statements as predictor variables. [14] compare a set of machine learning techniques, also including trees and the Random

Forest, to network- and regression-based approaches, showing that machine learning models mostly outperform the logistic regression in out-of-sample predictions and forecasting. The authors also offer a narrative for the predictions of machine learning models, based on the decomposition of the predicted crisis probability for each observation into a sum of contributions from each predictor. [67] implements BAGGING and the Random Forest to measure the risk of banking crises, using a long-run sample for 17 countries. He finds that tree ensembles yield a significantly better predictive performance compared to the logit. [20] use AdaBoost to identify the buildup of systemic banking crises, based on a dataset comprising 100 advanced and emerging economies. They also find that machine learning algorithms can have a better predictive performance than logit models. [13] is the only work, to our knowledge, finding an out-of-sample outperformance of conventional logit models over machine learning techniques, including decision trees and the Random Forest.

Also for sovereign crises several EWM have been developed based on tree ensemble techniques. The abundant literature on sovereign crises has documented the high complexity and the multidimensional nature of sovereign default, which often lead to predictive models characterized by irrelevant theory and poor/questionable conclusions. One of the first papers exploring machine learning methods in this literature is [49]. The authors compare the logit and the CART approach, concluding that the latter overperforms the logit with 89% of the crises correctly predicted; however, it issues more false alarms. [48] also use CART to investigate the roots of sovereign debt crises, finding that they differ depending on whether the country faces public debt sustainability issues, illiquidity, or various macroeconomic risks. [60] propose a procedure that mixes the CRAGGING and the probit approach. In particular, in the first step CRAGGING is used to detect the most important risk indicators with the corresponding threshold, while in a second step a simple pooled probit is used to parametrize the distances to the thresholds identified in the first step (so-called "Multidimensional Distance to Collapse Point"). [61] again use CRAGGING, to predict sovereign crises based on a sample of emerging markets together with Greece, Ireland, Portugal, and Spain. They show that this approach outperforms competing models, including the logit, while balancing in-sample goodness of fit and out-of-sample predictive performance. More recently, [5] use a recursive partitioning strategy to detect specific European sovereign risk zones, based on key predictors, including macroeconomic fundamentals and a contagion measure, and relevant thresholds.

Finally, decision trees have been used also for the prediction of currency crises. [36] first apply this methodology on a sample of 42 countries, covering 52 currency crises. Based on the binary classification tree they grow on this data, they identify two different sets of key predictors for advanced and emerging economies, respectively. The root node, associated with an index measuring the quality of public sector governance, essentially splits the sample into these two subsamples. [28] implement a set of methodological approaches, including regression trees, in their empirical investigation of macroeconomic crises in emerging markets. This approach allows each regressor to have a different effect on the dependent variable for different ranges of values, identified by the tree splits, and is thus able to capture nonlinear relationships and interactions. The regression tree analysis identifies three variables, namely, the ratio of external debt to GDP, the ratio of short-term external debt to reserve, and inflation, as the key predictors. [42] uses regression tree analysis to classify 96 currency crises in 20 countries, capturing the stylized characteristics of different types of crises. Finally, a recent paper using CART and the Random Forest to predict currency crises and banking crises is [41]. The authors identify the key predictors for each type of crisis, both in the short and in the long run, based on a sample of 36 industrialized economies, and show that different crises have different causes.

6.2 Sparse Models for Financial Stability

LASSO and Bayesian methods have so far been used in finance mostly for portfolio optimization. A vast literature starting with [8] uses a Bayesian approach to address the adverse effect due to the accumulation of estimation errors. The use of LASSO-based approaches to regularize the optimization problem, allowing for the stable construction of sparse portfolios, is far more recent (see, e.g., [19] and [24], among others).

Looking at financial stability applications of Bayesian techniques, [23] develop an early warning system where the dependent variable is an index of financial stress. They apply Bayesian Model Averaging to 30 candidate predictors, notably twice as many as those generally considered in the literature, and select the important ones by checking which predictors have the highest probability to be included in the most probable models. More recently, [55] investigate the determinants of the 2008 global financial crisis using a Bayesian hierarchical formulation that allows for the joint treatment of group and variable selection. Interestingly, the authors argue that the established results in the literature may be due to the use of different priors. [65] and [37] use Bayesian estimation to estimate the effects of the US subprime mortgage crisis. The first paper uses Bayesian panel data analysis for exploring its impact on the US stock market, while the latter uses time-varying Bayesian Vector AutoRegressions to estimate cross-asset contagion in the US financial market, using the subprime crisis as an exogenous shock.

Turning to the LASSO, not many authors have yet used this approach to predict financial crises. [45] use a logistic LASSO in combination with cross-validation to set the λ penalty parameter, and test their model in a real-time recursive out-of-sample exercise based on bank-level and macrofinancial data. The LASSO yields a parsimonious optimal early-warning model which contains the key risk-driver indicators and has good in-sample and out-of-sample signaling properties. More recently, [2] apply the LASSO in the context of sovereign crises prediction. In particular, they use it to identify the macro indicators that are relevant in explaining the cross-section of sovereign Credit Default Swaps (CDS) spreads in a recursive setting, thereby distilling time-varying market sensitivities to specific economic fundamentals. Based on these estimated sensitivities, the authors identify distinct crisis regimes characterized by different dynamics. Finally, [39] conduct a horse race of conventional statistical methods and more recent machine learning methods, including a logit LASSO as well as classification trees and the Random Forest, as early-warning models. Out of a dozen competing approaches, tree-based algorithms place in the middle of the ranking, just above the naive Bayes approach and the LASSO, which in turn does better than the standard logit. However, when using a different performance metric, the naive Bayes and logit outperform classification trees, and the standard logit slightly outperforms the logit LASSO.

6.3   Unsupervised Learning for Financial Stability

Networks have been extensively applied in financial stability. This stream of litera-ture is based on the notion that the financial system is ultimately a complex system, whose characteristics determining its resilience, robustness, and stability can be studied by means of traditional network approaches (see [12] for a discussion). In particular, network models have been successfully used to model contagion (see the seminal work by [3], as well as [35] for a review of the literature on contagion in financial networks)[19] and measure systemic risk (see, e.g., [11]). The literature applying network theory started to grow exponentially in the aftermath of the global financial crisis. DebtRank [10], e.g., is one of the first approaches put forward to identify systemically important nodes in a financial network. This work contributed to the debate on too-big-to-fail financial institutions in the USA by emphasizing that too-central-to-fail institutions deserve at least as much attention.[20] [51] explore the properties of the global banking network by modelling 184 countries as nodes of the network, linked through cross-border lending flows, using data over the 1978–2009 period. By today, countless papers use increasingly complex network approaches to make sense of the structure of the financial system. The tools they offer aim at enabling policymakers to monitor the evolution of the financial system and detect vulnerabilities, before a trigger event precipitates the whole system into a crisis state. Among the most recent ones, one may cite, e.g., [62], who study the type of systemic risk arising in a situation where it is impossible to decide which banks are in default.

Turning to artificial neural networks, while supervised ones have been used in a few works as early warning models for financial crises ([26] on sovereign debt crises, [27] and [54] on currency crises), unsupervised ones are even less common in the financial stability literature. In fact, we are only aware of one work, [59], using self-organizing maps. In particular, the authors develop a Self-Organizing Financial Stability Map where countries can be located based on whether they are

---

[19]Amini et al. [4], among others, also use financial networks to study contagion.

[20]On the issue of centrality, see also [44] who built a network based on co-movements in Credit Default Swaps (CDS) of major US and European banks.

in a pre-crisis, crisis, post-crisis, or tranquil state. They also show that this tool performs better than or equally well as a logit model in classifying in-sample data and predicting the global financial crisis out-of-sample.

7   Conclusions

Forecasting financial crises is essential to provide warnings to be used in preventing impending abnormalities, and taking action with a sufficient lead time to implement adequate policy measures. The global financial crisis that started with the Lehman collapse in 2008 and the subsequent Eurozone sovereign debt crisis over the years 2010–2013 have both profoundly changed economic thinking around machine learning. The ability to discover complex and nonlinear relationships, not fully biased by a priori theory/beliefs, has contributed to getting rid of the skepticism around machine learning. Ample evidence proved indeed the inconsistency of traditional models in predicting financial crisis, and the need to explore data-driven approaches. However, we should be aware about what machine learning can and cannot do, and how to handle these algorithms alone and/or in conjunction with common traditional approaches to make financial crisis predictions more statistically robust and theoretically consistent. Also, it would be important to work on improving the interpretability of the models, as there is a strong need to understand how decisions on financial stability issues are being made.

References

1. Alessi, L., & Detken, C. (2018). Identifying excessive credit growth and leverage. *Journal of Financial Stability, 35*, 215–225.
2. Alessi, L., Balduzzi, P., & Savona, R. (2019). *Anatomy of a Sovereign Debt Crisis: CDS Spreads and Real-Time Macroeconomic Data*. Brussels: European Commission—Joint Research Centre. Working Paper No. 2019-03.
3. Allen, F., & Gale, D. (2000). Financial contagion. *Journal of Political Economy 108*(1), 1–33.
4. Amini, H., Cont, R., & Minca, A. (2016). Resilience to contagion in financial networks. *Mathematical Finance, 26*, 329–365.
5. Arakelian, V., Dellaportas, P., Savona, R., & Vezzoli, M. (2019). Sovereign risk zones in Europe during and after the debt crisis. *Quantitative Finance, 19*(6), 961–980.
6. Athey, S. (2017). Beyond prediction: Using big data for policy problems. *Science, 355*(6324), 483–485.
7. Athey, S., & Imbens, G. W. (2019). Machine learning methods economists should know about. *Annual Review of Economics, 11*(1), 685–725.
8. Barry, C. B. (1974). Portfolio analysis under uncertain means, variances, and covariances. *Journal of Finance, 29*, 515–22.
9. Basel Committee on Banking Supervision (2018). *Implications of fintech developments for banks and bank supervisors, consultative document*. Accessed 12 March 2021. https://www.bis.org/bcbs/publ/d415.htm.

10. Battiston, S., Puliga, M., Kaushik, R., Tasca, P., & Caldarelli, G. (2012). DebtRank: Too Central to Fail? Financial Networks, the FED and Systemic Risk. *Scientific Reports, 2*, 541–541.
11. Battiston, S., Caldarelli, G., May, R. M., Roukny, T., & Stiglitz, J. E. (2016). The price of complexity in financial networks. *Proceedings of the National Academy of Sciences, 113*(36), 10031–10036.
12. Battiston, S., Farmer, J. D., Flache, A., Garlaschelli, D., Haldane, A. G., Heesterbeek, H., Hommes, C., Jaeger, C., May, R., Scheffer, M. (2016). Complexity theory and financial regulation. *Science, 351*(6275), 818–819.
13. Beutel, J., List, S., & von Schweinitz, G. (2018). *An evaluation of early warning models for systemic banking crises: Does machine learning improve predictions?*. Deutsche Bundesbank Discussion Paper Series, No. 48.
14. Bluwstein, K., Buckmann, M., Joseph, A., Kang, M., Kapadia, S., & Simsek, O. (2020). *Credit growth, the yield curve and financial crisis prediction: Evidence from a machine learning approach*. Bank of England Staff Working Paper No. 848.
15. Breiman, L. (1996). Bagging predictors. *Machine Learning, 24*, 123–140.
16. Breiman, L. (2001). Random forests. *Machine Learning, 45*(1), 5–32.
17. Breiman, L. (2001). Statistical Modelling: The Two Cultures. *Statistical Science, 16*(3), 199–215.
18. Breiman, L., Friedman, J., Olshen, R., & Stone, C. (1984). *Classification and regression trees*. Monterey, CA: Wadsworth and Brooks.
19. Brodie, J., Daubechies, I., De Mol, C., Giannone, D., & Loris, I. (2009). Sparse and stable Markowitz portfolios, *Proceedings of the National Academy of Sciences, 106*, 12267–12272.
20. Casabianca, E. J., Catalano, M., Forni, L., Giarda, E., & Passeri, S. (2019). An early warning system for banking crises: From regression-based analysis to machine learning techniques. In *Marco Fanno Working Papers 235, Dipartimento di Scienze Economiche "Marco Fanno"*.
21. Clements, M., & Hendry, D. (1998). *Forecasting economic time series*. Cambridge: Cambridge University.
22. Duttagupta, R., & Cashin, P. (2011). Anatomy of banking crises in developing and emerging market countries. *Journal of International Money and Finance, 30*(2), 354–376.
23. Eidenberger, J., Sigmund, M., Neudorfer, B., & Stein, I. (2014). *What Predicts Financial (In)Stability? A Bayesian Approach*. Bundesbank Discussion Paper No. 36/2014.
24. Fan, J., Zhang, J., & Yu, K. (2012). Vast portfolio selection with gross-exposure constraints. *Journal of the American Statistical Association, 107*, 592–606.
25. Financial Stability Board (2017). *Artificial intelligence and machine learning in financial services—Market developments and financial stability implications*. FSB Report, P011117. https://www.fsb.org/wp-content/uploads/P011117.pdf.
26. Fioramanti, M. (2008). Predicting sovereign debt crises using artificial neural networks: A comparative approach. *Journal of Financial Stability, 4*(2), 149–164.
27. Franck, R., & Schmied, A. (2003). *Predicting currency crisis contagion from East Asia to Russia and Brazil: An artificial neural network approach*. AMCB Working Paper No 2/2003, Aharon Meir Center for Banking.
28. Frankel, J., & Wei, S. J. (2004). *Managing macroeconomic crises: Policy lessons*. NBER Working Paper 10907.
29. Freund, Y., & Schapire, R. E. (1996). Experiments with a new boosting algorithm. *Proceedings of the International Conference on Machine Learning (ICML 96), 96*, 148–156.
30. Friedman, J. H. (2001). Greedy function approximation: A gradient boosting machine. *The Annals of Statistics, 29*(5), 1189–1232.
31. Friedman, J. H. (2002). Stochastic gradient boosting. *Computational Statistics & Data Analysis, 38*(4), 367–378.
32. Friedman, J. H., & Meulman, J. J. (2003). Multiple additive regression trees with application in epidemiology. *Statistics in Medicine, 22*(9),1365–1381.
33. Gabriele, C. (2019). *Learning from trees: A mixed approach to building early warning systems for systemic banking crises*. European Stability Mechanism Working Paper No. 40/2019.

34. Gelman, A., Carlin, J. B., Stern, H. S., Dunson, D. B., Vehtari, A., & Rubin, D. B. (2013). *Bayesian data analysis*. New York: CRC Press.
35. Glasserman, P., & Peyton Young, H. (2016). Contagion in financial networks. *Journal of Economic Literature, 54*(3), 779–831.
36. Gosh, S., & Gosh, A. (2002). *Structural vulnerabilities and currency crises*. IMF Working Paper 02/9.
37. Guidolin, M., Hansen, E., & Pedio, M. (2019). Cross-asset contagion in the financial crisis: A Bayesian time-varying parameter approach. *Journal of Financial Markets, 45*, 83–114.
38. Hastie, T., Tibshirani, R., & Friedman, J. (2009). *The elements of statistical learning*, 2 edn. New York: Springer.
39. Holopainen, M., & Sarlin, P. (2017). Toward robust early-warning models: A horse race, ensembles and model uncertainty. *Quantitative Finance, 17*(12), 1933–1963.
40. Huang, A., Xu, S., & Cai, X. (2015). Empirical Bayesian elastic net for multiple quantitative trait locus mapping. *Heredity, 114*, 107–115.
41. Joy, M., Rusnk, M., Smdkova, K., & Vacek, B. (2017). Banking and currency crises: Differential diagnostics for developed countries. *International Journal of Finance & Economics, 22*(1), 44–67.
42. Kaminsky, G. L. (2006). Currency crises: Are they all the same. *Journal of International Money and Finance, 25*, 503–527.
43. Kass, G. V. (1980). An exploratory technique for investigating large quantities of categorical data. *Journal of the Royal Statistical Society. Series C (Applied Statistics), 29*(2), 119–127.
44. Kaushik, R., & Battiston, S. (2013). Credit default swaps drawup networks: Too interconnected to be stable? *PLoS ONE, 8*(7), e61815.
45. Lang, J. H., Peltonen, T. A., & Sarlin, P. (2018). *A framework for early-warning modeling with an application to banks*. Working Paper Series 2182, European Central Bank.
46. Leamer, E. E. (1978). *Specification searches*. New York: Wiley.
47. Leng, C., Nott, D., & Minh-Ngoc, T. (2014). Bayesian adaptive lasso. *Annals of the Institute of Statistical Mathematics, 66*(2), 221–244.
48. Manasse, P., & Roubini, N. (2009). Rules of thumb for Sovereign debt crises. *Journal of International Economics, 78*, 192–205.
49. Manasse, P., Roubini, N., & Schimmelpfennig, A. (2003). *Predicting Sovereign debt crises*. IMF Working Paper WP 03/221.
50. Manasse, P., Savona, R., & Vezzoli, M. (2016). Danger zones for banking crises in emerging markets. *International Journal of Finance & Economics, 21*, 360–381.
51. Minoiu, C., & Reyes, J. (2011). *A Network Analysis of Global Banking:1978–2009*. IMF Working Papers. 11.
52. Mullainathan, S., & Spiess, J. (2017). Machine learning: An applied econometric approach. *Journal of Economic Perspectives, 31*(2), 87–106.
53. Park, T., & Casella, G. (2008). The Bayesian Lasso. *Journal of the American Statistical Association, 103*(482), 681–686.
54. Peltonen, T. (2006). *Are emerging market currency crises predictable?* A test. No 571, Working Paper Series, European Central Bank.
55. Ray-Bing, C., Kuo-Jung, L., Yi-Chi, C., & Chi-Hsiang, C. (2017). On the determinants of the 2008 financial crisis: A Bayesian approach to the selection of groups and variables. *Studies in Nonlinear Dynamics & Econometrics, 21*(5), 17.
56. Ren, Y., Zhang, L., & Suganthan, P. N. (2016). Ensemble classification and regression-recent developments, applications and future directions. *IEEE Computational Intelligence Magazine, 11*(1), 41–53.
57. Rojas, R. (1996). *Neural networks a systematic introduction* (p. 101). Berlin: Springer.
58. Santosa, F., & Symes, W. W. (1986). Linear inversion of band-limited reflection seismograms. *SIAM Journal on Scientific and Statistical Computing, 7*(4), 1307–1330.
59. Sarlin, P., & Peltonen, T. (2011). Mapping the state of financial stability. *Journal of International Financial Markets Institutions and Money, 26*, 46–76.

60. Savona, R., & Vezzoli, M. (2012). Multidimensional Distance-To-Collapse Point and sovereign default prediction. *Intelligent Systems in Accounting, Finance and Management, 19*(4), 205–228.
61. Savona, R., & Vezzoli, M. (2015). Fitting and forecasting sovereign defaults using multiple risk signals. *Oxford Bulletin of Economics and Statistics, 77*(1), 66–92.
62. Schuldenzucker, S., Seuken, S., & Battiston, S. (2020). Default ambiguity: Credit default swaps create new systemic risks in financial networks. *Management Science, 66*(5), 1981–1998.
63. Tanaka, K., Kinkyo, T., & Hamori, S. (2016). Random forests-based early warning system for bank failures. *Economics Letters, 148*, 118–121.
64. Tibshirani, R. (1996). Regression Shrinkage and Selection via the lasso. *Journal of the Royal Statistical Society. Series B (Methodological), 58*(1), 267–288.
65. Tsay, R., & Ando, T. (2012). Bayesian panel data analysis for exploring the impact of subprime financial crisis on the US stock market. *Computational Statistics & Data Analysis, 56*(11), 3345–3365.
66. Vezzoli, M., & Stone, C. J. (2007). CRAGGING. In *Book of short papers of CLADAG 2007, EUM*, pp. 363–366.
67. Ward, F. (2017). Spotting the danger zone: Forecasting financial crises with classification tree ensembles and many predictors. *Journal of Applied Econometrics, 32*(2), 359–378.
68. Zou, H. (2006). The adaptive lasso and its oracle properties. *Journal of the American Statistical Association, 101*(476), 1418–1429.
69. Zou, H., & Hastie, T. (2005). Regularization and variable selection via the elastic net. *Journal of the Royal Statistical Society. Series B (Statistical Methodology), 67*(2), 301–320.

Sharpening the Accuracy of Credit Scoring Models with Machine Learning Algorithms

Massimo Guidolin and Manuela Pedio

Abstract The big data revolution and recent advancements in computing power have increased the interest in credit scoring techniques based on artificial intelligence. This has found easy leverage in the fact that the accuracy of credit scoring models has a crucial impact on the profitability of lending institutions. In this chapter, we survey the most popular supervised credit scoring classification methods (and their combinations through ensemble methods) in an attempt to identify a superior classification technique in the light of the applied literature. There are at least three key insights that emerge from surveying the literature. First, as far as individual classifiers are concerned, linear classification methods often display a performance that is at least as good as that of machine learning methods. Second, ensemble methods tend to outperform individual classifiers. However, a dominant ensemble method cannot be easily identified in the empirical literature. Third, despite the possibility that machine learning techniques could fail to outperform linear classification methods when standard accuracy measures are considered, in the end they lead to significant cost savings compared to the financial implications of using different scoring models.

1 Introduction

Credit scoring consists of a set of risk management techniques that help lenders to decide whether to grant a loan to a given applicant [42]. More precisely, financial institutions use credit scoring models to make two types of credit decisions. First, a lender should decide whether to grant a loan to a new customer. The process

M. Guidolin
Bocconi University, Milan, Italy
e-mail: massimo.guidolin@unibocconi.it

M. Pedio (✉)
University of Bristol, Accounting and Finance Department, Bristol, UK
Bocconi University, Milan, Italy
e-mail: manuela.pedio@unibocconi.it

© The Author(s) 2021
S. Consoli et al. (eds.), *Data Science for Economics and Finance*,
https://doi.org/10.1007/978-3-030-66891-4_5

that leads to this decision is called *application scoring*. Second, a lender may want to monitor the risk associated with existing customers (the so-called behavioral scoring). In the field of retail lending, credit scoring typically consists of a binary classification problem, where the objective is to predict whether an applicant will be a "good" one (i.e., she will repay her liabilities within a certain period of time) or a "bad" one (i.e., she will default in part or fully on her obligations) based on a set of observed characteristics (*features*) of the borrower.[1] A feature can be of two types: continuous, when the value of the feature is a real number (an example can be the income of the applicant) or categorical, when the feature takes a value from a predefined set of categories (an example can be the rental status of the applicant, e.g., "owner," "living with parents," "renting," or "other"). Notably, besides traditional categories, new predictive variables, such as those based on "soft" information have been proposed in the literature to improve the accuracy of the credit score forecasts. For instance, Wang et al. [44] use text mining techniques to exploit the content of descriptive loan texts submitted by borrowers to support credit decisions in peer-to-peer lending.

Credit scoring plays a crucial role in lending decisions, considering that the cost of an error is relatively high. Starting in the 1990s, most financial institutions have been making lending decisions with the help of automated credit scoring models [17]. However, according to the Federal Reserve Board [15] the average delinquency rate on consumer loans has been increasing again since 2016 and has reached 2.28% in the first quarter of 2018, thus indicating that wide margins for improvement in the accuracy of credit scoring models remain. Given the size of the retail credit industry, even a small reduction in the hazard rate may yield significant savings for financial institutions in the future [45].

Credit scoring also carries considerable regulatory importance. Since the Basel Committee on Banking Supervision released the Basel Accords, especially the second accord in 2004, the use of credit scoring has grown considerably, not only for credit granting decisions but also for risk management purposes. Basel III, released in 2013, enforced increasingly accurate calculations of default risk, especially in consideration of the limitations that external rating agencies have shown during the 2008–2009 financial crisis [38]. As a result, over the past decades, the problem of developing superior credit scoring models has attracted significant attention in the academic literature. More recently, thanks to the increase in the availability of data and the progress in computing power the attention has moved towards the application of Artificial Intelligence (AI) and, in particular, Machine Learning (ML) algorithms to credit scoring, when machines may learn and make predictions without being explicitly assigned program instructions.

---

[1] There are also applications in which the outcome variable is not binary; for instance, multinomial models are used to predict the probability that an applicant will move from one class of risk to another. For example, Sirignano et al. [40] propose a nonlinear model of the performance of a pool of mortgage loans over their life; they use neural networks to model the conditional probability that a loan will transition to a different state (e.g., pre-payment or default).

There are four major ML paradigms: *supervised learning, semi-supervised learning, unsupervised learning,* and *reinforcement learning* [32]. In supervised learning, a training dataset should consist of both input data and their corresponding output target values (also called labels). Then, the algorithm is trained on the data to find relationships between the input variables and selected output labels. If only some target output values are available in a training dataset, then such a problem is known as semi-supervised learning. Unsupervised learning requires only the input data to be available. Finally, reinforcement learning does not need labelled inputs/outputs but focuses instead on agents making optimal decisions in a certain environment; a feedback is provided to the algorithm in terms of "reward" and "punishment" so that the final goal is to maximize the agent's cumulative reward. Typically, lending institutions store both the input characteristics and the output historical data concerning their customers. As a result, supervised learning is the main focus of this chapter.

Simple linear classification models remain a popular choice among financial institutions, mainly due to their adequate accuracy and straightforward implementation [29]. Furthermore, the majority of advanced ML techniques lack the necessary transparency and are regarded as "black boxes", which means that one is not able to easily explain how the decision to grant a loan is made and on which parameters it is based. In the financial industry, however, transparency and simplicity play a crucial role, and that is the main reason why advanced ML techniques have not yet become widely adopted for credit scoring purposes.[2] However, Chui et al. [12] emphasize that the financial industry is one of the leading sectors in terms of current and prospective ML adoption, especially in credit scoring and lending applications as they document that more than 25% of the companies that provide financial services have adopted at least one advanced ML solution in their day-to-day business processes.

Even though the number of papers on advanced scoring techniques has increased dramatically, a consensus regarding the best-performing models has not yet been reached. Therefore, in this chapter, besides providing an overview of the most common classification methods adopted in the context of credit scoring, we will also try to answer three key questions:

- Which individual classifiers show the best performance both in terms of accuracy and of transparency?
- Do ensemble classifiers consistently outperform individual classification models and which (if any) is the superior ensemble method?

---

[2]Casual interpretations of "black box" ML models have attracted considerable attention. Zhao and Hastie [50] provide a summary and propose partial dependence plots (PDP) and individual conditional expectations (ICE) as tools to enhance the interpretation of ML models. Dorie et al. [13] report interesting results of a data analysis competition where different strategies for causal inference—including "black box" models—are compared.

- Do one-class classification models score higher accuracy compared to the best individual classifiers when tested on imbalanced datasets (i.e., datasets where one class is underrepresented)?

Our survey shows that, despite that ML techniques rarely significantly outperform simple linear methods as far as individual classifiers are concerned, ensemble methods tend to show a considerably better classification performance than individual methods, especially when the financial costs of misclassification are accounted for.

2 Preliminaries and Linear Methods for Classification

A (supervised) learning problem is an attempt to predict a certain output using a set of variables (*features* in ML jargon) that are believed to exercise some influence on the output. More specifically, what we are trying to learn is the function $h(\mathbf{x})$ that best describes the relationship between the predictors (the features) and the output. Technically, we are looking for the function $h \in H$ that minimizes a *loss function*.

When the outcome is a categorical variable C (a label), the problem is said to be a classification problem and the function that maps the inputs $\mathbf{x}$ into the output is called *classifier*. The estimate $\hat{C}$ of C takes values in $\mathscr{C}$, the set of all possible classes. As discussed in Sect. 1, credit scoring is usually a classification problem where only two classes are possible, either the applicant is of the "good" (G) or of the "bad" (B) type. In a binary classification problem, the loss function can be represented by a 2×2 matrix L with zeros on the main diagonal and nonnegative values elsewhere. $L(k, l)$ is the cost of classifying an observation belonging to class $\mathscr{C}_k$ as $\mathscr{C}_l$. The expected prediction error (EPE) is

$$EPE = E[L(C, \hat{C}(X))] = E_X \sum_{k=1}^{2} L[\mathscr{C}_k, \hat{C}(X)] p(\mathscr{C}_k | X), \qquad (1)$$

where $\hat{C}(X)$ is the predicted class C based on X (the matrix of the observed features), $\mathscr{C}_k$ represents the class with label k, and $p(\mathscr{C}_k | X)$ is the probability that the actual class has label k conditional to the observed values of the features. Accordingly, the optimal prediction $\hat{C}(X)$ is the one that minimizes the EPE pointwise, i.e.,

$$\hat{C}(x) = \arg\min_{c \in \mathscr{C}} \sum_{k=1}^{2} L(\mathscr{C}_k, c) p(\mathscr{C}_k | X = x), \qquad (2)$$

where x is a realization of the features. Notably, when the loss function is of the 0–1 type, i.e., all misclassifications are charged a unit cost, the problem simplifies to

$$\hat{C}(x) = \arg\min_{c \in \mathscr{C}} [1 - p(c | X = x)], \qquad (3)$$

which means that

$$\hat{C}(x) = \mathscr{C}_k \quad \text{if} \quad p(\mathscr{C}_k | X = x) = \max_{c \in \mathscr{C}} p(c | X = x). \tag{4}$$

In this section, we shall discuss two popular classification approaches that result in linear *decision boundaries*: logistic regressions (LR) and linear discriminant analysis (LDA). In addition, we also introduce the Naïve Bayes method, which is related to LR and LDA as it also considers a *log-odds* scoring function.

2.1 Logistic Regression

Because of its simplicity, LR is still one of the most popular approaches used in the industry for the classification of applicants (see, e.g., [23]). This approach allows one to model the posterior probabilities of K different applicant classes using a linear function of the features, while at the same time ensuring that the probabilities sum to one and that their value ranges between zero and one. More specifically, when there are only two classes (coded via y, a dummy variable that takes a value of 0 if the applicant is "good" and of 1 if she is "bad"), the posterior probabilities are modeled as

$$p(C = G | X = x) = \frac{exp(\beta_0 + \boldsymbol{\beta}^T \mathbf{x})}{1 + exp(\beta_0 + \boldsymbol{\beta}^T \mathbf{x})}$$

$$p(C = B | X = x) = \frac{1}{1 + exp(\beta_0 + \boldsymbol{\beta}^T \mathbf{x})}. \tag{5}$$

Applying the *logit* transformation, one obtains the log of the probability odds (the log-odds ratio) as

$$\log \frac{p(C = G | X = x)}{p(C = B | X = x)} = \beta_0 + \boldsymbol{\beta}^T \mathbf{x}. \tag{6}$$

The input space is optimally divided by the set of points for which the log-odds ratio is zero, meaning that the posterior probability of being in one class or in the other is the same. Therefore, the decision boundary is the hyperplane defined by $\{x | \beta_0 + \boldsymbol{\beta}^T \mathbf{x} = 0\}$. Logistic regression models are usually estimated by maximum likelihood, assuming that all the observations in the sample are independently Bernoulli distributed, such that the log-likelihood functions is

$$\mathscr{L}(\theta | x) = p(y | x; \theta) = \sum_{i=1}^{T_0} \log p_{C_i}(\mathbf{x}_i; \theta), \tag{7}$$

where T_0 are the observations in the training sample, θ is the vector of parameters, and $p_k(\mathbf{x}_i; \theta) = p(C = k|X = \mathbf{x}_i; \theta)$. Because in our case there are only two classes coded via a binary response variable y_i that can take a value of either zero or one, $\hat{\beta}$ is found by maximizing

$$\mathscr{L}(\beta) = \sum_{i=1}^{T_0} (y_i \boldsymbol{\beta}^T \mathbf{x}_i - \log(1 + exp(\boldsymbol{\beta}^T \mathbf{x}_i))). \tag{8}$$

2.2   Linear Discriminant Analysis

A second popular approach used to separate "good" and "bad" applicants that lead to linear decision boundaries is LDA. The LDA method approaches the problem of separating two classes based on a set of observed characteristics $\mathbf{x}$ by modeling the class densities $f_G(\mathbf{x})$ and $f_B(\mathbf{x})$ as multivariate normal distributions with means $\boldsymbol{\mu}_G$, and $\boldsymbol{\mu}_B$ and the same covariance matrix Σ, i.e.,

$$f_G(\mathbf{x}) = (2\pi)^{-K/2} (|\Sigma|)^{-1/2} \exp\left(-\frac{1}{2}(\mathbf{x} - \boldsymbol{\mu}_\mathbf{G})^T \Sigma^{-1}(\mathbf{x} - \boldsymbol{\mu}_\mathbf{G})\right)$$

$$f_B(\mathbf{x}) = (2\pi)^{-K/2} (|\Sigma|)^{-1/2} \exp\left(-\frac{1}{2}(\mathbf{x} - \boldsymbol{\mu}_\mathbf{B})^T \Sigma^{-1}(\mathbf{x} - \boldsymbol{\mu}_\mathbf{B})\right). \tag{9}$$

To compare the two classes ("good" and "bad" applicants), one has then to compute and investigate the log-ratio

$$\log \frac{p(C = G|X = x)}{p(C = B|X = x)} = \log \frac{f_G(\mathbf{x})}{f_B(\mathbf{x})} + \log \frac{\pi_G}{\pi_B}$$

$$= \log \frac{\pi_G}{\pi_B} - \frac{1}{2}(\boldsymbol{\mu}_\mathbf{B} + \boldsymbol{\mu}_\mathbf{G})^T \Sigma^{-1}(\boldsymbol{\mu}_\mathbf{B} + \boldsymbol{\mu}_\mathbf{G}) + \mathbf{x}^T \Sigma^{-1}(\boldsymbol{\mu}_\mathbf{B} + \boldsymbol{\mu}_\mathbf{G}), \tag{10}$$

which is linear in $\mathbf{x}$. Therefore, the decision boundary, which is the set where $p(C = G|X = x) = p(C = B|X = x)$, is also linear in $\mathbf{x}$. Clearly the Gaussian parameters $\boldsymbol{\mu}_G, \boldsymbol{\mu}_B$, and Σ are not known and should be estimated using the training sample as well as the prior probabilities π_G and π_B (set to be equal to the proportions of good and bad applicants in the training sample). Rearranging Eq. (10), it appears evident that the Bayesian optimal solution is to predict a point to belong to the "bad" class if

$$\mathbf{x}^T \hat{\Sigma}^{-1}(\hat{\boldsymbol{\mu}}_B - \hat{\boldsymbol{\mu}}_G) > \frac{1}{2}\hat{\boldsymbol{\mu}}_B^T \hat{\Sigma}^{-1} \hat{\boldsymbol{\mu}}_B - \frac{1}{2}\hat{\boldsymbol{\mu}}_G^T \hat{\Sigma}^{-1} \hat{\boldsymbol{\mu}}_G + \log \hat{\pi}_G - log\hat{\pi}_G, \tag{11}$$

which can be rewritten as

$$\mathbf{x}^T \mathbf{w} > z \tag{12}$$

where $\mathbf{w} = \hat{\Sigma}^{-1}(\hat{\boldsymbol{\mu}}_B - \hat{\boldsymbol{\mu}}_G)$ and $z = \frac{1}{2}\hat{\boldsymbol{\mu}}_B^T \hat{\Sigma}^{-1} \hat{\boldsymbol{\mu}}_B - \frac{1}{2}\hat{\boldsymbol{\mu}}_G^T \hat{\Sigma}^{-1} \hat{\boldsymbol{\mu}}_G + \log \hat{\pi}_G - log\hat{\pi}_G$.

Another way to approach the problem, which leads to the same coefficients $\mathbf{w}$ is to look for the linear combination of the features that gives the maximum separation between the means of the classes and the minimum variation within the classes, which is equivalent to maximizing the separating distance M

$$M = \omega^T \frac{\hat{\mu}_G - \hat{\mu}_B}{(\omega^T \hat{\Sigma} \omega)^{1/2}}. \tag{13}$$

Notably, the derivation of the coefficients $\mathbf{w}$ does not require that $f_G(\mathbf{x})$ and $f_B(\mathbf{x})$ follow a multivariate normal as postulated in Eq. (9), but only that $\Sigma_G = \Sigma_B = \Sigma$. However, the choice of z as a cut-off point in Eq. (12) requires normality. An alternative is to use a cut-off point that minimizes the training error for a given dataset.

2.3 Naïve Bayes

The Naïve Bayes (NB) approach is a probabilistic classifier that assumes that given a class (G or B), the applicant's attributes are independent. Let π_G denote the prior probability that an applicant is "good" and π_B the prior probability that an applicant is "bad." Then, because of the assumption that each attribute x_i is conditionally independent from any other attribute x_j for $i \neq j$, the following holds:

$$p(\mathbf{x} \mid G) = p(x_1 \mid G) p(x_2 \mid G) \ldots p(x_n \mid G), \tag{14}$$

where $p(\mathbf{x} \mid G)$ is the probability that a "good" applicant has attributes $\mathbf{x}$. The probability of an applicant being "good" if she is characterized by the attributes $\mathbf{x}$ can now be found by applying Bayes' theorem:

$$p(G \mid \mathbf{x}) = \frac{p(\mathbf{x} \mid G) \pi_G}{p(\mathbf{x})}. \tag{15}$$

The probability of an applicant being "bad" if she is characterized by the attributes $\mathbf{x}$ is

$$p(B \mid \mathbf{x}) = \frac{p(\mathbf{x} \mid B) \pi_B}{p(\mathbf{x})}. \tag{16}$$

The attributes $\mathbf{x}$ are typically converted into a score, $s(\mathbf{x})$, which is such that $p(G \mid \mathbf{x}) = p(G \mid s(\mathbf{x}))$. A popular score function is the log-odds score [42]:

$$s(\mathbf{x}) = \log\left(\frac{p(G \mid \mathbf{x})}{p(B \mid \mathbf{x})}\right) = \log\left(\frac{\pi_G p(\mathbf{x} \mid G)}{\pi_B p(\mathbf{x} \mid B)}\right) =$$

$$= \log\left(\frac{\pi_G}{\pi_B}\right) + \log\left(\frac{p(\mathbf{x} \mid G)}{p(\mathbf{x} \mid B)}\right) = s_{pop} + woe(\mathbf{x}), \tag{17}$$

where s_{pop} is the log of the relative proportion of "good" and "bad" applicants in the population and $woe(\mathbf{x})$ is the weight of evidence of the attribute combination $\mathbf{x}$. Because of the conditional independence of the attributes, we can rewrite Eq. (17) as

$$s(\mathbf{x}) = \ln\left(\frac{\pi_G}{\pi_B}\right) + \ln\left(\frac{p(x_1|G)}{p(x_1|B)}\right) + \ldots + \ln\left(\frac{p(x_n|G)}{p(x_n|B)}\right)$$

$$= s_{pop} + woe(x_1) + woe(x_2) + \ldots + woe(x_n). \quad (18)$$

If $woe(x_i)$ is equal to 0, then this attribute does not affect the estimation of the status of an applicant. The prior probabilities π_G and π_B are estimated using the proportions of good and bad applicants in the training sample; the same applies to the weight of evidence of the attributes, as illustrated in the example below.

Example Let us assume that a bank makes a lending decision based on two attributes: the residential status and the monthly income of the applicant. The data belonging to the training sample are given in Fig. 1. An applicant who has a monthly income of USD 2000 and owns a flat, will receive a score of:

$$s(\mathbf{x}) = \ln\left(\frac{1300}{300}\right) + \ln\left(\frac{950/1300}{150/300}\right) + \ln\left(\frac{700/1300}{100/300}\right) = 2.32.$$

If $p(G \mid s(\mathbf{x}) = 2.32)$, the conditional probability of being "good" when the score is 2.32, is higher than $p(B \mid s(\mathbf{x}) = 2.32)$, i.e., the conditional probability of being "bad," this applicant is classified as "good" (and vice versa).

	Owner		Not owner		Total	
Income	G	B	G	B	G	B
$1000-	200	50	150	100	350	150
$1000+	500	50	450	100	950	150
Total	700	100	600	200	1300	300

Fig. 1 This figure provides the number of individuals in each cluster in a fictional training sample used to illustrate the NB approach. Two binary attributes are considered: the residential status (either "owner" or "not owner") and monthly income (either more than USD 1000 or less than USD 1000). Source: Thomas et al. [42]

A lender can therefore define a cutoff score, below which applicants are automatically rejected as "bad." Usually, the score $s(\mathbf{x})$ is linearly transformed so that its interpretation is more straightforward. The NB classifier performs relatively well in many applications but, according to Thomas et al. [42], it shows poor performance in the field of credit scoring. However, its most significant advantage is that it is easy to interpret, which is a property of growing importance in the industry.

3   Nonlinear Methods for Classification

Although simple linear methods are still fairly popular with practitioners, because of their simplicity and their satisfactory accuracy [29], more than 25% of the financial companies have recently adopted at least one advanced ML solution in their day-to-day business processes [12], as emphasized in Sect. 1. Indeed, these models have the advantage of being much more flexible and they may be able to uncover complex, nonlinear relationships in the data. For instance, the popular LDA approach postulates that an applicant will be "bad" if her/his score exceeds a given threshold; however, the path to default may be highly nonlinear in the mapping between scores and probability of default (see [39]).

Therefore, in this section, we review several popular ML techniques for classification, such as Decision Trees (DT), Neural Network (NN), Support Vector Machines (SVM), k-Nearest Neighbor (k-NN), and Genetic Algorithms (GA). Even if GA are not exactly classification methods, evolutionary computing techniques that help to find the "fittest" solution, we cover them in our chapter as this method is widely used in credit scoring applications (see, e.g., [49, 35, 1]). Finally, we discuss ensemble methods that combine different classifiers to obtain better classification accuracy. For the sake of brevity, we do not cover deep learning techniques, which are also employed for credit scoring purposes; the interested reader can find useful references in [36].

3.1   Decision Trees

Decision Trees (also known as Classification Trees) are a classification method that uses the training dataset to construct decision rules organized into tree-like structures, where each branch represents an association between the input values and the output label. Although different algorithms exist (such as classification and regression trees, also known as CART), we focus on the popular C4.5 algorithm developed by Quinlan [37]. At each node, the C4.5 algorithm splits the training dataset according to the most influential feature through an iterative process. The most influential feature is the one with the lowest entropy (or, similarly, with the highest information gain). Let $\hat{\pi}_G$ be the proportion of "good" applicants and $\hat{\pi}_B$ the proportion of "bad" applicants in the sample S. The entropy of S is then defined as in Baesens et al. [5]:

$$\text{Entropy}\,(S) = -\hat{\pi}_G \log_2\left(\hat{\pi}_G\right) - \hat{\pi}_B \log_2\left(\hat{\pi}_B\right). \tag{19}$$

According to this formula, the maximum value of the entropy is equal to 1 when $\hat{\pi}_G = \hat{\pi}_B = 0.5$ and it is minimal at 0, which happens when either $\hat{\pi}_G = 0$ or $\hat{\pi}_B = 0$. In other words, an entropy of 0 means that we have been able to identify the characteristics that lead to a group of good (bad) applicants. In order to split the sample, we compute the gain ratio:

$$\text{Gain ratio}\,(S, x_i) = \frac{\text{Gain}\,(S, x_i)}{\text{Split Information}(S, x_i)}. \tag{20}$$

$Gain\,(S, x_i)$ is the expected reduction in entropy due to splitting the sample according to feature x_i and it is calculated as

$$\text{Gain}\,(S, x_i) = \text{Entropy}\,(S) - \sum_{\upsilon} \frac{|S_{\upsilon}|}{|S|} \text{Entropy}\,(S_{\upsilon}), \tag{21}$$

where $\upsilon \in \text{values}(x_i)$, S_{υ} is a subset of the individuals in S that share the same value of the feature x_i, and

$$\text{Split Information}\,(S, x_i) = -\sum_{k} \frac{|S_k|}{|S|} \log_2 \frac{|S_k|}{|S|} \tag{22}$$

where $k \in \text{values}(x_i)$ and S_k is a subset of the individuals in S that share the same value of the feature x_i. The latter term represents the entropy of S relative to the feature x_i. Once such a tree has been constructed, we can predict the probability that a new applicant will be a "bad" one using the proportion of "bad" customers in the leaf that corresponds to the applicant's characteristics.

3.2 Neural Networks

NN models were initially inspired by studies of the human brain [8, 9]. A NN model consists of input, hidden, and output layers of interconnected neurons. Neurons in one layer are combined through a set of weights and fed to the next layer. In its simplest single-layer form, a NN consists of an input layer (containing the applicants' characteristics) and an output layer. More precisely, a single-layer NN is modeled as follows:

$$u_k = \omega_{k0} + \sum_{i=1}^{n} \omega_{ki} x_i$$

$$y_k = f\,(u_k), \tag{23}$$

where $x_1, \ldots, x_n$ are the applicant's characteristics, which in a NN are typically referred to as *signals*, $\omega_{k1}, \ldots, \omega_{kn}$ are the weights connecting characteristic i to the layer k (also called *synaptic weights*), and ω_{k0} is the "bias" (which plays a similar role to the intercept term in a linear regression). Eq. (23) describes a single-layer NN, so that $k = 1$. A positive weight is called *excitatory* because it increases the effect of the corresponding characteristic, while a negative weight is called *inhibitory* because it decreases the effect of a positive characteristic [42]. The function f that transform the inputs into the output is called *activation function* and may take a number of specifications. However, in binary classification problems, it may be convenient to use a logistic function, as it produces an output value in the range [0, 1]. A cut-off value is applied to y_k to decide whether the applicant should be classified as good or bad. Figure 2 illustrates how a single-layer NN works.

A single-layer NN model shows a satisfactory performance only if the classes can be linearly separated. However, if the classes are not linearly separable, a multilayer model could be used [33]. Therefore, in the rest of this section, we describe multilayer perceptron (MLP) models, which are the most popular NN models in classification problems [5]. According to Bishop [9], even though multiple hidden layers may be used, a considerable number of papers have shown that MLP NN models with one hidden layer are universal nonlinear discriminant functions that can approximate arbitrarily well any continuous function. An MLP model with one hidden layer, which is also called a three-layer NN, is shown in Fig. 3. This model can be represented algebraically as

$$y_k = f^{(1)}\left(\sum_{i=0}^{n} \omega_{ki} x_i\right), \tag{24}$$

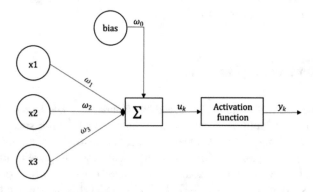

Fig. 2 The figure illustrates a single-layer NN with one output neuron. The applicant's attributes are denoted by $x_1, \ldots, x_n$, the weights are denoted by $\omega_1, \ldots, \omega_n$, and ω_0 is the "bias." The function f is called activation function and it transforms the sum of the weighted applicant's attributes to a final value. Source: Thomas et al. [42]

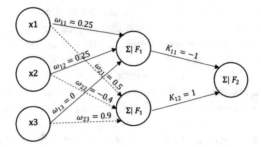

Fig. 3 The figure shows the weights of a three-layer MLP NN model, where the input characteristics are the following dummy variables: x_1 is equal to one if the monthly income is low; x_2 takes the value of one if the client has no credit history with the bank; x_3 represents the applicant's residential status

where $f^{(1)}$ is the activation function on the second (hidden) layer and y_k for $k = 1 \ldots, r$ are the outputs from the hidden layer that simultaneously represent the inputs to the third layer. Therefore, the final output values z_v can be written as

$$z_v = f^{(2)} \left(\sum_{k=1}^{r} K_{vk} y_k \right) = f^{(2)} \left(\sum_{k=1}^{r} K_{vk} f^{(1)} \left(\sum_{i=0}^{n} \omega_{ki} x_i \right) \right) \tag{25}$$

where $f^{(2)}$ is the activation function of the third (output) layer, z_v for $v = 1, \ldots, s$ are the final outputs, and K_{vk} are the weights applied to the y_k values. The estimation of the weights is called *training* of the model and to this purpose the most popular method is the back-propagation algorithm, in which the pairs of input values and output values are presented to the model many times with the goal of finding the weights that minimize an error function [42].

3.3   Support Vector Machines

The SVM method was initially developed by Vapnik [43]. The idea of this method is to transform the input space into a high-dimensional feature space by using a nonlinear function $\varphi(\bullet)$. Then, a linear classifier can be used to distinguish between "good" and "bad" applicants. Given a training dataset of N pairs of observations $(\mathbf{x}_i, y_i)_{i=1}^{N}$, where $\mathbf{x}_i$ are the attributes of customer i and y_i is the corresponding binary label, such that $y_i \in [-1, +1]$, the SVM model should satisfy the following conditions:

$$\begin{cases} \mathbf{w}^T \varphi(\mathbf{x}_i) + b \geq +1 \ \text{if} \ y_i = +1 \\ \mathbf{w}^T \varphi(\mathbf{x}_i) + b \leq -1 \ \text{if} \ y_i = -1, \end{cases}$$

Fig. 4 This figure illustrates the main concept of an SVM model. The idea is to maximize the perpendicular distance between the support vectors and the separating hyperplane. Source: Baesens et al. [5]

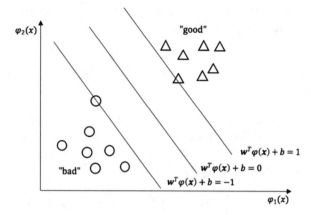

which is equivalent to

$$y_i \left[\mathbf{w}^T \varphi\left(\mathbf{x}_i\right) + b \right] \geq 1, \quad i = 1, \ldots, N. \tag{26}$$

The above inequalities construct a hyperplane in the feature space, defined by $\left\{x \mid \mathbf{w}^T \varphi\left(\mathbf{x}_i\right) + b = 0\right\}$, which distinguishes between two classes (see Fig. 4 for the illustration of a simple two-dimensional case). The observations on the lines $\mathbf{w}^T \varphi\left(x_i\right) + b = 1$ and $\mathbf{w}^T \varphi\left(x_i\right) + b = -1$ are called the *support vectors*. The parameters of the separating hyperplane are estimated by maximizing the perpendicular distance (called the *margin*), between the closest support vector and the separating hyperplane while at the same time minimizing the misclassification error.

The optimization problem is defined as:

$$\begin{cases} \min_{w,b,\xi} \; \mathscr{J}\left(\mathbf{w}, \, b, \, \xi\right) = \frac{1}{2}\mathbf{w}^T \mathbf{w} + C \sum_{i=1}^{N} \xi_i, \\ \text{subject to:} \\ y_i \left[\mathbf{w}^T \, \varphi\left(x_i\right) + b \right] \geq 1 - \xi_i, \quad i = 1, \ldots, N \\ \xi_i \geq 0, \quad i = 1, \ldots, N, \end{cases} \tag{27}$$

where the variables ξ_i are slack variables and C is a positive tuning parameter [5]. The Lagrangian to this optimization problem is defined as follows:

$$\mathscr{L}\left(\mathbf{w}, b, \xi; \boldsymbol{\alpha}, \, \boldsymbol{v}\right) = \mathscr{J}\left(\mathbf{w}, b, \xi\right) - \sum_{i=1}^{N} \alpha_i \left\{ y_i \left[\mathbf{w}^T \, \varphi\left(\mathbf{x_i}\right) + b \right] - 1 + \xi_i \right\} - \sum_{i=1}^{N} v_i \xi_i. \tag{28}$$

The classifier is obtained by minimizing $\mathscr{L}\left(\mathbf{w}, b, \xi; \boldsymbol{\alpha}, \boldsymbol{v}\right)$ with respect to $\mathbf{w}, b, \xi$ and maximizing it with respect to $\boldsymbol{\alpha}, \boldsymbol{v}$. In the first step, by taking the derivatives with respect to $\mathbf{w}, b, \xi$, setting them to zero, and exploiting the results, one may represent the classifier as

$$y(\mathbf{x}) = sign\left(\sum_{i=1}^{N} \alpha_i y_i K(\mathbf{x}_i, \mathbf{x}) + B\right) \tag{29}$$

where $K(\mathbf{x}_i, \mathbf{x}) = \varphi(\mathbf{x}_i)^T \varphi(\mathbf{x}_i)$ is computed using a positive-definite kernel function. Some possible kernel functions are the radial basis function $K(\mathbf{x}_i, \mathbf{x}) = exp(-\|x_i - x_j\|_2^2/\sigma^2)$ and the linear function $K(\mathbf{x}_i, \mathbf{x}) = \mathbf{x}_i^T \mathbf{x}_j$. At this point, the Lagrange multipliers α_i can be found by solving:

$$\begin{cases} \max_{\alpha_i} -\frac{1}{2} \sum_{i,j=1}^{N} y_i y_j K(\mathbf{x}_i, \mathbf{x}_j) \alpha_i \alpha_j + \sum_{i=1}^{N} \alpha_i \\ \text{subject to:} \\ \sum_{i=1}^{N} \alpha_i y_i = 0 \\ 0 \leq \alpha_i \leq C, \quad i = 1, \dots, N. \end{cases}$$

3.4 k-Nearest Neighbor

In the k-NN method, any new applicant is classified based on a comparison with the training sample using a distance metric. The approach consists of calculating the distances between the new instance that needs to be classified and each instance in the training sample that has been already classified and selecting the set of the k-nearest observations. Then, the class label is assigned according to the most common class among k-nearest neighbors using a majority voting scheme or a distance-weighted voting scheme [41]. One major drawback of the k-NN method is that it is extremely sensitive to the choice of the parameter k, as illustrated in Fig. 5. Given the same dataset, if k=1 the new instance is classified as "bad," while if k=3 the neighborhood contains one "bad" and two "good" applicants, thus, the new instance will be classified as "good." In general, using a small k leads to overfitting (i.e., excessive adaptation to the training dataset), while using a large k reduces accuracy by including data points that are too far from the new case [41].

The most common choice of a distance metric is the Euclidean distance, which can be computed as:

$$d(\mathbf{x}_i, \mathbf{x}_j) = \|\mathbf{x}_i - \mathbf{x}_j\| = \left[(\mathbf{x}_i - \mathbf{x}_j)^T (\mathbf{x}_i - \mathbf{x}_j)\right]^{\frac{1}{2}} \tag{30}$$

where $\mathbf{x}_i$ and $\mathbf{x}_j$ are the vectors of the input data of instances i and j, respectively. Once the distances between the newest and every instance in the training sample are calculated, the new instance can be classified based on the information available

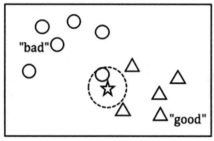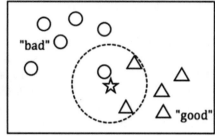

Fig. 5 The figure illustrates the main problem of a k-NN method with the majority voting approach: its sensitivity to the choice of k. On the left side of the figure, a model with k=1 is shown. Based on such a model, the new client (marked by a star symbol) would be classified as "bad." However, on the right side of the figure, a model with k=3 classifies the same new client as "good." Source: Tan et al. [41]

from its k-nearest neighbors. As seen above, the most common approach is to use the majority class of k-nearest examples, the so-called majority voting approach

$$y^{new} = \arg\max_{v} \sum_{(x_i, y_i) \in S_k} I(v = y_i), \tag{31}$$

where y^{new} is the class of the new instance, v is a class label, S_k is the set containing k-closest training instances, y_i is the class label of one of the k-nearest observations, and $I(\bullet)$ is a standard indicator function.

The major drawback of the majority voting approach is that it gives the same weight to every k-nearest neighbor. This makes the method very sensitive to the choice of k, as discussed previously. However, this problem might be overcome by attaching to each neighbor a weight based on its distance from the new instance, i.e.,

$$\omega_i = \frac{1}{d\left(\mathbf{x}_i, \mathbf{x}_j\right)^2} \tag{32}$$

This approach is known as the distance-weighted voting scheme, and the class label of the new instance can be found in the following way:

$$y^{new} = \arg\max_{v} \sum_{(x_i, y_i) \in S_k} \omega_i I(v = y_i), \tag{33}$$

One of the main advantages of k-NN is its simplicity. Indeed, its logic is similar to the process of traditional credit decisions, which were made by comparing a new applicant with similar applicants [10]. However, because estimation needs to be performed afresh when one is to classify a new instance, the classification speed may be slow, especially with large training samples.

3.5   Genetic Algorithms

GA are heuristic, combinatorial optimization search techniques employed to determine automatically the adequate discriminant functions and the valid attributes [35].
The search for the optimal solution to a problem with GA imitates the evolutionary
process of biological organisms, as in Darwin's natural selection theory. In order
to understand how a GA works in the context of credit scoring, let us suppose that
$(x_1, \ldots, x_n)$ is a set of attributes used to decide whether an applicant is good or bad
according to a simple linear function:

$$y = \beta_0 + \sum_{i=1}^{N} \beta_i x_i. \tag{34}$$

Each solution is represented by the vector $\boldsymbol{\beta} = (\beta_0, \beta_1, \ldots, \beta_N)$ whose elements
are the coefficients assigned to each attribute. The initial step of the process is the
generation of a random population of solutions $\boldsymbol{\beta}_j^0$ and the evaluation of their fitness
using a fitness function. Then, the following algorithms are applied:

1. **Selection**: a genetic operator selects the solutions that survive (the fittest
   solutions)
2. **Crossover**: a genetic operator recombines the survived solutions
3. **Mutation**: a genetic operator allows for random mutations in the survived
   solutions, with a low probability

The application of these algorithms results in the generation of a new population of
solutions $\boldsymbol{\beta}_j^0$. The algorithms selection-crossover-mutation are applied recursively
until an (approximate) optimal solution $\boldsymbol{\beta}_j^*$ is converged to.

Compared to traditional statistical approaches and NN, GA offers the advantage
of not being limited in its effectiveness by the form of functions and parameter
estimation [11]. Furthermore, GA is a nonparametric tool that can perform well
even in small datasets [34].

3.6   Ensemble Methods

In order to improve the accuracy of the individual (or *base*) classifiers illustrated
above, ensemble (or classifier combination) methods are often used [41]. Ensemble
methods are based on the idea of training multiple models to solve the same problem
and then combine them to get better results. The main hypothesis is that when
weak models are correctly combined, we can obtain more accurate and/or robust
models. In order to understand why ensemble classifiers may reduce the error rate
of individual models, it may be useful to consider the following example.

Example Suppose that an ensemble classifier is created by using 25 different base classifiers and that each classifier has an error rate $\epsilon_i = 0.25$. If the final credit decision is taken through a majority vote (i.e., if the majority of the classifiers suggests that the customer is a "good" one, then the credit is granted), the error rate of the ensemble model is

$$\epsilon_{ensemble} = \sum_{i=13}^{25} \binom{25}{i} \epsilon^i (1 - \epsilon)^{25-i} = 0.003, \tag{35}$$

where $i = 13, \ldots, 25$, which is much less than the individual rate of 0.25, because the ensemble model would make a wrong decision only if more than half of the base classifiers yield a wrong estimate.

It is easy to understand that ensemble classifiers perform especially well when they are uncorrelated. Although in real-world applications it is difficult to obtain base classifiers that are totally uncorrelated, considerable improvements in the performance of ensemble classifiers are observed even when some correlations exists but are low [17]. Ensemble models can be split into homogeneous and heterogeneous. Homogeneous ensemble models use only one type of classifier and rely on resampling techniques to generate k different classifiers that are then aggregated according to some rule (e.g., majority voting). Examples of homogeneous ensemble models are *bagging* and *boosting* methods. More precisely, the bagging algorithm creates k bootstrapped samples of the same size as the original one by drawing with replacement from the dataset. All the samples are created in parallel and the estimated classifiers are aggregated according to majority voting. Boosting algorithms work in the same spirit as bagging but the models are not fitted in parallel: a sequential approach is used and at each step of the algorithm the model is fitted, giving more importance to the observations in the training dataset that were badly handled in the previous iteration. Although different boosting algorithms are possible, one of the most popular is AdaBoost. AdaBoost was first introduced by Freund and Schapire [19]. This algorithm starts by calculating the error of a base classifier h_t:

$$\epsilon_t = \frac{1}{N} \left[\sum_{j=1}^{N} \omega_j I \left(h_t \left(x_j \right) \neq y_j \right) \right]. \tag{36}$$

Then, the importance of the base classifier h_t is calculated as:

$$\alpha_t = \frac{1}{2} \ln \left(\frac{1 - \epsilon_t}{\epsilon_t} \right). \tag{37}$$

The parameter α_t is used to update the weights assigned to the training instances. Let $\omega_i^{(t)}$ be the weight assigned to the training instance i in the tth boosting round. Then, the updated weight is calculated as:

$$\omega_i^{(t+1)} = \frac{\omega_i^{(t)}}{Z_t} \times \begin{cases} \exp(-\alpha_t) & \text{if } h_t(\mathbf{x}_i) = y_i \\ \exp(\alpha_t) & \text{if } h_t(\mathbf{x}_i) \neq y_i, \end{cases} \tag{38}$$

where Z_t is the normalization factor, such that $\sum_i \omega_i^{(t+1)} = 1$. Finally, the AdaBoost algorithm decision is based on

$$h(\mathbf{x}) = sign\left(\sum_{t=1}^{T} \alpha_t h_t(\mathbf{x})\right). \tag{39}$$

In contrast to homogeneous ensemble methods, heterogeneous ensemble methods combine different types of classifiers. The main idea behind these methods is that different algorithms might have different views on the data and thus combining them helps to achieve remarkable improvements in predictive performance [47]. An example of heterogeneous ensemble method can be the following:

1. Create a set of different classifiers $H = \{h_t, t = 1, \ldots, T\}$ that map an instance in the training sample to a class y_i: $h_t(\mathbf{x}_i) = y_i$.
2. Start with an empty ensemble ($S = \emptyset$).
3. Add to the ensemble the model from the initial set that maximizes the performance of the ensemble on the validation dataset according to the error metric.
4. Repeat Step 3 for k iterations, where k is usually less than the number of models in the initial set.

A comparative evaluation of alternative ensemble methods is provided in Sect. 4.2.

4   Comparison of Classifiers in Credit Scoring Applications

The selection of the best classification algorithm among all methods that have been proposed in the literature has always been a challenging research area. Although many studies have examined the performance of different classifiers, most of these papers have traditionally focused only on a few novel algorithms at the time and, thus, have generally failed to provide a comprehensive overview of pros and cons of alternative methods. Moreover, in most of these papers, a relatively small number of datasets were used, which limited the practical applicability of the empirical results reported. One of the most comprehensive studies that attempts to overcome these issues and to apply thorough statistical tests to compare different algorithms has been published by Stefan Lessmann and his coauthors [29]. By combining their results with other, earlier studies, this section seeks to isolate the best classification algorithms for the purposes of credit scoring.

4.1 Comparison of Individual Classifiers

In the first decade of the 2000s, the focus of most papers had been on performing comparisons among individual classifiers. Understandably, the question of whether advanced methods of classification, such as NN and SVM, might outperform LR and LDA had attracted much attention. While some authors have since then concluded that NN classifiers are superior to both LR and LDA (see, e.g., [2]), generally, it has been shown that simple linear classifiers lead to a satisfactory performance and, in most cases, that the differences between NN and LR are not statistically significant [5]. This section compares the findings of twelve papers concerning individual classifiers in the field of credit scoring. Papers were selected based on two features: first, the number of citations, and, second, the publishing date. The sample combines well-known papers (i.e., [45, 5]) with recent work (e.g., [29, 3]) in an attempt to provide a well-rounded overview.

One of the first comprehensive comparisons of linear methods with more advanced classifiers was West [45]. He tested five NN models, two parametric models (LR, LDA), and three nonparametric models (k-NN, kernel density, and DT) on two real-world datasets. He found that in the case of both datasets, LR led to the lowest credit scoring error, followed by the NN models. He also found that the differences in performance scores of the superior models (LR and three different way to implement NN) vs. the outperformed models were not statistically significant. Overall, he concluded that LR was the best choice among individual classifiers he tested. However, his methodology presented a few drawbacks that made some of his findings potentially questionable. First, West [45] used only one method of performance evaluation and ranking, namely, average scoring accuracy. Furthermore, the size of his datasets was small, containing approximately 1700 observations in total (1000 German credit applicants, 700 of which were creditworthy, and 690 Australian applicants, 307 of which were creditworthy).

Baesens et al. [5] remains one of the most comprehensive comparisons of different individual classification methods. This paper overcame the limitations in West [45] by using eight extensive datasets (for a total of 4875 observations) and multiple evaluation methods, such as the percentage of correctly classified cases, sensitivity, specificity, and the area under the receiver operating curve (henceforth, AUC, an accuracy metric that is widely used when evaluating different classifiers).[3] However, the results reported by Baesens et al. [5] were similar to West's [45]: NN

---

[3] A detailed description of the performance measurement metrics that are generally used to evaluate the accuracy of different classification methods can be found in the previous chapter by Bargagli-Stoffi et al. [6].

and SVM classifiers had the best average results; however, also LR and LDA showed a very good performance, suggesting that most of the credit datasets are only weakly nonlinear. These results have found further support in the work of Lessmann et al. [29], who updated the findings in [5] and showed that NN models perform better than LR model, but only slightly.[4]

These early papers did not contain any evidence on the performance of GA. One of the earliest papers comparing genetic algorithms with other credit scoring models is Yobas et al. [49], who compared the predictive performance of LDA with three computational intelligence techniques (a NN, a decision tree, and a genetic algorithm) using a small sample (1001 individuals) of credit scoring data. They found that LDA was superior to genetic algorithms and NN. Fritz and Hosemann [20] also reached a similar conclusion even though doubts existed on their use of the same training and test sets for different techniques. Recently, these early results have been overthrown. Ong et al. [35] compared the performance of genetic algorithms to MLP, decision trees (CART and C4.5), and LR using two real-world datasets, which included 1690 observations. Genetic algorithms turned out to outperform other methods, showing a solid performance even on relatively small datasets. Huang et al. [26] compared the performance of GA against NN, SVM, and decision tree models in a credit scoring application using the Australian and German benchmark data (for a total of almost 1700 credit applicants). Their study revealed superior classification accuracy from GA than under other techniques, although differences are marginal. Abdou [1] has investigated the relative performance of GA using data from Egyptian public sector banks, comparing this technique with probit analysis, reporting that GA achieved the highest accuracy rate and also the lowest type-I and type-II errors when compared with other techniques.

One more recent and comprehensive study is that of Finlay [16], who evaluated the performance of five alternative classifiers, namely, LR, LDA, CART, NN, and k-NN, using the rather large dataset of Experian UK on credit applications (including a total of 88,789 applications, 13,261 of which were classified as "bad"). He found that the individual model with the best performance is NN; however, he also showed that the overperformance of nonlinear models over their linear counterparts is rather limited (in line with [5]).

Starting in 2010, most papers have shifted their focus to comparisons of the performance of ensemble classifiers, which are covered in the next section. However, some recent studies exist that evaluate the performance of individual classifiers. For instance, Ala'raj and Abbod [2] (who used five real-world datasets for a total of 3620 credit applications) and Bequé and Lessmann [7] (who used three real-world credit datasets for a total of 2915 applications) have found that LR has the best performance among the range of individual classifiers they considered.

---

[4]Importantly, compared to Baesens et al. [5], Lessmann et al. [29] used the more robust H-measure instead of the AUC as a key performance indicator for their analysis. Indeed, as emphasized by Hand [21], the AUC has an important drawback as it uses different misclassification cost distributions for different classifiers (see also Hand and Anagnostopoulos [22]).

Although ML approaches are better at capturing nonlinear relationships, similarly to what is typical in credit risk applications (see [4]), it could be concluded that, in general, a simple LR model remains a solid choice among individual classifiers.

4.2   Comparison of Ensemble Classifiers

According to Lessmann et al. [29], the new methods that have appeared in ML have led to superior performance when compared to individual classifiers. However, only a few papers concerning credit scoring have examined the potential of ensemble methods, and most papers have focused on simple approaches. This section attempts to determine whether ensemble classifiers offer significant improvements in performance when compared to the best available individual classifiers and examines the issue of uncovering which ensemble methods may provide the most promising results. To succeed in this objective, we have selected and surveyed ten key papers concerning ensemble classifiers in the field of credit scoring.

West et al. [46] were among the first researchers to test the relative performance of ensemble methods in credit scoring. They selected three ensemble strategies, namely, cross-validation, bagging, and boosting, and compared them to the MLP NN as a base classifier on two datasets.[5] West and coauthors concluded that among the three chosen ensemble classifiers, boosting was the most unstable and had a mean error higher than their baseline model. The remaining two ensemble methods showed statistically significant improvements in performance compared to MLP NN; however, they were not able to single out which ensemble strategy performed the best since they obtained contrasting results on the two test datasets. One of the main limitations of this seminal study is that only one metric of performance evaluation was employed. Another extensive paper on the comparative performance of ensemble classifiers is Zhou et al.'s [51]. They compared six ensemble methods based on LS-SVM to 19 individual classifiers, with applications to two different real-world datasets (for a total of 1113 observations). The results were evaluated using three different performance measures, i.e., sensitivity, the percentage of correctly classified cases, and AUC. They reported that the ensemble methods assessed in their paper could not lead to results that would be statistically superior to an LR individual classifier. Even though the differences in performance were not large, the ensemble models based on the LS-SVM provided promising solutions to the classification problem that was not worse than linear methods. Similarly, Louzada et al. [30] have recently used three famous and publicly available datasets (the Australian, the German, and the Japanese credit data) to perform simulations under both balanced (p = 0:5, 50% of bad payers) and imbalanced cases (p = 0:1,

---

[5]While bagging and boosting methods work as described in Sect. 3, the cross-validation ensemble, also known as CV, has been introduced by Hansen and Salamon [24] and it consists of an ensemble of similar networks, trained on the same dataset.

10% of bad payers). They report that two methods, SVM and fuzzy complex systems offer a superior and statistically significant predictive performance. However, they also notice that in most cases there is a shift in predictive performance when the method is applied to imbalanced data. Huang and Wu [25] report that the use of boosted GA methods improves the performance of underlying classifiers and appears to be more robust than single prediction methods. Marqués et al. [31] have evaluated the performance of seven individual classifier techniques when used as members of five different ensemble methods (among them, bagging and AdaBoost) on six real-world credit datasets using a fivefold cross-validation method (each original dataset was randomly divided into five stratified parts of equal size; for each fold, four blocks were pooled as the training data, and the remaining part was employed as the hold out sample). Their statistical tests show that decision trees constitute the best solution for most ensemble methods, closely followed by the MLP NN and LR, whereas the k-NN and the NB classifiers appear to be significantly the worst.

All the papers discussed so far did not offer a comprehensive comparison of different ensemble methods, but rather they focused on a few techniques and compared them on a small number of datasets. Furthermore, they did not always adopt appropriate statistical tests of equal classification performance. The first comprehensive study that has attempted to overcome these issues is Lessmann et al. [29], who have compared 16 individual classifiers with 25 ensemble algorithms over 8 datasets. The selected classifiers include both homogeneous (including bagging and boosting) and heterogeneous ensembles. The models were evaluated using six different performance metrics. Their results show that the best individual classifiers, namely, NN and LR, had average ranks of 14 and 16 respectively, being systematically dominated by ensemble methods. Based on the modest performance of individual classifiers, Lessmann et al. [29] conclude that ML techniques have progressed notably since the first decade of the 2000s. Furthermore, they report that heterogeneous ensemble classifiers provide the best predictive performance.

Lessmann et al. [29] have also examined the potential financial implications of using ensemble scoring methods. They considered 25 different cost ratios based on the assumption that accepting a "bad" application always costs more than denying a "good" application [42]. After testing three models (NN, RF, and HCES-Bag) against LR, Lessmann et al. [29] conclude that for all cost ratios, the more advanced classifiers led to significant cost savings. However, the most accurate ensemble classifier, HCES-Bag, on average achieved lower cost savings than the radial basis function NN method, 4.8 percent and 5.7 percent, respectively. Based on these results, they suggested that the most statistically accurate classifier may not always be the best choice for improving the profitability of the credit lending business.

Two additional studies, Florez-Lopez and Ramon-Jeronimo [18] and Xia et al. [48], have focused on the interpretability of ensemble methods, constructing ensemble models that can be used to support managerial decisions. Their empirical results confirmed the findings of Lessmann et al. [29] that ensemble methods consistently lead to better performances than individual scoring. Furthermore, they concluded that it is possible to build an ensemble model that has both high interpretability and a high accuracy rate. Overall, based on the papers considered in this section, it is evident that ensemble models offer higher accuracy compared to the best individual models. However, it is impossible to select one ensemble approach that will have the best performance over all datasets and error costs. We expect that scores of future papers will appear with new, more advanced methods and that the search for "the silver bullet" in the field of credit scoring will not end soon.

4.3   One-Class Classification Methods

Another promising development in credit scoring concerns one-class classification methods (OCC), i.e., ML methods that try to learn from one class only. One of the biggest practical obstacles to applying scoring methods is the class imbalance feature of most (all) datasets, the so-called low-default portfolio problem. Because financial institutions only store historical data concerning the accepted applicants, the characteristics of "bad" applicants present in their data bases may not be statistically reliable to provide a basis for future predictions ([27]. Empirical and theoretical work has demonstrated that the accuracy rate may be strongly biased with respect to imbalance in class distribution and that it may ignore a range of misclassification costs [14], as in applied work it is generally believed that the costs associated with type-II errors (bad customers misclassified as good) are much higher than the misclassification costs associated with type-I errors (good customers mispredicted as bad). OCC attempts to differentiate a set of target instances from all the others. The distinguishing feature of OCC is that it requires labeled instances in the training sample for the target class only, which in the case of credit scoring are "good" applicants (as the number of "good" applicants is larger than that of "bad" applicants). This section surveys whether OCC methods can offer a comparable performance to the best two-class classifiers in the presence of imbalanced data features.

The literature on this topic is still limited. One of the most comprehensive studies is a paper by Kennedy [27], in which he compared eight OCC methods, in which models are separately trained over different classes of datasets, with eight two-class individual classifiers (e.g., k-NN, NB, LR) over three datasets. Two important conclusions emerged. First, the performance of two-class classifiers deteriorates significantly with an increasing class imbalance. However, the performance of some classifiers, namely, LR and NB, remains relatively robust even for imbalanced datasets, while the performance of NN, SVM, and k-NN deteriorates rapidly. Second, one-class classifiers show superior performance compared to two-class classifiers only at high levels of imbalance (starting at 99% of "good" and 1% of "bad" applicants). However, the differences in performance between OCC models and LR model were not statistically significant in most cases. Kennedy [27] concluded that OCC methods failed to show statistically significant improvements in performance compared to the best two-class classification methods. Using a proprietary dataset from a major US commercial bank from January 2005 to April

2009, Khandani et al. [28] showed that conditioning on certain changes in a consumer's bank account activity can lead to considerably more accurate forecasts of credit card delinquencies by analyzing subtle nonlinear patterns in consumer expenditures, savings, and debt payments using CART and SVM compared to simple regression and logit approaches. Importantly, their trees are "boosted" to deal with the imbalanced features of the data: instead of equally weighting all the observations in the training set, they weight the scarcer observations more heavily than the more populous ones.

These findings are in line with studies in other fields. Overall, the conclusion that can be drawn is that OCC methods should not be used for classification problems in credit scoring. Two-class individual classifiers show superior or comparable performance for all cases, except for cases of extreme imbalances.

5 Conclusion

The field of credit scoring represents an excellent example of how the application of novel ML techniques (including deep learning and GA) is in the process of revolutionizing both the computational landscape and the perception by practitioners and end-users of the relative merits of traditional vs. new, advanced techniques. On the one hand, in spite of their logical appeal, the available empirical evidence shows that ML methods often struggle to outperform simpler, traditional methods, such as LDA, especially when adequate tests of equal predictive accuracy are deployed. Although some of these findings may be driven by the fact that some of the datasets used by the researchers (especially in early studies) were rather small (as in the case, for instance, of West [45]), linear methods show a performance that is often comparable to that of ML methods also when larger datasets are employed (see, e.g., Finlay [17]). On the other hand, there is mounting experimental and on-the-field evidence that ensemble methods, especially those that involve ML-based individual classifiers, perform well, especially when realistic cost functions of erroneous classifications are taken into account. In fact, it appears that the issues of ranking and assessing alternative methods under adequate loss functions, and the dependence of such rankings on the cost structure specifications, may turn into a fertile ground for research development.

References

1. Abdou, H. A. (2009). Genetic programming for credit scoring: The case of Egyptian public sector banks. *Expert Systems with Applications, 36*(9), 11402–11417.
2. Abdou, H., Pointon, J., & El-Masry, A. (2008). Neural nets versus conventional techniques in credit scoring in Egyptian banking. *Expert Systems with Applications, 35*(3), 1275–1292.
3. Ala'raj, M., & Abbod, M. F. (2016). Classifiers consensus system approach for credit scoring. *Knowledge-Based Systems, 104*, 89–105.

4. Bacham, D., & Zhao, J. (2017). Machine learning: challenges, lessons, and opportunities in credit risk modelling. *Moody's Analytics Risk Perspectives/Managing Disruptions, IX*, 1–5.
5. Baesens, B., Gestel, T. V., Viaene, S., Stepanova, M., Suykens, J., & Vanthienen, J. (2003) Benchmarking state-of-the-art classification algorithms for credit scoring. *Journal of the Operational Research Society, 54*, 627–635.
6. Bargagli-Stoffi, F. J., Niederreiter, J., & Riccaboni, M. (2021). Supervised learning for the prediction of firm dynamics. In S. Consoli, D. Reforgiato Recupero, & M. Saisana (Eds.) *Data Science for Economics and Finance: Methodologies and Applications* (pp. 19–41). Switzerland: Springer-Nature.
7. Bequé, A., & Lessmann, S. (2017). Extreme learning machines for credit scoring: An empirical evaluation. *Expert Systems with Applications, 86*, 42–53.
8. Bishop, C. (1994). Novelty detection and neural network validation. *IEE Proceedings on Vision, Image and Signal Processing, 141*, 217–222.
9. Bishop, C. M. (1999). *Neural Networks for Pattern Recognition*. Oxford, United Kingdom: Oxford University.
10. Bunker, R., Naeem, A., & Zhang, W. (2016). *Improving a credit scoring model by incorporating bank statement derived features*. Working paper, Auckland University of Technology. arXiv, CoRR abs/1611.00252.
11. Chi, B., & Hsu, C. (2011). A hybrid approach to integrate genetic algorithm into dual scoring model in enhancing the performance of credit scoring model. *Expert Systems with Applications, 39*, 2650–2661.
12. Chui, M., Manyika, J., & Miremadi, M. (2018). *What AI can and can't do (yet) for your business*. https://www.mckinsey.com/business-functions/mckinsey-analytics/our-insights/what-ai-can-and-cant-do-yet-for-your-business.
13. Dorie, V., Hill, J., Shalit, U., Scott, M., & Cervone, D. (2019). Automated versus do-it-yourself methods for causal inference: Lessons learned from a data analysis competition? *Statistical Science, 34*, 43–68.
14. Fawcett, T. & Provost, F. (1997). Adaptive fraud detection. *Data Mining and Knowledge Discovery, 1*(3), 291–316.
15. Federal Reserve Bank of New York (2020). *Household debt and credit report (Q4 2020), Center FRO Microeconomic data*. https://www.newyorkfed.org/microeconomics/hhdc.
16. Finlay, S. M. (2009). Are we modelling the right thing? The impact of incorrect problem specification in credit scoring. *Expert Systems with Applications, 36*(5), 9065–9071.
17. Finlay, S. (2011). Multiple classifier architectures and their application to credit risk assessment. *European Journal of Operational Research, 210*, 368–378.
18. Florez-Lopez, R., & Ramon-Jeronimo, J. M. (2015). Enhancing accuracy and interpretability of ensemble strategies in credit risk assessment. A correlated-adjusted decision forest proposal. *Expert Systems with Applications, 42*, 5737–5753.
19. Freund, Y., & Schapire, R. E. (1997). A decision-theoretic generalization of on-line learning and an application to boosting. *Journal of Computer and System Sciences, 55*, 119–139.
20. Fritz, S., & Hosemann, D. (2000). Restructuring the credit process: Behaviour scoring for German corporates. *Intelligent Systems in Accounting, Finance & Management, 9*(1), 9–21.
21. Hand, D. J. (2009). Measuring classifier performance: A coherent alternative to the area under the roc curve. *Machine Learning, 77*(1), 103–123.
22. Hand, D. J., & Anagnostopoulos, C. (2014). A better beta for the h measure of classification performance. *Pattern Recognition Letters, 40*, 41–46.
23. Hand, D. J., & Zhou, F. (2010). Evaluating models for classifying customers in retail banking collections. *Journal of the Operational Research Society, 61*, 1540–1547.
24. Hansen, L. K., & Salamon, P. (1990). Neural network ensembles. *IEEE Transactions on Pattern Analysis and Machine Intelligence, 12*, 993–1001.
25. Huang, S. C. & Wu, C. F. (2011). Customer credit quality assessments using data mining methods for banking industries. *African Journal of Business Management, 5*(11), 4438–4445.
26. Huang, C. L., Chen, M. C., & Wang, C. J. (2007). Credit scoring with a data mining approach based on support vector machines. *Expert Systems with Applications, 33*(4), 847–856.

27. Kennedy, K. (2013). Credit scoring using machine learning. Doctoral thesis, Technological University Dublin. https://doi.org/10.21427/D7NC7J.
28. Khandani, A. E., Kim, A. J., & Lo, A. W. (2010). Consumer credit-risk models via machine-learning algorithms. *Journal of Banking & Finance, 34*(11), 2767–2787.
29. Lessmann, S., Baesens, B., Seow, H., & Thomas, L. C. (2015). Benchmarking state-of-the-art classification algorithms for credit scoring: An update of research. *European Journal of Operational Research, 247*(1), 124–136.
30. Louzada, F., Ara, A., & Fernandes, G. B. (2016). Classification methods applied to credit scoring: Systematic review & overall comparison. *Surveys in Operations Research and Management Science, 21*(2), 117–134.
31. Marqués, A. I., García, V., & Sánchez, J. S. (2012). Exploring the behaviour of base classifiers in credit scoring ensembles. *Expert Systems with Applications, 39*(11), 10244–10250.
32. McCarthy, B., Chui, M., & Kamalnath, V. (2018). *An executive's guide to AI.* https://www.mckinsey.com/business-functions/mckinsey-analytics/our-insights/an-executives-guide-to-ai.
33. Minsky, M., & Papert, S. (1969). *Perceptrons: An introduction to computational geometry.* Cambridge, MA: MIT Press.
34. Nath, R., Rajagopalan, B., & Ryker, R. (1997). Determining the saliency of input variables in neural network classifiers. *Computers and Operations Researches, 24*, 767–773.
35. Ong, C., Huang, J., & Tzeng, G. (2005). Building credit scoring models using genetic programming. *Expert Systems with Applications, 29*, 41–47.
36. Ozbayoglu, A. M., Gudelek, M. U., & Sezer, O. B. (2020). Deep learning for financial applications: A survey. *Applied Soft Computing, 93*, 106384.
37. Quinlan, J. R. (1993) *C4.5—Programs for machine learning.* San Francisco, CA, United States: Morgan Kaufmann Publishers.
38. Rohit, V. M., Kumar, S., Kumar, J. (2013). Basel II to Basel III the way forward. In *Infosys White Paper.* https://srinath-keshavan-naj7.squarespace.com/s/Basel-III_Basel-II-to-III.pdf.
39. Saunders, A., Allen, L. (2002). *Credit risk measurement: New approaches to value at risk and other paradigms.* New York: Wiley.
40. Sirignano, J., Sadhwani, A., Giesecke, K. (2018). Deep learning for mortgage risk. Technical report, Working paper available at SSRN. https://ssrn.com/abstract=2799443.
41. Tan, P., Steinbach, M., & Kumar, V. (2006). *Introduction to Data Mining.* New York, US: Pearson Educatio.
42. Thomas, L., Crook, J., & Edelman, D. (2017). Credit scoring and its applications. In *Society for Industrial and Applied Mathematics (SIAM), Philadelphia, US.* https://doi.org/10.1137/1.9781611974560.
43. Vapnik, N. (1998). *Statistical learning theory.* New York: Wiley.
44. Wang, Z., Jiang, C., Zhao, H., & Ding, Y. (2020). Mining semantic soft factors for credit risk evaluation in Peer-to-Peer lending. *Journal of Management Information Systems, 37*(1), 282–308.
45. West, D. (2000). Neural network credit scoring models. *Computers and Operations Research, 27*, 1131–1152.
46. West, D., Dellana, S., & Qian, J. (2005). Neural network ensemble strategies for financial decision applications. *Computers and Operations Research, 32*, 2543–2559.
47. Whalen, S., & Pandey, G. (2013). A comparative analysis of ensemble classifiers: Case studies in genomics. In *Data Mining (ICDM), 2013 IEEE 13th International Conference* (pp. 807–816). New Jersey: IEEE.
48. Xia, Y., Liu, C., Li, Y., & Liu, N. (2017). A boosted decision tree approach using Bayesian hyper-parameter optimization for credit scoring. *Expert Systems with Applications, 78*, 225–241.
49. Yobas, M. B., Crook, J. N. & Ross, P. (2000). Credit scoring using neural and evolutionary techniques. *IMA Journal of Mathematics Applied in Business and Industry, 11*(4), 111–125.

50. Zhao, Q., & Hastie, T. (2019). Causal interpretations of black-box models. *Journal of Business & Economic Statistics, 39*(1), 1–10.
51. Zhou, L., Lai, K. K., & Yu, L. (2010). Least squares support vector machines ensemble models for credit scoring. *Expert Systems with Applications, 37*, 127–133.

Classifying Counterparty Sector in EMIR Data

Francesca D. Lenoci and Elisa Letizia

Abstract The data collected under the European Market Infrastructure Regulation ("EMIR data") provide authorities with voluminous transaction-by-transaction details on derivatives but their use poses numerous challenges. To overcome one major challenge, this chapter draws from eight different data sources and develops a greedy algorithm to obtain a new counterparty sector classification. We classify counterparties' sector for 96% of the notional value of outstanding contracts in the euro area derivatives market. Our classification is also detailed, comprehensive, and well suited for the analysis of the derivatives market, which we illustrate in four case studies. Overall, we show that our algorithm can become a key building block for a wide range of research- and policy-oriented studies with EMIR data.

1 Introduction

During the Pittsburgh Summit in 2009, G20 leaders agreed to reform the derivatives markets to increase transparency, mitigate systemic risk, and limit market abuse [14]. As a result of this internationally coordinated effort, counterparties trading derivatives in 21 jurisdictions are now required to daily report their transactions to trade repositories (TR) [16]. To accomplish the G20's reform agenda, the EU introduced in 2012 the European Market Infrastructure Regulation (EMIR, hereafter).

Authors are listed in alphabetic order since their contributions have been equally distributed.
This work was completed while Elisa Letizia was at the European Central Bank.

F. D. Lenoci (✉)
European Central Bank, Frankfurt am Main, Germany
e-mail: francesca_daniela.lenoci@ecb.europa.eu

E. Letizia
Single Resolution Board, Brussels, Belgium
e-mail: elisa.letizia@srb.europa.eu

However, the use of these data poses numerous challenges, especially when it comes to data aggregation [15, 16]. To enhance data quality and usability, over the past years public institutions and private entities have jointly worked to harmonize critical data fields [27]. The harmonization effort has focused on key variables, one of which is the legal entity identifier (LEI). The LEI uniquely identifies legally distinct entities that engage in financial transactions based on their domicile.[1] The LEI was introduced in 2012, and currently covers 1.4 million entities in 200 countries. It identifies entities reporting over-the-counter (OTC) derivatives with a coverage close to 100% of the gross notional outstanding, and debt and equity issuers for 78% of the outstanding amount, across all FSB jurisdiction [17]. LEIs are linked to reference data which provide basic information on the legal entity itself, such as the name and address, and its ownership (direct and ultimate parent entities). However, the counterparties' sector is not included in the reference data. This information is crucial to derive the sectoral risk allocation in this global and diverse market, especially if the aim is to identify potential concentration of risk in specific sectors of the financial system. In EMIR data, even though counterparties are obliged to report their sector using a classification given in the regulation, the available information suffers from several conceptual and data quality limitations. In particular, the sector breakdown is not detailed enough to obtain a comprehensive view of the sectoral allocation of risk. For example, central clearing counterparties (CCPs), which play a key role in the market, are not readily identifiable, as they do not need to report to any sector. To fill this gap, we propose an algorithm to enrich the current classification and uniquely assign a sector to each counterparty trading derivatives, identified by its LEI. We employ a greedy algorithm [7] based on eight different data sources. Firstly we use lists of institutions available from relevant EU public authorities competent for various sectors. Even though comprehensive at EU level, these lists are not sufficient to gain the whole picture because of the global scale of the derivatives market, where many entities outside EU interact with EU investors. Therefore we complement the official lists with sector-specialized commercial data providers. Our work contributes to the existing body of published research dealing with the problem of assigning sectors to individual institutions. In [13] this is done by grouping firms according to their Standard Industrial Classification code in a way to have similar exposure to risk factors within the same group. Despite the popularity of this method in the academic literature, [5] showed that the Global Industry Classifications Standard (GICS) system, jointly developed by Standard & Poor's and Morgan Stanley Capital International (MSCI), is significantly better at explaining stock return co-movements with respect to [13]. The GICS, however, is not very detailed for the financial sector, so not suitable

---

[1] The LEI is a 20-digit alpha-numeric code based on ISO standards provided by the Global Legal Entity Identifier Foundation (GLEIF). It excludes natural persons, but includes governmental organizations and supranationals.

to fairly describe the derivatives market. More recent works [32] have used deep learning to predict the sector of companies[2] from the database of business contacts.

The methodology presented in this chapter has a proven track record, as it has been used by several studies. It has been effectively employed to support analysis in the areas of financial stability [19, 12, 23] and monetary policy [6].

Our approach has three main advantages with respect to existing research: it is comprehensive and detailed, flexible, and helps reproducibility and comparability.

We use a multilayered taxonomy to allow a wide range of applications and granularity. The final classification allows classifying entities trading 96% of notional outstanding in the euro area at the end of 2018Q2 and is tailored for the derivatives market, recognizing entities having crucial roles (like market makers, large dealers, and CCPs).

The algorithm is flexible and can easily accommodate future changes in regulation regarding institutional sectors and can be used in other markets.

Lastly, by choosing to give prominence to publicly available official lists, our method makes the aggregates produced from transactional data *comparable* with other aggregates published by the same authorities we use as sources. At the same time, the data being public and easily available to any researcher helps produce stable and reproducible results, which is of paramount importance in many policy and research applications. Reproducibility is dependent on the researcher having access to EMIR data, which is currently available to a number of public authorities in the EU. However, the core of the algorithm is based on publicly available data, while commercial data sources can be easily excluded or replaced depending on what is available to the researcher or policy officer. The reproducibility also depends on the fact that the algorithm can be adapted to other datasets of transactional data, such as those collected under SFTR.

To this regard, our methodology contributes to the growing body of research using TR data [1, 29, 20, 15, 6, 10] by providing a stable building block to conduct a wide range of analyses. To show this potential, we present four cases studies where we use our classification on the sample of EMIR data available to the ECB.[3] In the first case we describe, for the first time to our knowledge, the derivatives portfolios of euro area investment funds, with emphasis on their overall investment strategy. In the second, we disentangle the role of investment and commercial banks in the market. In the third, we measure how large dealers provide liquidity in the Credit Default Swaps (CDS) market. In the last, we show how relying only on the sector reported in EMIR data can lead to a very different picture of the euro area insurance companies activity in the market.

The rest of the chapter is structured as follows: Sect. 2 describes reporting under EMIR, Sect. 3 describes the methodology, Sect. 4 discusses the performance of the algorithm, Sect. 5 includes the four case studies.

---

[2]More specifically the North American Industry Classification System code.

[3]This includes trades where at least one counterparty is domiciled in the euro area, or the reference entity is in the euro area.

2 Reporting Under EMIR

EMIR enabled authorities in the EU to improve their oversight of the derivatives market by requiring European counterparties to report their derivatives transactions to TRs.[4] The reporting obligation applies to both OTC and exchange-traded derivatives in all five main asset classes, i.e., commodity, equity, foreign exchange, credit and interest rate derivatives.

Since 2014 all EU-located entities that enter a derivatives contract must report the details of the contract, within one day from its execution to one of the TRs authorized by the European Securities and Markets Authority (ESMA).[5] Each opening of a new contract should be reported by the counterparties to the trade repository as a new entry, and all life-cycle events must be reported as well (modification, early termination, compression, and valuation update of contracts). Intragroup transactions are not exempt from the obligation, and trades with nonfinancial counterparties must be reported alike.[6]

The EU implemented the reform with double reporting, i.e., counterparties of the trade have to be compliant in reporting the details of the transaction to one of the trade repositories active in the jurisdiction.

Daily transaction-by-transaction derivatives data are made available by the TRs to over one hundred authorities in the EU, depending on their mandate and jurisdiction. The ECB has access to trades where at least one of the counterparties is located in the euro area, the reference entity is resident in the euro area, to euro-denominated contracts or when the derivatives contract is written on sovereigns domiciled in the euro area: these trades constitute the sample for the implementation of the algorithm presented in this chapter.

With more than 2000 entities reporting every day roughly 30 million outstanding derivatives contracts, with an overall value of slightly less than €300 trillion, EMIR data can be classified as "big data." On a daily basis, counterparties report roughly 250 fields, of which 85 are subject to mandatory reporting.[7] These include information on entities involved in the transactions, the characteristics and terms of the contract, which are static and common across asset classes, and the value of the contract, which may change over the life cycle of a trade.

The regulation requires counterparties to report their own sector choosing from a specific list of codes as reported in EMIR.[8] For nonfinancial corporations, a single

---

[4]The reporting obligation extends to non-European counterparties when the reference entity of the contract is resident in the EU and when they trade CDS written on EU-domiciled sovereigns.

[5]Currently there are seven TRs authorized by ESMA in the EU.

[6]Only individuals not carrying out an economic activity are exempt from the reporting obligation.

[7]All fields included in the Annex of the Commission Delegated Regulation (EU) No 148/2013 are subject to mandatory reporting, except those not relevant for the specific asset class.

[8]Commission Implementing Regulation (EU) 2017/105 of October 19, 2016, amending Implementing Regulation (EU) No 1247/2012 laying down implementing technical standards with regard to the format and frequency of trade reports to trade repositories according to Regulation

Table 1 Sector classification reported in EMIR

Code	Description	Directive
A:U	Nonfinancial corporations (NACE)	
AIFMD	Alternative investment fund and their management companies	2011/61/EU
ASSU	Assurance undertaking	2002/83/EC
CDTI	Credit institution	2006/48/EC
INUN	Insurance undertaking	73/239/EC
INVF	Investment firm	2004/39/EC
ORPI	Institution for occupational retirement provision	2003/41/EC
REIN	Reinsurance undertaking	2005/68/EC
UCIT	Undertaking for the Collective Investment in transferable Securities and its management company	2009/65/EC
OTHR		

letter distinguishes the sector each firm belongs to, while for others the relevant regulation assigns entities to a specific sector (as shown in Table 1).

The existing reporting requirements present five main drawbacks related either to data quality or to the level of granularity:

i. The sector breakdown is not sufficiently detailed, or at least not for all industries. For example, it distinguishes between Alternative Investment Funds (AIF) and Undertakings for Collective Investments in Transferable Securities (UCITS) in the investment fund sector but does not allow to distinguish between commercial and investment banks.

ii. The granularity for the banking sector is not sufficiently detailed. For example, banks belonging to the G16 group of dealers[9] and entities acting as clearing members[10] cannot be identified through a dedicated field.

iii. Does not recognize Central Clearing Counterparties (CCPs) as a separate sector, even though they play an important role in efficiently reallocating counterparty credit risks and liquidity risks. In recent years, derivatives and repo markets have become heavily reliant on CCPs for the clearing of transactions either on voluntary basis or because traders are obliged to use a CCP to clear their trades.

(EU) No 648/2012 of the European Parliament and of the Council on OTC derivatives, central counterparties, and trade repositories.

[9]G16 dealers are defined as such by NY Fed as the group of banks which originally acted as primary dealers in the US Treasury bond market but nowadays happens to be also the group of largest derivatives dealers. The sample, which has world coverage, changed over time, and originally comprised: Bank of America, Barclays, BNP Paribas, Citigroup, Crédit Agricole, Credit Suisse, Deutsche Bank, Goldman Sachs, HSBC, JPMorgan Chase, Morgan Stanley, Nomura, Royal Bank of Scotland, Société Générale, UBS, and Wells Fargo. In 2019, the list is made up of 24 entities and is available at https://www.newyorkfed.org/markets/primarydealers.

[10]All G16 dealers are usually member of one or more CCPs with the role of clearing members. Only clearing members can clear trades on behalf of their clients.

In such cases, CCP interposes itself between the original buyer and seller, acting as the buyer to each seller and the seller to each buyer.

iv. Although the sector definition of each entity is in line with the one provided by either the European System of National and Regional accounts (ESA)[11] or the European Classification of Economic Activities (NACE),[12] the classifications do not overlap consistently, making comparisons difficult. For example, nonfinancial corporations are classified using a one-digit NACE, while for other sectors there is no explicit mapping.

v. It happens that the same counterparty reports to belong to different sectors over time even if other data sources do not suggest a material change in its activity.

3 Methodology

To overcome the limitations of the sectors available in EMIR, we define a greedy algorithm in order to uniquely identify to which sector each counterparty belongs to. As shown in Fig. 1 the algorithm comprises three parts:

- Data collection from several data sources
- Data harmonization
- Greedy assignment of a unique sector to each LEI

Our algorithm is greedy as the "local" optimal is determined by looking at a single (ordered) source at the time, without considering whether the same LEI appears in another source later in the hierarchy.

3.1 First Step: The Selection of Data Sources

In the first step we collect information from different data sources using both publicly available official lists and commercial data providers. The choice of sources is crucial, therefore in what follows we explain the reasons for choosing each of them.

As counterparties are identified by LEI in EMIR data, we opt for sources which include this identifier systematically. The final set of sources used is a trade-off between completeness and parsimony: we aim at assigning a sector to as many LEIs as possible, but also keeping a simple and updatable procedure for data collection.

---

[11]For details on the ESA, see https://ec.europa.eu/eurostat/cache/metadata/Annexes/nasa_10_f_esms_an1.pdf.

[12]For details on NACE, see https://ec.europa.eu/eurostat/documents/3859598/5902521/KS-RA-07-015-EN.PDF.

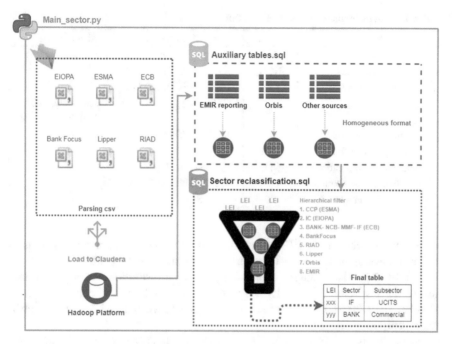

Fig. 1 A schematic overview of the algorithm

The list of Central Clearing Counterparties (CCP) is published officially by ESMA and includes authorized EU CCPs, recognized third-country CCPs, and CCPs established in non-EEA countries which have applied for recognition.[13] At the last update in January, July, and September 2019, these lists comprised 17, 34, and 54 CCPs, respectively.

The list of Insurance Undertakings (IC) rely on the public Register provided by the European Insurance and Occupational Pensions Authority (EIOPA).[14] The Register of Insurance undertakings is a representation of the information provided by the respective National Competent Authorities that are responsible for authorization and/or registration of the reported insurance undertakings activities. It comprises roughly 30,000 institutions operating in the EU, which are either domestic undertakings or EEA/3rd country branches or insurers domiciled in EEA

---

[13]The list is disclosed in accordance with Article 88 of EMIR and is updated on a nonregular frequency, when changes occur. Furthermore, under Article 25 of EMIR, non-EEA CCPs have to expressly agreed to have their name mentioned publicly; therefore the list is not necessarily exhaustive for this category. For the latest update see https://www.esma.europa.eu/sites/default/files/library/ccps_authorised_under_emir.pdf.

[14]In accordance with Article 8 of EIOPA Regulation (Regulation EU No 1094/2010). For the latest update see https://register.eiopa.europa.eu/registers/register-of-insurance-undertakings.

Table 2 Sector classification based on ESA 2010

Sector	Label	ESA code(s)	Other attributes
Banks	BANK	S122A	
Central Banks	NCB	S121	
Government	GOVT	S1311, S1312, S1313	
Insurance undertakings	IC	S128	Composite, Life Nonlife, Reinsurer
Investment funds	IF	S124	Asset manager
Money market funds	MMF	S123	
Nonfinancial corporations	NFC	S11, S15, S14	
Other financial institutions	OFI	S122B, S125, S126, S127	
Pension Funds	PF	S129, S1314	Private/Public

or having branches in the EEA using the internet or other communication tools to sell insurance in the EU under Freedom of Providing Services (FPS).

The ECB publishes the list of monetary financial institutions (MFIs) according to several regulations.[15] The list is updated on a daily basis and comprises, as of October 2019, 20 NCBs, 4526 credit institutions, 455 MMFs, and 224 other deposit taking corporations.

The ECB also publishes a list of EU investment funds on a quarterly basis.[16] The list included 63427 institutions as of 2019 Q2 and allows to distinguish between Exchange Trade Funds (ETF), Private Equity Funds (PEF), and Mutual funds; it provides further details in terms of capital variability (open-ended vs. closed mutual funds), UCITS compliance, investment policy (mixed, equity, bond, hedge, real estate), and the legal setup.

Furthermore, we use the Register of Institutions and Affiliated Database (RIAD). RIAD is the European System of Central Banks registry and is compiled by National Central Banks, National Competent Authorities, international organizations, and commercial data providers. RIAD collects information on institutions, financial and nonfinancial companies, including granular relationship data on eight million individual entities. From RIAD we take the information on the ESA 2010 sector code associated with LEIs, as detailed in Table 2.

---

[15]Regulation ECB/2013/33 as resident undertakings belonging to central banks (NCB), credit institutions according to Art. 4 575/2013 (BANK), and other resident financial institutions whose business is to receive deposits or close substitutes for deposits from institutional units, to grant credit, and/or to make investments in securities for their own account, electronic money institutions (Art.2 2009/110/EC), and money market funds (MMF). For the latest update see https://www.ecb.europa.eu/stats/financial_corporations/list_of_financial_institutions/html/index.en.html.

[16]Under Regulation EC No 1073/2013 concerning statistics on the assets and liabilities of investment funds (ECB/2013/38), collects information on investment fund undertakings (IF) to provide a comprehensive picture of the financial activities of the sector and to ensure that the statistical reporting population is complete and homogeneous. For the latest update see https://www.ecb.europa.eu/stats/financial_corporations/list_of_financial_institutions/html/index.en.html.

To facilitate the reproducibility of the final classification, the algorithm would ideally rely only on publicly available lists. However ESMA, ECB, and EIOPA collect information for different purposes and their registers do not cover institutions domiciled outside the EU. For this reason it is crucial to identify entities not operating or domiciled in the EU but trading derivatives referencing euro area underlying and subject to the reporting mandate under EMIR. Consequently, the algorithm enriches the pool of sources using commercial data providers as well. These additional sources are used to classify entities which are not in the public lists.

Data on investment firms and commercial banks are complemented using Bank-Focus from Moody's Analytics. These data include information on *specialization* on more than 138,000 institutions active worldwide (see also Sect. 3.3.1 below).

To enlarge the set of investment funds, asset managers, and pension funds, the algorithm relies also on 768,000 undertakings reported in Lipper Fund Research Data from Refinitiv.

Orbis is used to assign a sector to LEIs not classified using any of the previous publicly and commercial sources, it represents the main database to identify pension funds via NACE codes. Orbis is the most comprehensive database, not being specialized in any particular sector, and provides cross reference for all the industry classification codes (NACE, NAICS, and SIC) for 310 million entities[17] including banks, insurance companies, and non-bank financial institutions covering all countries.

Finally, we rely on the EMIR reported sector for entities not reporting with LEI or not classified using any official or commercial data source.

3.2 Second Step: Data Harmonisation

In the second stage, data from each source is harmonized and made compatible with the EMIR data structure. In the harmonization phase, the algorithm rearranges information from several data providers in a functional way with respect to the final classification. For example, from the ESMA list it treats in the same way euro area CCPs and third-country CCPs with rights to provide their services in the euro area; from the EIOPA list, as well as for other lists, it excludes insurance companies which do not have the LEI. From ECB Investment Fund and Lipper lists, the algorithm makes uniform the breakdowns provided by each source to the ones provided by our classification: e.g., by merging government and corporate fixed income funds from Lipper in one category like "bond-funds," by merging closed-ended funds and funds with no redemption rights from Lipper in "closed funds" and so on. The algorithm also uniforms the itemization provided by BankFocus in saving, cooperative, and universal banks by creating only one category, like "commercial bank." For each

---

[17]Only 1.3 million entities have LEIs.

Table 3 Sector classification based on EMIR. NACE code K indicates nonfinancial corporations specialized in financial activities

Sector	Label	EMIR sector
Banks	BANK	CDTI
Insurance undertakings	IC	REIN, ASSU, INUN
Investment funds	IF	UCITS and AIFMD
Nonfinancial corporations	NFC	all entities reporting with a single digit, except K
Other financial institutions	OFI	INVF, OFI, K
Pension funds	PF	ORPI

public and commercial data provider, the algorithm creates a table storing relevant fields in a uniform way.

To extract a stable information from the sector reported in EMIR we proceed as follows. We extract the reported sector from EMIR data, keeping only consistently reported classification. That is, an auxiliary table tracks, for each reporting counterparty, the number of times, starting from November 2017, it declares to belong to one of the six sectors in Table 3.

For each reporting counterparty, the procedure assigns to each LEI the sector corresponding to the mode values, only when no ties occur. For example, if entity i reports to be a credit institution in 500 reports and an insurance company in 499 reports, the procedure assigns to the LEI of entity i the sector "CDTI."[18] This step tackles the fifth drawback of the existing reporting requirements presented in Sect. 2, i.e., the same counterparty reporting different sectors. As of 2019Q2, 10.9% of reporting entities reported two sectors, and around 0.3% reported at least three different sectors for the same LEI. In this way, the algorithm cleans the reported sector information, and, hereafter, we refer to the outcome of this procedure as source "EMIR sector." A description of the algorithm performing this procedure is presented in Sect. 3.4.

3.3 Third Step: The Classification

In the third stage, the final classification is performed in a greedy way: an entity is classified by looking at one source at a time, establishing a hierarchy of importance among sources.

With the exception of Orbis and RIAD, which are useful to classify several sectors, the majority of sources are specialized to classify one sector. Table 4 summarizes the sectors in our classification and its sources in order, reflecting our ranking which prioritizes official lists followed by commercial data providers.

---

[18]We exclude entities reporting with sector "OTHR."

Table 4 Hierarchy of sources for each sector. The ECB publishes several lists, so we indicate in parentheses the specific one we use for each sector in our classification. For pension funds we use the NACE code available in Orbis (6530)

Sector	label	Sources			
		1st	2nd	3rd	4th
Banks	BANK	ECB (MFI)	BankFocus	RIAD	Orbis
Central banks	NCB	ECB (MFI)	BankFocus	RIAD	Orbis
CCPs	CCP	ESMA			
Government	GOVT	RIAD	Orbis		
Insurance undertakings	IC	EIOPA	RIAD	Orbis	
Investment funds	IF	ECB (IF)	Lipper	Orbis	
Money market funds	MMF	ECB (MFI)	Lipper	RIAD	Orbis
Nonfinancial corporations	NFC	RIAD	Orbis		
Other financial institutions	OFI	ECB (MFI)	Bank Focus	RIAD	Orbis
Pension funds	PF	Orbis (NACE)	Lipper	RIAD	

The final classification recognizes ten sectors and includes a more granular subsector, when available (see Table 5). The following sections describe the subsector granularity for banks and investment funds. For the latter we also provide a further set of dedicated dimensions in terms of structure, vehicle, and strategy (see Sect. 3.3.2).

Entities acting as clearing members and banks within the group of G16 dealers are identified by the algorithm with a proper flag.

We complement sector classification with information on geographical dispersion by providing the country of domicile[19] from GLEIF. In addition to that, we add three dummy variables for entities domiciled in the euro area, in Europe and in the European Economic Area.

For reproducibility purposes, the final table includes a column indicating the source used for the classification. The algorithm is implemented for regular updates and we keep track of historical classification to account for new or inactive players.

Even though our classification shares some features of EU industry classifications (like ESA and NACE which we use as sources), we chose not to rely solely on them to make our classification more tailored to the derivatives market.

On one side, we inherit the concepts of assigning a sector to legally independent entities, and the use of multilayered classification, which allows different levels of detail depending on the analysis to be carried out. On the other side, ESA classification is aimed at describing the whole economies of Member States and the EU in a consistent and statistically comparable way. For this reason ESA classification covers all aspects of the economy, of which the derivatives market is a marginal part. As a result, entities which play key roles in the derivatives market, but not in other segments of the economy, do not necessarily have a dedicated code

---

[19]ISO 3166 country code.

Table 5 Our sector classification: taxonomy. *Other* indicates the residual categories for unclassified entities, both for the sector and the subsector. Blank in the subsector indicates no further granularity

Sector	Sub-sector
Bank	Investment
	Commercial
	Other
Central bank	
CCP	
Government	
Insurance undertaking	Life
	Nonlife
	Composite
	Reinsurance
	Assurance
	Other
Investment fund	UCITS
	AIF
	Management company
	Other
Money market mutual fund	
Nonfinancial corporation	
Other financial institution	Other MFI
	Other
Pension Fund	Private
	Public
	Other
Other	

in ESA. For example, CCPs may be classified under different sectors and not have a specific one[20] and the banking sector is all grouped under one category, without clear distinction for dealers. As these two categories are crucial for the market, we provide a clear distinction for them. Similarly, not much granularity is available in ESA and NACE for the investment fund sector, while we provide several dimensions to map this sector which is of growing importance in the derivatives market. Other sectors, like households, nonprofit institutions, government and nonfinancial corporations, play a marginal role in the derivatives market; therefore we do not provide further breakdown, even though they are more prominent in ESA (and NACE). Finally, ESA and NACE only refer to EU domiciled entities, therefore we needed to go beyond their scope because of the global scale of the derivatives market.

---

[20]Some CCPs are classified in ESA with the code S125, which includes also other types of institutions, e.g., financial vehicle corporations. Others, with a banking license, have as ESA sector S122.

3.3.1 Classifying Commercial and Investment Banks

For entities classified as banks, we disentangle those performing commercial banking from those performing investment banking activities. This is important because of the different role they might have in the derivatives market. Due to the exposure of commercial banks towards particular sectors/borrowers via their lending activity, they might need to enter the derivatives market to hedge their position via credit derivatives, to transform their investments or liabilities' flows from fixed to floating rate or from one currency to another via interest rate or currency swaps respectively. Moreover, commercial banks might use credit derivatives to lower the risk-weighted assets of their exposures for capital reliefs [22, 4, 26]. On the contrary, investment banks typically enter the derivatives market with the role of market makers. Leveraging on their inventories from large turnovers in the derivatives and repo market, their offsetting positions result in a matched book [24]. The distinction between commercial and investment banks is based on two sources: the list of large dealers and BankFocus. The list of large dealers is provided by ESMA and includes roughly one hundred LEIs and BIC codes referring to G16 dealers and institutions belonging to their group. The classification of investment and commercial banks using BankFocus relies on the field *specialization*. We classify as *commercial banks* those reporting in the *specialization* field as commercial banks as well as cooperative, Islamic, savings, and specialized governmental credit institutions. *Investment banks* includes entities specialized both as investment banks and securities firms.[21]

Combining the two sources above, the algorithm defines firstly as *investment banks* all entities flagged as G16 dealers in the ESMA list and all banks classified as such from BankFocus, secondly as *commercial banks* all banks defined as such by BankFocus. As residuals, when LEIs are not filtered in any of the two, entities can still be classified as *banks* using the ECB official list of Monetary Financial Institutions, RIAD when reported with ESA code S122A, Orbis, or EMIR. In these cases it is not possible to distinguish between commercial and investment banks and we leave the subsector field blank.

3.3.2 Classifying Investment Funds

Since EMIR requires reporting at the fund level and not at the fund manager level, the investment fund sector in EMIR comprises a very high number of entities and it is very heterogeneous. For this reason, we include dedicated dimensions for this sector which allows to better characterize entities broadly classified as investment

---

[21]When preparing the reference data from BankFocus the algorithm disregards some specializations. They are: bank holding companies, clearing institutions, group finance companies, multilateral government bank, other non-banking credit institutions, real estate, group finance company, private banking, and microfinancing institutions.

funds. We focus on three aspects, namely, their compliance to the UCITS and AIFM directives,[22] their capital variability, their strategy, and the vehicle according to which they run their business in order to define the following dimensions: subsector, structure, vehicle, strategy.

We recognize as subsectors UCITS, AIF, and Asset Managers. We identify *Asset Managers* when the trade is reported with the LEI of the Asset Manager and not at the fund level, as it should be reported. This might occur when the trade refers to proprietary trading of the asset manager or when the transaction refers to more than one fund. To disentangle UCITS from AIFs,[23] we rely first on the ECB official list of investment funds which includes a dummy for UCITS compliance and secondly on Lipper, which also has separated fields for funds compliant with one or the other regulation. Both sources assign to each fund the LEI of the fund manager allowing to create a list of asset managers and define the subsector as *AM* when the trade is reported by the asset manager.

Using the ECB list of investment funds and Lipper, we filter investment funds according to their capital variability.[24] The algorithm leaves the field blank when the source does not provide information on the structure for a specific mutual fund.

The *vehicle* defines the legal structure according to which the fund operates. We distinguish exchange trade funds (vehicles in the form of investment funds that usually replicate a benchmark index and whose shares are traded on stock exchanges), private equity funds, and we leave the field blank for all mutual funds.

Strategy defines the investment profile of the fund in terms of asset allocation. Relying on the investment policy reported in ECB's official list, on the asset type field as well as the corporate and government dummies reported in Lipper, we define the fund investment strategy encompassing bond, real estate, hedge, mixed, and equity. Those investing mainly in corporate and government bonds are identified as bond funds.

---

[22] Alternative investment fund—AIFD—are authorized or registered in accordance with Directive 2011/61/EU while UCITS and its management company—UCIT—are authorized in accordance with Directive 2009/65/EC.

[23] UCITS-compliant funds are open-ended European and non-EU funds compliant with the EU regulation which raise capital freely between European Union members. Alternative investment funds (AIF) are funds that are not regulated at EU level by the UCITS directive. The directive on AIF applies to (i) EU AIFMs which manage one or more AIFs irrespective of whether such AIFs are EU AIFs or non-EU AIFs; (ii) non-EU AIFMs which manage one or more EU AIFs; (iii) and non-EU AIFMs which market one or more AIFs in the Union irrespective of whether such AIFs are EU AIFs or non-EU AIFs.

[24] We define as *closed-ended* those non-MMMFs which do not allow investors to redeem their shares in any moment or which can suspend the issue of their shares, while as *open-ended* all funds which allow investors ongoing withdrawals and can issue an unlimited number of shares.

Table 6 Sector
classification: dedicated
dimensions for investment
funds

Structure	Vehicle	Strategy
ETF	Open	Bond
Mutual	Closed	Equity
Private Equity Funds		Mixed
Investment trust		Hedge
		Money Market
		Real Estate
		Commodity
		Other

3.4 Description of the Algorithm

The classification consists of an SQL-code and it is made up by eight intermediate
tables which could be grouped into the stages below:

1. Data preparation. The first five tables aim at defining the sample on which the
   algorithm will be applied and identifying the most frequently reported sector
   among those allowed by the Regulation (see Table 6).[25] The first table creates a
   list of distinct LEI-sector tuples. In this stage, LEIs which report different sectors
   over time are included. A mapping from EMIR sector to our classification is done
   based on Table 3. We count how many times each LEI report belongs to each
   sector. Each reporting counterparty is then assigned the sector which is reported
   more often. In case of ties, OFI and NFC are not considered as options. Finally,
   LEIs only available as other counterparty in trades but not reporting are added to
   the sample.
2. Data enrichment. The list of distinct LEIs is joined with various sources. In
   particular, the list is complemented with the NACE, RIAD, or Orbis sectors,
   and the algorithm adds Boolean flags for CCPs, G16 dealers, IC, banks, and
   MMF funds if the LEI is classified from an official list and adds other attributes
   according to each data provider.
3. The third stage creates the final table via a greedy process implemented in SQL
   through the case condition. Thus, while stage 2 enriches the list of LEIs with
   information from different data providers, here LEIs pass through a bottleneck
   for the final classification. Using the enrichment performed in stage 2, this stage
   scans each LEI and assigns a unique sector according to the first source which
   includes that LEI, in the order of hierarchy as per Table 4. In practice this
   means checking each Boolean flag added in the enrichment phase, and have an
   assignment if one is TRUE. The sources are ordered, so the first TRUE value
   causes the clause to terminate, without further checks if the LEI exists in other
   sources, in a greedy approach. If no list includes the LEI, i.e., all Boolean flags

---

[25] For each trade, EMIR prescribes that the reporting counterparty report only its sector and not the
sector of the other counterparty involved in the trade.

are FALSE and additional classification from RIAD, Orbis, and EMIR are empty, it is assigned to the residual class "Other." For example, to classify an LEI as BANK, the algorithm first looks for that LEI in the ECB list of MFIs, then in the list of G16 dealers, then in RIAD if that LEI is reported with ESA sector "S122A," then in BankFocus, then in Orbis, and finally in the EMIR reported sector. The same process is used for the identification of the subsector and for the investment funds' strategy, vehicle, and structure.

4 Results

In this section we test our algorithm on the ECB's sample of EMIR data, including outstanding contracts as of $2018Q2$, and we demonstrate its added value with respect to the EMIR sector classification, both as reported and processed to avoid ambiguous classification.[26]

We first show in Table 7 how our sector classification (rows) compares to the sector reported in EMIR data (columns). To this aim, aggregation is based on the sector of the reporting counterparty.[27] By increasing the overall granularity from ten to seventeen categories (including subsectors), there is not only a reshuffling among existing categories but also a transition towards other sectors. As expected, the most significant transitions occur towards the sectors of CCP and investment bank, which are known to play a very important role in the market, but do not have a dedicated sector in EMIR classification. 88% of gross notional outstanding which was in the residual group (NULL) is now classified as traded by CCPs.[28] Furthermore, 69% and 73% of gross notional traded by credit institutions (CDTI) and investment firms (INVF), respectively, is allocated to investment banks according to our classification.

The sectors of insurance companies, pension funds, and nonfinancial corporations are also deeply affected. Forty-four percent (7%) of gross notional allocated to assurance companies (ASSU) are reclassified as investment funds (nonfinancial corporations) once we apply our classification.[29] Only 62% of gross notional outstanding reported by pension funds under EMIR remains as such, while the remaining 23% of gross notional is found to be traded by insurance companies, investment funds, other financial institutions, or nonfinancial corporations.

---

[26]See Sect. 3 for details on how we process the sector reported in EMIR data to avoid ambiguous cases.

[27]As mentioned in Sect. 2 this is the only information mandated to be reported.

[28]The remaining part of the residual group is traded by banks (4%), nonfinancial corporations (3%), other financial institutions (2%), and governments or alternative investment funds (1% each).

[29]A similar finding applies to insurance companies (INUN) where 10% of gross notional outstanding refers either to investment funds, pension funds, or nonfinancial corporations, and reinsurance companies where 4% refers to investment funds or nonfinancial corporations.

Table 7 Comparison between reported sector and our classification, 2018Q2. The table indicates the percentage of notional allocated from the EMIR reported sector (column) across our classification (rows, for acronyms see Table 4). Sub-sector is left as blank for sectors for which we do not provide a sub-sector classification or for entities which are not assigned to any sub-sector according to our set of sources. Each column sums up 100%, only allocations larger than 1% are reported

SECTOR	SUBSECTOR	CDTI	ASSU	INUN	REIN	AIFD	UCIT	INVF	ORPI	NFC	OTHR	Null
Bank	Commercial	26						4		1	28	2
	Investment	69					3	73			66	1
NCB		4						13			1	1
CCP											1	88
GOVT									1		1	0
IC	Composite		5	2					1		1	
	Life		15	64	1							
	Nonlife		7	12	6				2			
	Reinsurance		18	11	90				1			
IF	AIFD		3	2		41	9	1	3	2	1	1
	UCIT			3	1	2	84	3				
	AM		44			55	2		4	3	1	
MMF							1					
NFC			7	2	3					38		3
OFI		1						4	10	54		2
PF	Private		2	2		1			62	1		
	Public								16			
Other				1								

Our method shows its value also when compared to EMIR data as source for the sector of both counterparties. In this case, aggregation is based on the two sectors, and in order to assign a sector also to the other counterparty, EMIR data needs to be processed to avoid ambiguity.[30] Our algorithm reaches a coverage of 96% of notional amount outstanding, for which it successfully classifies both counterparties. For the remaining 4%, entities' domicile is either located outside EU or not available.[31] This compares with 80% when using only EMIR data as source, but this figure is inflated by the fact that one CCP is wrongly identified as a credit institution.[32]

On top of the improved coverage, the detailed granularity of our classification enhances the understanding of the market structure (see Fig. 2). It allows to recognize that CCPs and investment banks play a key role in the market, being a counterparty in 76% of outstanding trades in terms of gross notional.

Specifically, trades between CCP and investment banks represent 32% notional (blue bubble CCP—Investment Bank in Fig. 2), while 14% is interdealer activity (yellow bubble Investment Bank—Investment Bank). Among CCPs, the volume of notional is concentrated in a few large players, with seven players clearing 98% of the market. The largest player covers 60% of the outstanding notional among cleared contracts, the second 15% and the third 14%, each specialized in some segments of the market: interest rate, equity, and credit derivatives, respectively. Some asset classes are characterized by a monopoly-oriented market in the provision of clearing services, where the first player clears more than 50% of cleared contracts in interest rate, commodity, and equity derivatives. While credit and currency derivatives show a sort of duopoly. Finally, two major European CCPs seem to benefit from economies of scope providing clearing services in the commodity and credit derivatives market, and currency and interest rate derivatives market, respectively. For further details on the CCPs' business model, and their role in the derivatives market after the reforms, see, e.g., [28, 9, 25, 18].

Commercial banks trade mainly with CCPs and investment banks, with notional amounts of similar magnitude (9% each pair). On the other hand investment banks interact with all the other sectors in the market, owing to their market making and dealer activities. Notably, we find that 7% of notional outstanding is represented by trades between investment funds and investment banks (three red-labeled bubbles at the bottom).

When RIAD, and hence ESA classification, is employed instead of the official lists, results for some sectors change considerably. Most notably, 86% of notional allocated to CCPs according to our classification is allocated to OFIs (S125) with ESA classification. Furthermore, 14% of notional allocated to banks in our

---

[30]See footnote 26.

[31]The fact that there is no domicile is indication of missing or misreported LEI.

[32]Since CCPs do not report any sector according to the regulation, a single mis-reported trade alters greatly the final classification. Some euro area CCPs have a banking license to facilitate their role in the market, but they cannot provide credit and are exempted from some capital requirements.

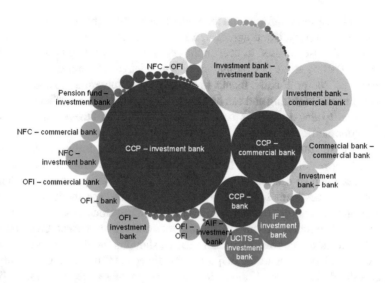

Fig. 2 Notional breakdown by sector based on outstanding contracts, 2018Q2. The size of the circles is proportional to the notional amounts. The colors indicate the pair of sectors, e.g., blue indicates trades between CCPs and banks, and when available we present further breakdown by subsector

classification is allocated as OFI (S125), financial auxiliaries (S126), and captive financial institutions (S127), and 1% is not classified at all. Five percent of notional allocated to the insurance sector is not allocated in ESA while 8% is classified as nonfinancial corporations (S11) or pension funds (S129). Finally, using only ESA classification does not allow to classify 15%, 23%, and 22% of entities classified as nonfinancial corporations, OFI, and pension funds, respectively ,according to our classification.

Overall, the results show several advantages of our sector classification with respect to the reported EMIR sector classification. Firstly, it improves the coverage, allowing for a more comprehensive market description. Secondly, it introduces separate categories for key players in the market, CCPs and investment banks, providing a fairer representation of the market. Lastly, its detailed and multilayered granularity allows to better characterize the market structure.

5 Applications

This section presents four case studies that demonstrate our new classification effectiveness and robustness. At the same time, this section shows the potential of our method as a building block for economic and financial econometric research on the derivatives market. For example, it can be used to investigate market microstructure implications and price formation in these markets, to indicate whether a specific sector would bear more information than others or to study the pricing strategies of derivatives market participants aggregated at the sector level. The application of this algorithm could also be used to deepen the research on monetary economics, e.g., by studying trading strategies on underlyings subject to QE with a breakdown by counterparties' sector. Finally, thanks to the level of automation the algorithm can support a time series setting and can be used to analyze the number of counterparties active in the euro area derivatives market, with a breakdown of the sector they belong to, or in econometric modeling and forecasting.

In some case studies the enhanced granularity provides further insight on the market or on investors' behavior, in others, the extended coverage allows for more precise assessment of sectoral exposures. Case study I leverages on the dedicated taxonomy for investment funds, to show how their strategy significantly affects their portfolio allocation in the derivatives market; Case study II shows the role of investment and commercial banks in the euro area derivatives market; Case study III focuses on the euro area sovereign CDS market, showing the liquidity provisioning role of G16 dealers in one of the major intermediated OTC markets; Case study IV compares the derivatives portfolio of insurance companies as reported in EMIR to previous published reports.

5.1 Case Study I: Use of Derivatives by EA Investment Funds

In this case study, we present, for the first time to our knowledge, a detailed breakdown of euro area investment funds portfolio composition. Furthermore we take full advantage of the detailed level of information on investment fund strategy to investigate whether some asset classes are more or less used by some investment funds depending on their strategy. Data refers to a snapshot at $2019Q3$. We select only funds in ECB's publicly available list.

Funds can opt for different products in the derivatives market according to their mandate. Like other counterparts, they can use derivatives both for hedging balance sheet exposures or to take position; in the second case they are building the so-called synthetic leverage.

Overall we find $20,494$ funds trading derivatives in the euro area,[33] of which 61% are UCITS. For 83% of them, we are able to assign a strategy, with a clear abundance of Mixed (33%), Bond (23%), and Equity (20%) funds. They trade a notional amount of €14 tr, of which 59% is traded by UCITS funds. The most commonly used derivatives are currency derivatives (39%) followed by interest rate (37%) and equity (27%).

There is, however, a large heterogeneity in the portfolio composition when grouping funds by their strategy. Figure 3 provides a summary of funds portfolios according to their strategy. Bond funds largely use interest rate derivatives (47%

---

[33]They represent 35% of active EA funds.

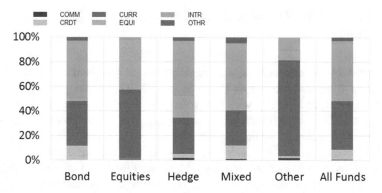

Fig. 3 Notional breakdown of investment funds derivatives portfolio by asset class of the underlying and strategy of the fund. Data refer to 2019Q3

of their portfolio in terms of notional). They are also the largest users of credit derivatives. Equity funds almost exclusively use currency (56%) and equity (41%) derivatives. Hedge and Mixed funds have similar portfolios, with a large share of interest rate (around 40% for each) and currency derivatives (around 28% for each).

To assess whether these differences are statistically significant, we perform a multinomial test on the portfolio allocation of the groups of investment funds with the same strategy, using the overall portfolio allocation as the null distribution (see [31] for details on the methodology). The idea is that for every billion of notional, the fund can decide how to allocate across the six asset classes according to its strategy. If the fraction of notional allocated to a certain asset class is greater (smaller) than the percentage in the overall sample, we will say that it is over-(under-)represented.

The significance is assessed by computing the p-value for the observed fraction in each subgroup using as null a multinomial distribution with parameters inferred from the whole sample. To control for the fact that we are performing multiple tests on the same sample, we apply the Bonferroni correction to the threshold values, which we set at 1% and 5%.

We find that the differences in strategy are generally statistically significant. Bond funds use significantly less currency, commodity, and equity derivatives than average, while they use significantly more credit and interest rate. Equity funds use significantly less interest rate derivatives, while they use significantly more equity, and to a lesser extent currency derivatives. Hedge funds use less credit and currency derivatives, while they significantly use all other asset classes. Real estate funds use significantly less credit and equity derivatives than average, while they use significantly more currency derivatives.

For robustness, we repeat the test on the subsamples of UCIT and non-UCIT and we find very similar results. The only discrepancy is in the use of equity and interest rate derivatives by funds with hedge strategy, which are concentrated in UCIT and non-UCIT funds, respectively.

5.2 Case Study II: The Role of Commercial and Investment Banks

As proved by several studies, the participation of the banking sector in the derivatives market is overriding [8, 3, 26, 2, 30, 21]. Banks participate in the derivatives market typically with two roles: (i) as liquidity providers or (ii) as clearing members. In their liquidity provisioning role, a few dealers intermediate large notional amounts acting as potential sellers and buyers to facilitate the conclusion of the contract. Dealers are willing to take the other side of the trade, allowing clients to buy or sell quickly without waiting for an offsetting customer trade. As a consequence, dealers accumulate net exposures, sometimes long and sometimes short, depending on the direction of the imbalances. Thus, their matched book typically results in large gross exposures.

Given their predominance, the aim of this case study is to analyze the participation of commercial and investment banks in the euro area derivatives market (see Fig. 2). EMIR classification (Table 1) mandates counterparties to report their sector as Credit Institutions or Investment firms as defined by the regulation. However, the classification proposed by our algorithm (Table 5) categorizes banks based on their activity and operating perspective. The reason behind this choice refers to the business model and the domicile of banks operating in the euro area derivatives market. The UK, US, Japanese, and Switzerland counterparties are active in the euro area derivatives markets as much as euro area banks are. Due to the different banking models with which they operate in their home jurisdiction this might affect the final classification and, more importantly, the *role* they play in the market. Using information from several data sources, we define as investment banks those entities performing investment banking activities other than providing credit, while as commercial banks entities which are involved only in the intermediation of credit. Figure 4 shows a comparison between the notional traded by Credit Institutions (CDTI) and Investment Firms (INVF) according to EMIR (LHS) and our classification (RHS). For interest rate derivatives, according to EMIR classification, 68 €trillion is traded by credit institutions and 30 €trillion by investment banks while, applying our classification, these amounts swap. At the same time the breakdown by contract type remains fairly the same across the two groups. The amount traded in currency derivatives by investment banks is the same applying EMIR and our classification, but the breakdown by contract type shows different results: 9% and 52% are the shares of the notional traded in forwards and options according to EMIR reporting which become 79% and 19% according to our classification. For credit and equity derivatives, the gross notional traded by commercial banks double when passing from EMIR to our classification, although the breakdown by contract types remains fairly the same.

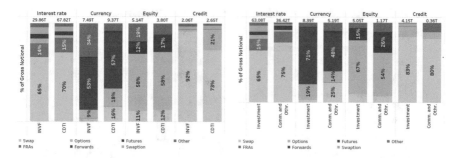

Fig. 4 Banks classified according to EMIR reporting vs. our reclassification, with a breakdown by asset classes. On top of each bar the gross notional reported at the end of the third quarter 2019

5.3   Case Study III: The Role of G16 Dealers in the EA Sovereign CDS Market

The flag *G16* allows to identify entities belonging to the group of G16 dealers. These are investment banks that provide liquidity in the market by buying and selling derivatives on request of the other counterparties. Figure 5 shows the role of these players in the euro area sovereign CDS market as of 2019Q2. The protection traded on euro area government bonds amounts to 600 billion euro in terms of gross notional outstanding. Almost 67% of the gross notional outstanding is traded on Italian government bonds, while the remaining is traded on French, Spanish, German, Portuguese, Irish, Dutch, and Greek government bonds. The position of G16 banks in the market is characterized by a large notional outstanding but a very tiny net notional, because a lot of buying and selling positions offset each other. Although the market making activity implies that the net positions of entities making the market is close to zero, banks may temporarily or persistently have a directional exposure in one market. Hence, the *G16* flag helps to identify which institutions are providing liquidity on specific segments, whether they are specialized or operate across several segments, and how long they maintain their positions. If this might seem irrelevant during calm periods, it might have financial stability implications when liquidity in the derivatives market dries up.

Figure 5 shows G16 net exposures in sovereign CDS aggregated at country level (left) and at *solo* level (right). Overall, UK dealers have the largest net exposures in the euro area sovereign CDS market. G16 domiciled in the UK and US do not have a homogeneous exposure on EA countries: net buying positions result in net buying/selling when passing from exposures aggregated at country level to exposures at solo level. On the contrary, G16 banks domiciled in France or Germany have a directional exposure as net sellers at country level, which is reflected when banks' positions are shown at solo level.

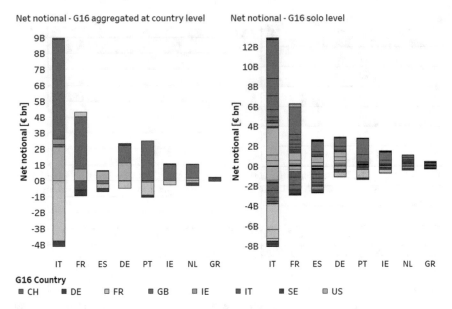

Fig. 5 Net notional exposure on EA sovereign bonds. (**a**) Country level. (**b**) Solo level

5.4 Case Study IV: The Use of Derivatives by EA Insurance Companies

In this application we show how our classification significantly improves assessing euro area insurance companies derivatives portfolio.

In [12], the authors presented the first evidence of insurance companies activity in the market, by employing our proposed classification. The authors considered as insurers only those companies listed in the publicly available register of insurance undertaking companies published by EIOPA. They could easily select those companies from our sector classification, owing to the dedicated column which indicates the data source. The choice to disregard other sources was linked to the intent to make results comparable to those published by EIOPA.[34]

To assess the quality of our classification, we compute the same statistics as presented in [12] but using a sample filtered by the categories *INSU*, *ASSU*, or *REIN* as reported in EMIR data (see again Table 1).

Using only reported information, the total notional outstanding for the insurance sector amounts to €784bn, e.g., 51% of the gross notional of €1.3tr presented in

---

[34]EIOPA has access to the central repository of the quantitative reporting under Solvency II. The data collection includes a template on derivatives positions, see, e.g., [11]

Fig. 6 Percentage of notional associated with LEIs in the EIOPA list allocated to each reported sector in the case it differs from INUN, ASSU, or REIN. Class *Null* includes trades for which the reporting is one sided from the noninsurance counterparts.

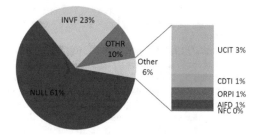

[12], and considerably lower than the figures published by EIOPA.[35] The reason for this discrepancy is largely due to several trades that are reported only by the other counterparty in the contract, represented as *null* (in blue) in Fig. 6. To this extent, our classification efficiently exploits the double reporting implementation of EMIR.[36] For those with a misreported sector, a significant share identify themselves as investment firms (23% of misclassified notional) or in the residual class Other (10% of misclassified notional).

Acknowledgments This chapter should not be reported as representing the views of the European Central Bank (ECB) or the Single Resolution Board (SRB). The views expressed are those of the authors and do not necessarily reflect those of the European Central Bank, the Single Resolution Board, or the Eurosystem. We are kindly grateful for comments and suggestions received by Linda Fache Rousová. We also thank P. Antilici, A. Kharos, G. Nicoletti, G. Skrzypczynski, C. Weistroffer, and participants at the ESRB EMIR Data Workshop (Frankfurt, December 2018), at the ESCoE Conference on Economic Measurement (London, May 2019), and at the European Commission/Joint Research Centre Workshop on Big Data (Ispra, May 2019).

References

1. Abad, J., Aldasoro, I., Aymanns, C., D'Errico, M., Rousová, L., Hoffmann, P., Langfield, S., Neychev, M., & Roukny, T. (2016). Shedding light on dark markets: First insights from the new EU-wide OTC derivatives dataset. Technical report, *ESRB Occasional Paper Series 11*. European Systemic Risk Board. https://www.esrb.europa.eu/pub/pdf/occasional/20160922_occasional_paper_11.en.pdf.
2. Adkins, L. C., Carter, D. A., & Simpson, W. G. (2007). Managerial incentives and the use of foreign-exchange derivatives by banks. *Journal of Financial Research, 30*(3), 399–413.

---

[35][11] reports €2.4tr of notional outstanding. This figure refers to derivatives portfolios of all EU insurers, while [12] only present figures for the portfolio of euro area insurers.

[36] As mentioned in Sect. 2, EMIR is implemented with double reporting. This means that the ECB sample should include two reports for any trade between euro area counterparties, each declaring its own sector. If this is not the case, the information on the sector of the entity failing to report is lost, and therefore the sector aggregates based only on the sector reported in EMIR may not be accurate.

3. Akhigbe, A., Makar, S., Wang, L., & Whyte, A. M. (2018). Interest rate derivatives use in banking: Market pricing implications of cash flow hedges. *Journal of Banking & Finance, 86,* 113–126.
4. Ashraf, D., Altunbas, Y., & Goddard, J. (2007). Who transfers credit risk? Determinants of the use of credit derivatives by large US banks. *The European Journal of Finance, 13*(5), 483–500.
5. Bhojraj, S., Lee, C. M. C., & Oler, D. K. (2003). What's my line? A comparison of industry classification schemes for capital market research. *Journal of Accounting Research, 41*(5), 745–774. https://doi.org/10.1046/j.1475-679X.2003.00122.x.
6. Boninghausen, B., Boneva, L., Fache Rousová, L., & Letizia, E. (2019). Derivatives transactions data and their use in central bank analysis. *ECB Economic Bulletin, 6.* https://www.ecb.europa.eu/pub/economic-bulletin/articles/2019/html/ecb.ebart201906_01~dd0cd7f942.en.html.
7. Cormen, T. H., Leiserson, C. E., Rivest, R. L., & Stein, C. (2009). *Introduction to algorithms.* Cambridge, MA: MIT Press.
8. Duffee, G. R., & Zhou, C.: Credit derivatives in banking: Useful tools for managing risk? *Journal of Monetary Economics, 48*(1), 25–54 (2001).
9. Duffie, D., & Zhu, H.: Does a central clearing counterparty reduce counterparty risk? *The Review of Asset Pricing Studies, 1*(1), 74–95 (2011).
10. Duffie, D., Scheicher, M., & Vuillemey, G.: Central clearing and collateral demand. *Journal of Financial Economics, 116*(2), 237–256 (2015).
11. *EIOPA: EIOPA Financial Stability Report June 2018* (2018). https://www.eiopa.europa.eu/content/financial-stability-report-june-2018_en.
12. Fache Rousová, L., & Letizia, E. (2018). Insurance companies and derivatives exposures: Evidence from EMIR data. In *ECB Financial Stability Review—Box 8.* https://www.ecb.europa.eu/pub/pdf/fsr/ecb.fsr201811.en.pdf.
13. Fama, E. F., & French, K. R. (1997). Industry costs of equity. *Journal of Financial Economics, 43*(2), 153–193. https://doi.org/10.1016/S0304-405X(96)00896-3.
14. Financial Stability Board (2010). Implementing OTC derivatives market reforms. In *Financial Stability Board Report.* https://www.fsb.org/wp-content/uploads/r_101025.pdf.
15. Financial Stability Board (2017). Review of OTC derivatives market reforms: Effectiveness and broader effects of the reforms. In *Financial Stability Board Report.* https://www.fsb.org/wp-content/uploads/P290617-1.pdf.
16. Financial Stability Board (2019). OTC derivatives market reforms thirteenth: Progress report on implementation. In *Financial Stability Board Report.* https://www.fsb.org/wp-content/uploads/P151019.pdf.
17. Financial Stability Board (2019). Thematic review on implementation of the legal entity identifier. *Financial Stability Board Report.* https://www.fsb.org/wp-content/uploads/P280519-2.pdf.
18. Ghamami, S., & Glasserman, P. (2017). Does OTC derivatives reform incentivize central clearing? *Journal of Financial Intermediation, 32,* 76–87.
19. Guagliano, C., Lenoci, F., Mazzacurati, J., & Weistroffer, C. (2019). Use of CDS by non-bank financial institutions in the EU. In *Financial Stability Board (FSB) Global Monitoring Report on Non-Bank Financial Intermediation* (2018). https://www.fsb.org/wp-content/uploads/P040219.pdf.
20. Hau, H., Hoffmann, P., Langfield, S., & Timmer, M. Y. (2019). Discriminatory pricing of over-the-counter derivatives. In *International Monetary Fund Working Paper, No. 19/100.* https://www.imf.org/-/media/Files/Publications/WP/2019/WPIEA2019100.ashx.
21. Hirtle, B.: Credit derivatives and bank credit supply. *Journal of Financial Intermediation, 18*(2), 125–150 (2009).
22. Jones, D. (2000). Emerging problems with the Basel Capital Accord: Regulatory capital arbitrage and related issues. *Journal of Banking & Finance, 24*(1–2), 35–58.

23. Jukonis, A., Cominetta, M., & Grill, M. (2019). Investigating initial margin procyclicality and corrective, tools using EMIR data. *ECB Macroprudential Bulletin, 9.* https://econpapers.repec.org/scripts/redir.pf?u=https%3A%2F%2Fwww.ecb.europa.eu%2F%2Fpub%2Ffinancial-stability%2Fmacroprudential-bulletin%2Fhtml%2Fecb.mpbu201910_5~6c579ba94e.en.html; h=repec:ecb:ecbmbu:2019:0009:5.
24. Kirk, A., McAndrews, J., Sastry, P., & Weed, P. (2014). Matching collateral supply and financing demands in dealer banks. In *Federal Reserve Bank of New York, Economic Policy Review* (pp. 127–151). https://www.newyorkfed.org/medialibrary/media/research/epr/2014/1412kirk.pdf.
25. Loon, Y. C., & Zhong, Z. K. (2014). The impact of central clearing on counterparty risk, liquidity, and trading: Evidence from the credit default swap market. *Journal of Financial Economics, 112*(1), 91–115.
26. Minton, B. A., Stulz, R., & Williamson, R. (2009). How much do banks use credit derivatives to hedge loans?. *Journal of Financial Services Research, 35*(1), 1–31.
27. OICV-IOSCO (2012). Report on OTC derivatives data reporting and aggregation requirements. In *Committee on Payment and Settlement Systems and Technical Committee of the International Organization of Securities Commissions.* https://www.iosco.org/library/pubdocs/pdf/IOSCOPD366.pdf.
28. Pirrong, C. (2011). The economics of central clearing: Theory and practice. In *ISDA Discussion Papers, number 1-May-2011.* https://www.isda.org/a/yiEDE/isdadiscussion-ccp-pirrong.pdf.
29. Rosati, S., & Vacirca, F. (2019). Interdependencies in the euro area derivatives clearing network: A multilayer network approach. *Journal of Network Theory in Finance, 5*(2). https://doi.org/10.21314/JNTF.2019.051.
30. Sinkey Jr, J. F., & Carter, D. A. (2000). Evidence on the financial characteristics of banks that do and do not use derivatives. *The Quarterly Review of Economics and Finance, 40*(4), 431–449.
31. Tumminello, M., Miccichè, S., Lillo, F., Varho, J., Piilo, J., & Mantegna, R. N. (2011). Community characterization of heterogeneous complex systems. *Journal of Statistical Mechanics: Theory and Experiment, 2011*(01), P01019.
32. Wood, S., Muthyala, R., Jin, Y., Qin, Y., Rukadikar, N., Rai, A., & Gao, H. (2017). Automated industry classification with deep learning. In *Proceedings of the 2017 IEEE International Conference on Big Data (Big Data)* (pp. 122–129). https://doi.org/10.1109/BigData.2017.8257920.

Massive Data Analytics for Macroeconomic Nowcasting

Peng Cheng, Laurent Ferrara, Alice Froidevaux, and Thanh-Long Huynh

Abstract Nowcasting macroeconomic aggregates have proved extremely useful for policy-makers or financial investors, in order to get real-time, reliable information to monitor a given economy or sector. Recently, we have witnessed the arrival of new large databases of alternative data, stemming from the Internet, social media, satellites, fixed sensors, or texts. By correctly accounting for those data, especially by using appropriate statistical and econometric approaches, the empirical literature has shown evidence of some gain in nowcasting ability. In this chapter, we propose to review recent advances of the literature on the topic, and we put forward innovative alternative indicators to monitor the Chinese and US economies.

1 Introduction

Real-time assessment of the economic activity in a country or a sector has proved extremely useful for policy-makers in order to implement contra-cyclical monetary or fiscal policies or for financial investors in order to rapidly shift portfolios. Indeed in advanced economies, Quarterly National Accounts are generally published on

All views expressed in this paper are those of the authors and do not represent the views of JPMorgan Chase or any of its affiliates.

P. Cheng (✉)
JPMorgan Chase, New York, NY, USA
e-mail: peng.cheng@jpmorgan.com

L. Ferrara
QuantCube Technology, Paris, France

SKEMA Business School, Lille, France
e-mail: laurent.ferrara@skema.edu

A. Froidevaux · T.-L. Huynh
QuantCube Technology, Lille, France
e-mail: af@q3-technology.com; thanh-long.huynh@q3-technology.com

145

a quarterly basis by Statistical Institutes, with a delay of release of about 1 or 2 months, especially the benchmark macroeconomic indicator, that is, the gross domestic product (GDP). For example, to be aware of the economic activity in the first quarter of the year (from beginning of January to end of March), we need sometimes to wait until mid-May, depending on the country considered. This is also especially true when dealing with emerging economies, where sometimes only low-frequency macroeconomic aggregates are available (e.g., annual aggregates). In this context, macroeconomic nowcasting has become extremely popular in both theoretical and empirical economic literatures. Giannoné et al. [35] were the first to develop econometric models, namely, dynamic factor models, in order to propose high-frequency nowcast of US GDP growth. Currently, the Federal Reserve Bank of Atlanta (GDPNow) and the Federal Reserve Bank of New York have developed their own nowcasting tools for US GDP, available in real time. Beyond the US economy, many nowcasting tools have been proposed to monitor macroeconomic aggregates either for advanced economies (see among others [2] for the euro area and [13] for Japan) or for emerging economies (see among others [14] for Brazil or [42] for Turkey). At the global level, some papers are also trying to assess world economic conditions in real time by nowcasting the world GDP that is computed on a regular basis by the IMF when updating the *World Economic Outlook* report, four times per year (see, e.g., [28]).

Assessing economic conditions in real time is generally done by using standard official economic information such as production data, sales, opinion surveys, or high-frequency financial data. However, we recently witnessed the arrival of massive datasets that we will refer to as *alternative datasets* in opposition to *official datasets*, stemming from various sources of information. The multiplication in recent years of the number of accessible alternative data and the development of methods based on machine learning and artificial intelligence capable of handling them constitute a break in the way of following and predicting the evolution of the economy. Moreover, the power of digital data sources is the real-time access to valuable information stemming from, for example, multi-lingual social media, satellite imagery, localization data, or textual databases.

The availability of those new alternative datasets raises important questions for practitioners about their possible use. One central question is to know whether and when those alternative data can be useful in modeling and nowcasting/forecasting macroeconomic aggregates, once we control for official data. From our reading of the recent literature, it seems that the gain from using alternative data depends on the country under consideration. If the statistical system of the country is well developed, as it is generally the case in advanced economies, then alternative data are able to generate proxies that can be computed on a high-frequency basis, well in advance of the release of official data, with a high degree of reliability (see, e.g., [29], as regards the euro area). In some cases, alternative data allows a high-frequency tracking of some specific sectors (e.g., tourism or labor market). If the statistical system is weak, as it may be in some emerging or low-income economies where national accounts are only annual and where some sectors are not covered, alternative data are likely to fill, or at least narrow, some information gaps and efficiently complement the statistical system in monitoring economic activity (see, e.g., [44]).

In this chapter, we first propose to review some empirical issues for practitioners when dealing with massive datasets for macroeconomic nowcasting. Then we give some examples of tracking for some sectors/countries, based on recent methodologies for massive datasets. The last section contains two real-time proxies for US and Chinese economic growth that have been developed by QuantCube Technology in order to track GDP growth rate on a high-frequency basis. The last section concludes by proposing some applications of macroeconomic nowcasting tools.

2   Review of the Recent Literature

This section presents a short review of the recent empirical literature on nowcasting with massive datasets of alternative data. We don't pretend to do an exhaustive review as this literature is quite large, but rather to give a flavor of recent trends. We first present the various types of alternative data that have been recently considered, then we describe econometric approaches able to deal with this kind of data.

2.1   Various Types of Massive Data

Macroeconomic nowcasting using alternative data involves the use of various types of massive data.

Internet data that can be obtained from webscraping techniques constitute a broad source of information, especially Google search data. Those data have been put forward by Varian [53] and Choi and Varian [19] and have been widely and successfully used in the empirical literature to forecast and nowcast various macroeconomic aggregates.[1] Forecasting prices with Google data has been also considered, for example, by Seabold and Coppola [48] who focus on a set of Latin American countries for which publication delays are quite large. Besides Google data, crowd-sourced data from online platforms, such as Yelp, provide accurate real-time geographical information. Glaeser et al. [37] present evidence that Yelp data can complement government surveys by measuring economic activity in real time at a granular level and at almost any geographic scale in the USA.

The availability of high-resolution satellite imagery has led to numerous applications in economics such as urban development, building type, roads, pollution, or agricultural productivity (for a review, see, e.g., [24]). However, as regards high-frequency nowcasting of macroeconomic aggregates, applications are more

---

[1]Examples of applications include household consumption [19], unemployment rate [23], building permits [21], car sales [45], or GDP growth [29].

scarce. For example, Clarck et al. [20] propose to use data on satellite-recorded nighttime lights as a benchmark for comparing various published indicators of the state of the Chinese economy. Their results are consistent with the rate of Chinese growth being higher than is reported in the official statistics. Satellites can be considered as mobile sensors, but information can also be taken from fixed sensors such as weather/pollution sensors or traffic sensors/webcams. For example, Askitas and Zimmermann [5] show that toll data in Germany, which measure monthly transportation activity performed by heavy transport vehicles, are a good early indicator of German production and are thus able to predict in advance German GDP. Recently, Arslanalp et al. [4] put forward vessel traffic data from automatic identification system (AIS) as a massive data source for nowcasting trade activity in real time. They show that vessel data are good complements of existing official data sources on trade and can be used to create a real-time indicator of global trade activity.

Textual data have been also recently used for nowcasting purposes in order to compute various indexes of sentiment that are then put into standard econometric models. In general, textual analyses are useful to estimate unobserved variables that are not directly available or measured by official sources. A well-known example is economic policy uncertainty that has been estimated for various countries by Baker et al. [8] starting from a large dataset of newspapers by identifying some specific keywords. Those economic policy uncertainty (EPU) indexes have proved useful to anticipate business cycle fluctuations, as recently shown by Rogers and Xu [47], though their real-time performance has to be taken with caution. Various extensions of this approach have been proposed in the literature, such as the geopolitical risk index by Caldara and Iacovello [17] that can be used to forecast business investment. Kalamara et al. [41] recently proposed to extract sentiment from various newspapers using different machine learning methods based on dictionaries and showed that they get some improvement in terms of UK GDP forecasting accuracy. In the same vein, Fraiberger et al. [32] estimate a media sentiment index using more than 4.5 million Reuters articles published worldwide between 1991 and 2015 and show that it can be used to forecast asset prices.

Payment data by credit cards have been shown to be a valuable source of information to nowcast household consumption. These card payment data are generally free of sampling errors and are available without delays, providing thus leading and reliable information on household spending. Aastveit et al. [1] show that credit card transaction data improve both point and density forecasts for Norway and underline the usefulness of getting such information during the Covid-19 period. Other examples of application of payment data for nowcasting economic activity include among others Galbraith and Tkacz [33], who nowcast Canadian GDP and retail sales using electronic payment data, or Aprigliano et al. [3], who assess the ability of a wide range of retail payment data to accurately forecast Italian GDP and its main domestic components.

Those massive alternative data have the great advantage of being available at a very high frequency, thus leading to signals that can be delivered well ahead of official data. Also, those data are not revised, avoiding thus a major issue for forecasters. However, there is no such thing as a free lunch. An important aspect that is not often considered in empirical works is about the cleaning of raw data. Indeed, it turns out that unstructured raw data are often polluted by outliers, seasonal patterns, or breaks, temporary or permanent. For example, when dealing with daily data, they can present two or more seasonalities (e.g., weekly and annual). In such a case, seasonal adjustment is not an easy task and should be carefully considered. An exhaustive review of various types of alternative data that can be considered for nowcasting issues is presented in [16].

2.2 Econometric Methods to Deal with Massive Datasets

Assume we have access to a massive dataset ready to be put into an econometric model. Generally, those datasets present two stylized facts: (1) a large number n of variables compared to the sample size T and (2) a frequency mismatch between the targeted variable (quarterly in general) and explanatory variables (monthly, weekly, or daily) .

Most of the time, massive datasets have an extremely large dimension, with the number of variables much larger than the number of observations (i.e., $n >> T$, sometimes referred to as *fat datasets*). The basic equation for nowcasting a target variable y_t using a set of variables

$$y_t = \beta_1 x_{1t} + \ldots + \beta_n x_{nt} + \varepsilon_t, \tag{1}$$

where $\varepsilon_t \sim N(0, \sigma^2)$. To account for dynamics, x_{jt} can also be a lagged value of the target variable or of other explanatory variables. In such a situation, usual least-squares estimates are not necessarily a good idea as there are too many parameters to estimate, leading to a high degree of uncertainty in estimates, as well as a strong risk of over-fitting in-sample associated to poor out-of-sample performances. There are some econometric approaches to address the curse of dimensionality. Borrowing from Giannone et al. [36], we can classify those approaches in two categories: *sparse* and *dense* models. Sparse methods assume that some β_j coefficients in Eq. (1) are equal to zero. This means that only few variables have an impact on the target variable. Zeros can be imposed ex ante by the practitioners based on specific a priori information. Alternatively, zeros can be estimated using an appropriate estimation method such as the LASSO (*least absolute shrinkage and selection operator*) regularization approach [51] or some Bayesian techniques that impose some coefficients to take null values during the estimation step (see, e.g., Smith et al. [49] who develop a Bayesian approach that can shrink some coefficients to zero and allows coefficients that are shrunk to zero to vary through regimes).

In opposition, dense methods assume that all the explanatory variables have a role to play. A typical example is the dynamic factor model (DFM) that tries to estimate a common factor from all the explanatory variables in the following way:

$$x_t = \Lambda f_t + \xi_t, \tag{2}$$

where $x_t = (x_{1t}, \ldots, x_{nt})'$ is a vector of n stationary time series and x_t is decomposed into a common component Λf_t where $f_t = (f_{1t}, \ldots, f_{rt})'$ and Λ is the loading matrix such that $\Lambda = (\lambda_1, \ldots, \lambda_n)'$ and an idiosyncratic component $\xi_t = (\xi_{1t}, \ldots, \xi_{nt})'$ a vector of n mutually uncorrelated components. A VAR(p) dynamics is sometimes allowed for the vector f_t. Estimation is carried out using the diffusion index approach of Stock and Watson [50] or the generalized DFM of Forni et al. [30]. As the number r of estimated factors $\hat{f}_t$ is generally small, they can be directly put in a second step into the regression equation to explain y_t in the following way:

$$y_t = \gamma_1 \hat{f}_{1t} + \ldots + \gamma_r \hat{f}_{rt} + \varepsilon_t. \tag{3}$$

We refer, for example, to [7, 10, 9], for examples of application of this approach.

Another well-known issue when nowcasting a target macroeconomic variable with massive alternative data is the frequency mismatch as y_t is generally a low-frequency variable (e.g., quarterly), while explanatory variables x_t are generally high frequency (e.g., daily). A standard approach is to first aggregate the high-frequency variables to the low frequency by averaging and then to estimate Eq. (1) at the lowest frequency. Alternatively, mixed-data sampling (MIDAS hereafter) models have been put forward by Ghysels et al. [34] in order to avoid systematically aggregating high-frequency variables. As an example, let's consider the following MIDAS bivariate equation:

$$y_t = \beta_0 + \beta_1 \times B(\theta) x_t^{(m)} + \varepsilon_t \tag{4}$$

where $(x_t^{(m)})$ is an exogenous stationary variable sampled at a frequency higher than (y_t) such that we observe m times $(x_t^{(m)})$ over the period $[t-1, t]$. The term $B(\theta)$ controls the polynomial weights that allows the frequency mixing. Indeed, the MIDAS specification consists in smoothing the past values of $(x_t^{(m)})$ by using the polynomial $B(\theta)$ of the form:

$$B(\theta) = \sum_{k=1}^{K} b_k(\theta) L^{(k-1)/m} \tag{5}$$

where K is the number of data points on which the regression is based, L is the lag operator such that $L^{s/m} x_t^{(m)} = x_{t-s/m}^{(m)}$, and $b_K(.)$ is the weight function that can take various shapes. For example, as in [34], a two-parameter exponential Almon lag polynomial can be implemented such as $\theta = (\theta_1, \theta_2)$,

$$b_k(\theta) = b_k(\theta_1, \theta_2) = \frac{\exp\left(\theta_1 k + \theta_2 k^2\right)}{\sum_{k=1}^{K} \exp\left(\theta_1 k + \theta_2 k^2\right)} \tag{6}$$

The parameter θ is part of the estimation problem. It is only influenced by the information conveyed by the last K values of the high-frequency variable $(x_t^{(m)})$, the window size K being an exogenous specification.

A useful alternative is the unrestricted specification (U-MIDAS) put forward by Foroni et al. [31] that does not consider any specific function $b_k(.)$ but assume a linear relationship of the following form:

$$y_t = \beta_0 + c_0 x_t^{(m)} + c_1 x_{t-1/m}^{(m)} + \ldots + c_{mK} x_{t-(K-1)/m}^{(m)} + \varepsilon_t \tag{7}$$

The advantage of the U-MIDAS specification is that it is linear and can be easily estimated by ordinary least-squares under some reasonable assumption. However, to avoid a proliferation of parameters ($2+mK$ parameters have to be estimated), m and $\tilde{K}$ have to be relatively small. Another possibility is to impose some parameters c_j in Eq. (7) to be equal to zero. We will use this strategy in our applications (see details in Sect. 4.1).

3 Example of Macroeconomic Applications Using Massive Alternative Data

In this section, we present three examples using the methodology that we have developed in order to nowcast growth rates of macroeconomic aggregates using the flow of information coming from alternative massive data sources. Nowcasts for current-quarter growth rates are in this way updated each time new data are published. It turns out that those macroeconomic nowcasts have the great advantage of being available well ahead of the publication of official data, sometimes several months, while being extremely reliable. In some countries, where official statistical systems are weak, such macroeconomic nowcasts can efficiently complement the standard macroeconomic indicators to monitor economic activity.

3.1 A Real-Time Proxy for Exports and Imports

3.1.1 International Trade

There are three various modes of transportation for international trade: ocean, air, and land. Each mode of transportation possesses its own advantages and drawbacks based on services, delivery schedules, costs, and inventory levels. According to *Transport and Logistics of France*, maritime market represents about 90% of the world market of imports and exports of raw materials, with a total of more than 10 million tonnes of goods traded per year, according to UNCTAD [52]. Indeed, maritime transport remains the cheapest way to carry raw materials and products. For example, raw materials in the energy sector dominate shipments by sea with 45% of total shipments. They are followed by those in the metal industry, which represents 25% in total, then by agriculture, which accounts for 13%.

Other productions such as textiles, machines, or vehicles represent only 3% of the sea transport but constitute around 50% of the value of raw materials transported because of their high value. Depending on their nature, raw materials are transported on cargo ships or tankers. Indeed, we generally refer to four main types of vessels: fishing vessels, cargo (dry cargo), tankers (liquid cargo), and offshore vessels (urgent parts and small parcels). In our study, we will only focus on the cargo ships and tankers as they represent the largest part of the volume traded by sea.

In the following of this section, we develop the methodology to analyze the ship movements and to create a proxy of imports and exports for various countries and commodities.

3.1.2 Localization Data

We get our data from the automatic identification system (AIS), the primary method of collision avoidance for water transport. AIS integrates a standardized VHF transceiver with a positioning system, such as GPS receiver, as well as other electronic navigation sensors, such as a gyro-compass. Vessels fitted with AIS transceivers can be tracked by AIS base stations located along coast lines or, when out of range of terrestrial networks, through a growing number of satellites that are fitted with special AIS receivers which are capable of de-conflicting a large number of signatures. In this respect, we are able to track more than 70,000 ships with a daily update since 2010.

3.1.3 QuantCube International Trade Index: The Case of China

The QuantCube International Trade Index that we have developed tracks the evolution of official external trade numbers in real time by analyzing shipping data from ports located all over the world and taking into account the characteristics of the ships. As an example, we will focus here on international trade exchanges of China, but the methodology of the international trade index can be extended to various countries and adapted for specific commodities (crude oil, coal, and iron ore).

First of all, we carry out an analysis of variance of Chinese official exports and imports by products (see Trade Map, monthly data 2005–2019). It turns out that (1) "electrical machinery and equipment" and "machinery" mainly explain the variance of Chinese exports and (2) "mineral fuels, oils, and products", "electrical machinery and equipment," and "commodities" mainly explain the variance of Chinese imports.

As those products are transported by ships, we count the number of various ships arriving in all Chinese ports. In fact, we are interested in three various types of ships: (1) bulk cargo ships that transport commodities, (2) container cargo ships transporting electrical machinery as well as equipment and machinery, and (3) tankers transporting petroleum products. For example, the total number of container cargo ships arriving in Chinese ports for each day, from July 2012 to July 2019, is presented in Fig. 1. Similar daily series are available for bulk cargo ships and tankers.

In order to avoid too much volatility present in the daily data, we compute the rolling average over 30 days of the daily arrivals of the three selected types of ships in all the Chinese ports, such as:

$$Ship_{(i,j)}^{q3}(t) = \frac{1}{30} \sum_{m=1}^{30} X_{i,j}(t-m) \tag{8}$$

with $X_{i,j}$ the number of ship arrivals of type i [container cargo, tanker, bulk cargo] in a given Chinese port j.

Finally, we compute our final QuantCube International Trade Index for China using Eq. (8) by summing up the three types of shipping and by computing its year-over-year changes. This index is presented in Fig. 2. We get a correlation of 80%

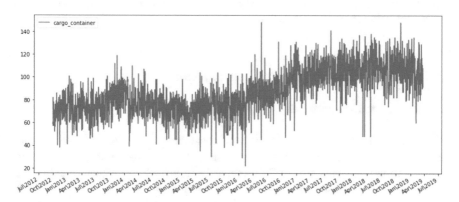

Fig. 1 Sum of cargo container arrivals in all Chinese ports

Fig. 2 China global trade index (year-over-year growth in %)

between the real-time QuantCube International Trade Index and Chinese official trade numbers (imports + exports). It is a 2-month leading index as the official numbers of imports and exports of goods are published with a delay of 2 months after the end of the reference month. We notice that our indicator clearly shows the slowing pace of total Chinese trade, mainly impacted by the increasing number of US trade sanctions since mid-2018.

For countries depending strongly on maritime exchanges, this index can reach a correlation with total country external trade numbers up to 95%. For countries relying mostly on terrestrial exchanges, it turns out that the index is still a good proxy of overseas exchanges. However, in this latter case, proxies of aerial and land exchanges can be computed to complement the information, by using cargo flights, tolls, and train schedule information.

3.2 A Real-Time Proxy for Consumption

3.2.1 Private Consumption

When tracking economic activity, private consumption is a key macroeconomic aggregate that we need to evaluate in real time. For example, in the USA, private consumption represents around 70% of the GDP growth. As official numbers of private consumption are available on a monthly basis (e.g., in the USA) or a quarterly basis (e.g., in China) and with delays of publication ranging from 1 to 3 months, alternative data sources, such as Google Trends, can convey useful information when official information is lacking.

As personal expenditures fall under the durable goods, non-durable goods, and services, we first carry out a variance analysis of the consumption for the studied countries, to highlight the key components of the consumption we have to track. For example, for the Chinese consumption, we have identified the following categories : Luxury (bags, watches, wine, jewelry), Retail sales (food, beverage, clothes, tobacco, smartphones, PC, electronics), Vehicles, Services (hotel, credit loan, transportation), and Leisure (tourism, sport, cinema, gaming). In this section, we focus on one sub-indicator of the QuantCube Chinese consumption proxy, namely, *Tourism* (Leisure category). The same methodology is developed to track the other main components of household consumption.

3.2.2 Alternative Data Sources

The touristic sub-component of the Chinese consumption is supposed to track the spending of the Chinese population for tourist trips, inside and outside the country. To create this sub-component of the consumption index, we have used touristic-related search queries retrieved by means of the Google Trends and Baidu applications. In fact, Internet search queries available through Google Trends and Baidu allow us to build a proxy of the private consumption for tourist trips per country as search queries done by tourists reflect the trends of their traveling preferences but also a prediction of their future travel destination. Google and Baidu Trends have search trends features that show how frequently a given search term is entered into Google's or Baidu's search engine relative to the site's total search volume over a given period of time. From these search queries, we built two different indexes: the tourist number per destination using the region filter "All country" and the tourist number from a specific country per destination by selecting the country in the region filter.

3.2.3 QuantCube Chinese Tourism Index

The QuantCube Chinese Tourism Index is a proxy of the tourist number from China per destination. To create this index, we first identified the 15 most visited countries by Chinese tourists that represent 60% of the total volume of Chinese tourists. We create a Chinese tourism index per country by identifying the relevant categories based on various aspects of trip planning, including transportation, touristic activities, weather, lodging, and shopping. As an example, to create our Chinese tourism index in South Korea, we identified the following relevant categories: Korea Tourism, South Korea Visa, South Korea Maps, Korea Tourism Map, South Korea Attractions, Seoul Airport, and South Korea Shopping (Fig. 3).

Finally, by summing the search query trends of those identified keywords, our Chinese Tourism Index in South Korea tracks in real time the evolution of official

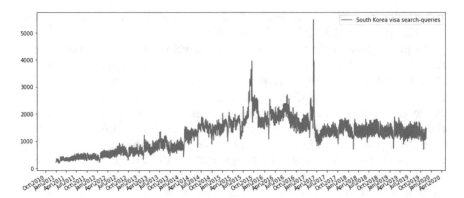

Fig. 3 Baidu: "South Korea Visa" search queries

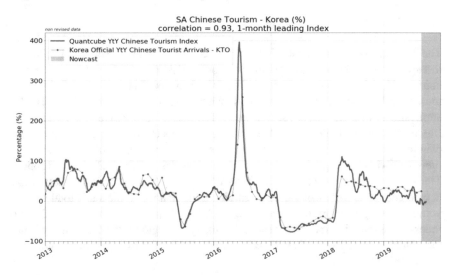

Fig. 4 South Korea Chinese tourism index short term (year-over-year in %)

Chinese tourist arrivals in Korea. We calculate the year-over-year variation of this index and validate it using official numbers of Chinese tourists in South Korea (see Fig. 4).

From Fig. 4, we observe that the QuantCube Chinese Tourism Index correctly tracks the arrivals of Chinese tourists in South Korea. As, for example, the index caught the first drop in June 2015 due to the MERS outbreak. Furthermore, in 2017, after the announcement of a future installation of the powerful Terminal High Altitude Area Defense (THAAD) system at the end of 2016, the Chinese government banned the tour groups to South Korea as economic retaliation. For 2017 as a whole, South Korea had 4.2 million Chinese visitors—down 48.3% from the previous year. The decrease in Chinese tourists leads to a 36% drop in tourist entries. Therefore, this real-time Chinese Tourism indicator is also useful to estimate in real time the trend of the South Korea Tourism industry.

It tracks Chinese tourist arrivals with a correlation up to 95%.

Finally, we developed similar indexes to track in real time the arrivals of Chinese tourists in the most 15 visited countries (USA, Europe, etc.); we get an average correlation of 80% for the most visited countries. By aggregating those indexes we are able to construct an index for tracking the arrival of Chinese tourists over the world that provides a nice proxy of Chinese households' consumption in this specific sector.

3.3 A Real-Time Proxy for Activity Level

QuantCube Technology has developed a methodology based on the analytic of the satellite Sentinel-2 images to detect new infrastructures (commercial, logistics, industrial, or residential) and measure the evolution of the shape and the size of urban areas. But, the level of activities or exploitation of these sites is hardly determined by building inspection and could be inferred from vehicle presence from nearby streets and parking lots. For this purpose, QuantCube Technology in partnership with IRISA developed a deep learning model for vehicles counting from satellite images coming from the Pleiades sensor at 50-cm spatial resolution. In fact, we select the satellite depending on the pixel resolution needed per application.

3.3.1 Satellite Images

Satellite imagery has become more and more accessible in recent years. In particular, some public satellites provide an easy and cost-free access to their image archives, with a spatial resolution high enough for many applications concerning land characterization. For example, the ESA (European Space Agency) satellite family Sentinel-2, launched on June 23, 2015, provides 10-meter resolution multi-spectral images covering the entire world. We analyze those images for infrastructure detection. To detect and count cars we use higher-resolution VHR (very high resolution) images acquired by the Pleiades satellite (PHR-1A and PHR-1B), launched by the French Space Agency (CNES), Distribution Airbus DS. These images are pan-sharpened products obtained by the fusion of 50-cm panchromatic data (70 cm at nadir, resampled at 50 cm) and 2-m multispectral images (visible RGB (red, green, blue) and infrared bands). They cover a large region of heterogeneous environments including rural, forest, residential, as well as industrial areas, where the appearance of vehicles is influenced by shadow and occlusion effects. On the one hand, one of the advantages of satellite imaging-based applications is their natural world scalability. And on the other hand, the evolution and improvements of artificial intelligence algorithms enable us to process the huge amount of information contained in satellite images in a straightforward way, giving a standardized and automatic solution working on real-time data.

3.3.2 Pre-processing and Modeling

Vehicle detection from satellite images is a particular case of object detection as objects are uniform and very small (around 5×8 pixels/vehicle in Pleiades images) and do not overlap. To tackle this task, we use a model called 100 layers Tiramisu (see [40]). It is a quite economical model since it has only 9 million parameters, compared to around 130 million for early deep learning network referred to as $VGG19$. The goal of the model is to exploit the feature reuse by extending DenseNet architecture while avoiding the feature explosion. To train our deep learning model, we created a training dataset using an interactive labeling tool that enables to label 60% of vehicles with one click using flood-fill methods adapted for this application. The created dataset contains 87,000 annotated vehicles from different environments depending on the level of urbanization. From the segmentation, an estimation of the number of vehicles is computed based on the size and the shape of the predictions.

3.3.3 QuantCube Activity Level Index

Finally, the model realizes a satisfying prediction for vehicle detection and counting application since the precision reached more than 85% on a validation set of 2673 vehicles belonging to urban and industrial zones. The algorithm is currently able to deal with different urban environments. As can be seen in Fig. 5, showing a view of Orly area near Paris including the prediction for the detection and counting of cars in yellow, the code is able to accurately count vehicles in some identified areas.

The example in Fig. 5 shows the number of vehicles for every identified bounding box corresponding to parking of hospitality, commercial, or logistics sites. Starting

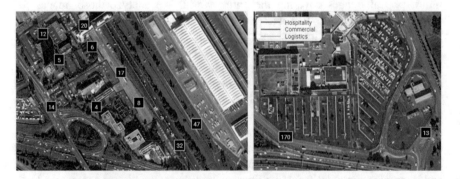

Fig. 5 Number of vehicles per zone in Orly

from this satellite-based information, we are able to compute an index to detect and count vehicles in identified sites and to track their level of activities or exploitation looking at the evolution of the index which is correlated to sales indexes. Using satellite images, it enables to create a methodology and normalized measures of the level of activities that enable financial institutions and corporate groups to anticipate new investment trends before the release of official economics numbers.

4 High-Frequency GDP Nowcasting

When dealing with the most important macroeconomic aggregate, that is, GDP, our approach relies on the expenditure approach that computes GDP by evaluating the sum of all goods and services purchased in the economy. That is, we decompose GDP into its main components, namely, consumption (C), investment (I), government spending (G), and net exports (X-M) such as:

$$GDP = C + I + G + (X - M) \tag{9}$$

Starting from this previous decomposition, our idea is to provide high-frequency nowcasts for each GDP component given in Eq. (9), using the indexes based on alternative data that we previously computed. However, depending on the country considered, we do not necessarily cover all the GDP components with our indexes. Thus, the approach that we developed within QuantCube consists in mixing in-house indexes based on alternative data and official data stemming from opinion surveys, production, or consumption. This is a way to have a high-frequency index that covers a large variety of economic activities. In this section, we present results that we get on the two largest economies in the world, that is, the USA and China.

4.1 Nowcasting US GDP

The US economy ranks as the largest economy by nominal GDP; they are the world's most technologically powerful economy and the world's largest importer and second largest exporter. In spite of some already existing nowcasting tools on the market, provided by the Atlanta Fed and the New York Fed, it seems useful to us to develop a US GDP nowcast on a daily basis.

To nowcast US GDP, we are mixing official information on household consumption (personal consumption expenditures) and on consumer sentiment (University of Michigan) with in-house indexes based on alternative data. In this respect, we use the *QuantCube International Trade Index* and *QuantCube Crude Oil Index*, developed using the methodology presented in the Sect. 3.1 of this chapter, and the *QuantCube Job Opening Index* that is a proxy of the job market and nonfarm payroll created by aggregating job offers per sector. The two official variables that we use are published with 1-month delay and are available on a monthly frequency. However, the three QuantCube indexes are daily and are available in real time without any publication lags.

Daily US GDP nowcasts are computed using the U-MIDAS model given in Eq. (7) by imposing some constraints. Indeed we assume that only the latest values of the indexes enter the U-MIDAS equation. As those values are the averages of the last 30 days, we thus account for the recent dynamics by imposing MIDAS weights to be uniform. The US QuantCube Economic Growth Index aiming at tracking year-over-year changes in US GDP is presented in Fig. 6. We clearly see that this index is able to efficiently track US GDP growth, especially as regards peaks and troughs in the cycles. For example, focusing on the year 2016, we observe that the index anticipates the slowing pace of the US economy for this specific year which was the worst year in terms of GDP growth since 2011, at 1.6% annually. The lowest point of index was reached on October 12, 2016, giving a leading signal of a decelerating fourth quarter in 2016. As a matter of fact, the US economy lost momentum in the final 3 months of 2016.

Then, the indicator managed to catch the strong economic trend in 2017 (+2.3% annually, acceleration from the 1.6% logged in 2016). It even reflected the unexpected slowdown in the fourth quarter of 2017 two months in advance, because of surging imports, a component that is tracked in real time. Focusing on the recent Covid-19 crisis, official US GDP data shows a decline to a value slightly above zero in year-over-year growth for 2020q1, while our index reflects a large drop in subsequent months close to −6% on July 2, 2020, indicating a very negative growth in 2020q2. As regards the US economy, the Atlanta Fed and New York Fed release on a regular basis estimates of current and future quarter-over-quarter GDP

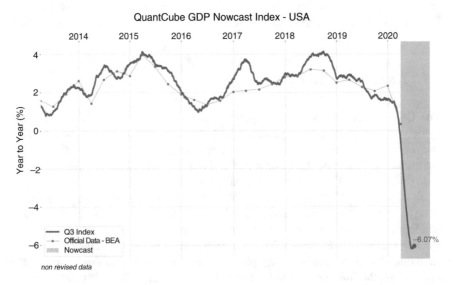

Fig. 6 US economic growth index

growth rate, expressed in annualized terms. Surprisingly, as of March 25, 2020, the nowcast of Atlanta Fed for 2020q1 was at 3.1%, and the one of New York Fed was a bit lower at 1.49% as of March 20, 2020, but still quite high. How is that? In fact, all those nowcasting tools have been extremely well built, but they only integrate official information, such as production, sales, and surveys, that is released by official sources with a lag. Some price variables, such as stock prices that are reacting more rapidly to news, are also included in the nowcasting tools, but they do not strongly contribute to the indicator. So how can we improve nowcasting tools to reflect high-frequency evolutions of economic activity, especially in times of crises? A solution is to investigate alternative data that are available on a high-frequency basis as we do with our indicator. It turns out that at the same date, our US nowcast for 2020q1 was close to zero in year-over-year terms, consistent with a quarter-over-quarter GDP growth of about -6.0% in annualized terms, perfectly in line with official figures from the BEA. This real-time economic growth indicator appears as a useful proxy to estimate in real time the state of the US economy.

4.2 Nowcasting Chinese GDP

China ranks as the second largest economy in the world by nominal GDP. It has the world's fastest growing major economy, with a growth rate of 6% in average over 30 years. It is the world's largest manufacturing economy and exporter of goods and the world's largest fastest growing consumer market and second largest importer of goods.

Yet, despite the importance for the world economy and the region, there are few studies on nowcasting Chinese economic activity (see [27]). Official GDP data are available only with a 2-month lag and are subject to several revisions.

To nowcast the Chinese GDP in real time, we use the *QuantCube International Trade Index* and the *QuantCube Commodity Trade Index* developed in the Sect. 3.1 of this chapter; the *QuantCube Job Opening Index*, a proxy of the job market created by aggregating the job offers per sector; and the *QuantCube Consumption Index* developed in Sect. 3.2. It turns out that all the variables have been developed in-house based on alternative massive datasets and are thus available on a daily frequency without any publication lags.

Daily GDP nowcasts are computed using the U-MIDAS model given in Eq. (7) by imposing the same constraints as for the USA (see previous sub-section). The China Economic Growth Index, aiming at tracking China GDP year-over-year growth, is presented in Fig. 7. First of all, we observe that our index is much more volatile than official Chinese GDP, which seems more consistent with expectations about fluctuations in GDP growth. Our measure thus reveals a bias, but it is not systematic. In fact, most of the time the *true* Chinese growth is likely to be lower than the official GDP, but for some periods of time, the estimated GDP can also be higher as, for example, in 2016–2017. The Chinese GDP index captured the deceleration of the Chinese economy from the middle of 2011. The index showed a sharp drop in Q2

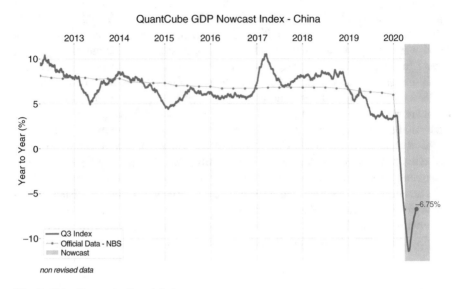

Fig. 7 China Economic Growth Index

2013, when according to several analysts, the Chinese economy actually shrank. The indicator shows the onset of the deceleration period beginning in 2014, in line with the drop in oil and commodity prices. According to our index, the Chinese economy is currently experiencing a deceleration that started beginning of 2017. This deceleration is not as smooth as in the official data disclosed by the Chinese government. In particular, a marked drop occurred in Q2 2018, amid escalating trade tensions with the USA. The year 2019 begins with a sharp drop of the index, showing that the China economy still did not reach a steady growth period. As regards the recent Covid-19 episode, QuantCube GDP Nowcast Index for China shows a sharp year-over-year decline starting at the end of January 2020 from 3.0% to a low of about −11.5% beginning of May 2020, ending at −6.7% on July 2, 2020. This drop is larger than the official data from the National Bureau of Statistics which stated a negative yearly GDP growth of −6.8% in 2020q1. Overall, this indicator is a unique valuable source of information about the state of the economy since very few economic numbers are released in China.

5 Applications in Finance

There is a long recognized intricate relationship between real macroeconomy and financial markets. Among the various academic works, Engel et al. [25] show evidence of predictive power of inflation and output gap on foreign exchange rates, while Cooper and Priestly [22] show that the output gap is a strong predictor of US government bond returns. Such studies are not limited to the fixed income market.

As early as 1967, Brown and Ball [15] show that a large portion of the variation in firm-level earnings is explained by contemporaneous macroeconomic conditions. Rangvid [46] also shows that the ratio of share price to GDP is a good predictor of stock market returns in the USA and international developed countries.

However, economic and financial market data have a substantial mismatch in the observation frequency. This presents a major challenge to analyzing the predictive power of economic data on financial asset returns, given the low signal-to-noise ratio embedded in financial assets. With the increasing accessibility of high-frequency data and computing power, real-time, high-frequency economic forecasts have become more widely available. The Federal Reserve Bank of Atlanta and New York produce nowcasting models of US GDP figures which are available at least on a weekly basis and are closely followed by the media and the financial markets. Various market participants have also developed their own economic nowcasting models. As previously pointed out in this chapter, QuantCube produces US GDP nowcasts available at daily frequency. A number of asset management firms and investment banks have also made their GDP nowcasts public. Together these publicly available and proprietary nowcast information are commonly used by discretionary portfolio managers and traders to assess investment prospects. For instance, Blackrock [39] uses their recession probability models for macroeconomic regime detection, in order to inform asset allocation decisions. Putnam Investments [6] uses global and country GDP nowcasts as key signals in their interest rates and foreign exchange strategies. While the investment industry has embraced nowcasting as an important tool in the decision-making process, evaluating the effectiveness of real-time, high-frequency economic nowcasts on financial market returns is not without its own challenges. Most of the economic nowcasts have short history and evolving methodology. Take the two publicly available US GDP nowcasts mentioned above as examples. The Atlanta Fed GDPNow was first released in 2014 and introduced a methodology change in 2017, whereas the NY Fed GDP nowcast was first released in 2016. Although longer in-sample historical time series are available, the out-of-sample historical periods would be considered relatively short by financial data standards. As a result, the literature evaluating the out-of-sample predictive power of nowcasting models is relatively sparse. Most studies have used point-in-time data to reconstruct historical economic nowcasts for backtesting purposes. We survey some of the available literature below.

Blin et al. [12] used nowcasts for timing alternative risk premia (ARP) which are investment strategies providing systematic exposures to risk factors such as value, momentum, and carry across asset classes. They showed that macroeconomic regimes based on nowcast indicators are effective in predicting ARP returns. Molodtsova and Papell [43] use real-time forecasts of Taylor rule model and show outperformance over random walk models on exchange rates during certain time periods. Carabias [18] shows that using macroeconomic nowcasts is a leading indicator of firm-level end-of-quarter realized earnings, which translates into risk-adjusted returns around earnings announcements. Beber et al. [11] developed latent factors representing economic growth and its dispersion, which together explain almost one third of the implied stock return volatility index (VIX). The results are encouraging since modeling stock market volatility is of paramount importance for financial risk management, but historically, financial economists have struggled to identify the relationship between the macroeconomy and the stock market volatility [26]. More recently, Gu et al. [38] have shown that machine learning approaches, based on neural networks and trees, lead to a significant gain for investors, basically doubling standard approaches based on linear regressions. Obviously, more research is needed about the high-frequency relationship between macroeconomic aggregates and financial assets, but this line of research looks promising.

6 Conclusions

The methodology reported in this chapter highlights the use of large and alternative datasets to estimate the current situation in systemic countries such as China and the USA. We show that massive alternative datasets are able to account for real-time information available worldwide on a daily frequency (AIS position, flight traffic, hotel prices, satellite images, etc.). By correctly handling those data, we can create worldwide indicators calculated in a systematic way. In countries where the statistical system is weak or non-credible, we can thus rely more on alternative data sources than on official ones. In addition, the recent Covid-19 episode highlights the gain in timeliness from using alternative datasets for nowcasting macroeconomic aggregates, in comparison with standard official information. When large shifts in GDP occur, generating thus a large amount of uncertainty, it turns out that alternative data are an efficient way to assess economic conditions in real time. The challenge for practitioners is to be able to deal with massive non-structured datasets, often affected by noise, outliers, and seasonal patterns . . . and to extract the pertinent and accurate information.

References

1. Aastveit, K. A., Albuquerque, B., & Anundsen, A. K. (2020). Changing supply elasticities and regional housing booms. Bank of England Working Paper No. 844. https://www.bankofengland.co.uk/working-paper/2020/changing-supply-elasticities-and-regional-housing-booms.
2. Angelini, E., Camba-Mendez, G., Giannone, D., Reichlin, L., & Ruenstler, G. (2011). Short-term forecasts of euro area GDP growth. *Economic Journal, 14*, C25–C44.
3. Aprigliano, V., Ardizzi, G. & Monteforte, L. (2019). Using the payment system data to forecast the economic activity. *International Journal of Central Banking 15*, 4.
4. Arslanalp, S., Marini, M., & Tumbarello, P. (2019). *Big data on vessel traffic: Nowcasting trade flows in real time*. IMF Working Paper No. 19/275.
5. Askitas, N., & Zimmermann, K. (2013). Nowcasting business cycles using toll data. *Journal of Forecasting, 32*(4), 299–306.

6. Atkin, M., Chan, A., Embre, O., Hornder, S., Solyanik, I., & Yildiz, I. (2020). *Waiting for the next wave of growth*. Macro Report, Putnam Investments. https://www.putnam.com/ institutional/content/macroReports/434-waiting-for-the-next-wave-of-growth.
7. Baffigi, A., Golinelli, R., & Parigi, G. (2004). Bridge models to forecast the euro area GDP. *International Journal of Forecasting, 20*(3), 447460.
8. Baker, S. R., Bloom, N., & Davis, S. J. (2016). Measuring economic policy uncertainty. *The Quarterly Journal of Economics, 131*(4), 1593–1636. https://doi.org/10.1093/qje/qjw024.
9. Banbura, M., Giannone, D., Modugno, M., & Reichlin, L. (2013). Now-casting and the real-time data flow. In *Handbook of Economic Forecasting*, vol. 2. Part A (pp. 195–237). Amsterdam: Elsevier.
10. Barhoumi, K., Darne, O., & Ferrara, L. (2013). Testing the number of factors: An empirical assessment for a forecasting purpose. *Oxford Bulletin of Economics and Statistics, 75*(1), 64–79.
11. Beber, A., Brandt, M., & Luizi, M. (2015). Distilling the macroeconomic news flow. *Journal of Financial Economics, 117*, 489–507.
12. Blin, Ol., Ielpo, F., Lee, J., & Teiletche, J. (2020). Alternative risk premia timing: A point-in-time macro, sentiment, valuation analysis. *Forthcoming in Journal of Systematic Investing*. http://dx.doi.org/10.2139/ssrn.3247010.
13. Bragoli, D. (2017). Nowcasting the Japanese economy. *International Journal of Forecasting, 33*(2), 390–402.
14. Bragoli, D., Metelli, L., & Modugno, M. (2015). The importance of updating: Evidence from a Brazilian nowcasting model. *OECD Journal: Journal of Business Cycle Measurement and Analysis, 1*, 5–22.
15. Brown, P., & Ball, R. (1967). Some preliminary findings on the association between the earnings of a firm, its industry and the economy. *Journal of Accounting Research, 5*, 55–77.
16. Buono, D., Kapetanios, G., Marcellino, M., Mazzi, G. L., & Papailias, F. (2018). *Big data econometrics: Nowcasting and early estimates*. Italy: Universita Bocconi. Technical Report 82, Working Paper Series.
17. Caldara, D., & Iacovello, M. (2019). Measuring geopolitical risk. In *Federal Reserve System, International Finance Discussion Papers, number 1222*. https://www.federalreserve.gov/ econres/ifdp/files/ifdp1222.pdf.
18. Carabias, J. M. (2018). The real-time information content of macroeconomic news: Implications for firm-level earnings expectations. *Review of Accounting Studies, 23*, 136–166.
19. Choi, H., & Varian, H. (2012). Predicting the present with Google Trends. *Economic Record, 88*(s1), 2–9. https://doi.org/10.1111/j.1475-4932.2012.00809.x.
20. Clarck, H, Pinkovskiy, M., & Sala-i-Martin, X. (2017). China's GDP growth may be understated. In *NBER Working Paper No. 23323*.
21. Coble, D., & Pincheira, P. (2017). Nowcasting building permits with Google Trends. In *MPRA Paper 76514*. Germany: University Library of Munich.
22. Cooper, I., & Priestley, R. (2009). Time-varying risk premiums and the output gap. *Review of Financial Studies, 22*(7), 2801–2833.
23. D'Amuri, F., & Marcucci, J. (2017). The predictive power of Google searches in forecasting unemployment. *International Journal of Forecasting, 33*, 801–816.
24. Donaldson, D., & Storeygard, A. (2016). The view from above: Applications from satellite data in economics. *Journal of Economic Perspectives, 30*(4), 171–198.
25. Engel, C., Mark, N. C., & West, K. D. (2008). Exchange rate models are not as bad as you think. In *NBER Macroeconomics Annual 2007* (pp. 381–441). Chicago: University of Chicago.
26. Engle, R. F., & Rangel, J. G. (2008). The Spline-GARCH model for low-frequency volatility and its global macroeconomic causes. *Review of Financial Studies, 21*, 1187–1222.
27. Fernald, J., Hsu, E., & Spiegel, M. (2019). Is China fudging its GDP figures? In *Evidence from Trading Partner Data*. FRB San Francisco Working Paper 2019–19.
28. Ferrara, L., & Marsilli, C. (2018). Nowcasting global economic growth: A factor-augmented mixed-frequency approach. *The World Economy, 42*(3), 846–875.

29. Ferrara, L. & Simoni, A. (2019). When are Google data useful to nowcast GDP? In *An Approach via Pre-selection and Shrinkage*. Banque de France Working Paper No. 717.
30. Forni, M., Gambetti, L., Lippi, M., & Sala, L. (2017). Noisy News in business cycles. *American Economic Journal: Macroeconomics, 9*(4): 122–52. https://doi.org/10.1257/mac.20150359.
31. Foroni, C., Marcellino, M., & Schumacher, C. (2015). Unrestricted mixed data sampling (MIDAS): MIDAS regressions with unrestricted lag polynomials. *Journal of the Royal Statistical Society A, 178*, 57–82. https://doi.org/10.1111/rssa.12043.
32. Fraiberger, S., Lee, D., Puy, D., & Ranciere, R. (2018). *Media sentiment and international asset prices*. IMF Working Paper No. WP/18/274.
33. Galbraith, J. W., & Tkacz, G. (2018). Nowcasting with payments system data. *International Journal of Forecasting, 34*(2), 366376.
34. Ghysels, E., Sinko, A., & Valkanov, R. (2007). MIDAS regressions: Further results and new directions. *Econometric Reviews, 26*(1), 53–90. https://doi.org/10.1080/07474930600972467.
35. Giannone, D., Reichlin, L., & Small, D. (2008). Nowcasting: The real-time informational content of macroeconomic data. *Journal of Monetary Economics, 55*(4), 665–676.
36. Giannone, D., Lenza, M., & Primiceri, G. E. (2017). Economic predictions with big data: The Illusion of Sparsity. In *Centre for Economic Policy Research*. Discussion Paper No. DP12256. https://cepr.org/active/publications/discussion_papers/dp.php?dpno=12256.
37. Glaeser E., Kim, H., & Luca, M. (2017). *Nowcasting the local economy: Using Yelp data to measure economic activity*. NBER Working Paper No. 24010.
38. Gu, S., Kelly, B., & Xiu, D. (2020). Empirical asset pricing via machine learning. *Review of Financial Studies, 33*(5), 2223–2273.
39. Hildebrand, P., Boivin, J., & Bartsch, E. (2020). On the risk of regime shifts. In *Macro and Market Perspectives*. New York: Blackrock Investment Institute. https://www.blackrock.com/americas-offshore/en/insights/blackrock-investment-institute/global-macro-outlook.
40. Jégou, S., Drozdzal, M., Vazquez, D., Romero, A., & Bengio, Y. (2017). The one hundred layers Tiramisu: Fully convolutional DenseNets for semantic segmentation. In *Proceedings of the 2017 IEEE Conference on Computer Vision and Pattern Recognition Workshops (CVPRW), Honolulu, HI, USA* (pp. 1175–1183). https://doi.org/10.1109/CVPRW.2017.156.
41. Kalamara, E., Turrell, A., Redl, C., Kapetanios, G., & Kapadia, S. (2020). Making text count: Economic forecasting using newspaper text. In *Bank of England*. Working paper No. 865. http://dx.doi.org/10.2139/ssrn.3610770.
42. Modugno, M., Soybilgen, B., & Yazgan, E. (2016). Nowcasting Turkish GDP and news decomposition. *International Journal of Forecasting, 32*(4),1369–1384.
43. Molodtsova, T., & Papell, D. (2012). Taylor rule exchange rate forecasting during the financial crisis. In *National Bureau of Economic Research (NBER)*. Working paper 18330. https://doi.org/10.3386/w18330.
44. Narita, F., & Yin, R. (2018). In search for information: Use of Google Trends' data to narrow information gaps for low-income developing countries. Technical Report WP/18/286. IMF Working Paper.
45. Nymand-Andersen, P., & Pantelidis, E. (2018). Google econometrics: Nowcasting Euro area car sales and big data quality requirements. In *European Central Bank*. Statistics Paper Series, No. 30. https://www.ecb.europa.eu/pub/pdf/scpsps/ecb.sps30.en.pdf?21f8c889572bc4448f92acbfe4d486af.
46. Rangvid, J. (2006). Output and expected returns. *Journal of Financial Economics, 81*(3), 595–624.
47. Rogers, J. H., & Xu, J. (2019). How well does economic uncertainty forecast economic activity? *Finance and Economics Discussion Series* (pp. 2019–085). Washington: Board of Governors of the Federal Reserve System.
48. Seabold, S., & Coppola, A. (2015). Nowcasting prices using Google Trends: An application to Latin America. Policy Research Working Paper No. 7398.
49. Smith, S., Timmermann, A., & Zhu, Y. (2019). Variable selection in panel models with breaks. *Journal of Econometrics. 212*(1), 323–344.

50. Stock, J. H., & Watson, M. W. (2012). Disentangling the channels of the 2007–2009 recession. In *National Bureau of Economic Research (NBER)*. Working Paper No. w18094. https://www.nber.org/papers/w18094

51. Tibshirani, R. (1996). Regression shrinkage and selection via the Lasso. *Journal of the Royal Statistical Society. Series B (Methodological), 58*(1), 267–88. http://www.jstor.org/stable/2346178.

52. UNCTAD (2016). Trade and development report 2016: Structural transformation for inclusive and sustained growth. In *United Nations Conference on Trade And Development (UNCTAD)*. https://unctad.org/system/files/official-document/tdr2016_en.pdf.

53. Varian, H. (2014). Big data: New tricks for econometrics. *Journal of Economic Perspectives, 28*(2), 3–28.

New Data Sources for Central Banks

Corinna Ghirelli, Samuel Hurtado, Javier J. Pérez, and Alberto Urtasun

Abstract Central banks use structured data (micro and macro) to monitor and forecast economic activity. Recent technological developments have unveiled the potential of exploiting new sources of data to enhance the economic and statistical analyses of central banks (CBs). These sources are typically more granular and available at a higher frequency than traditional ones and cover structured (e.g., credit card transactions) and unstructured (e.g., newspaper articles, social media posts, or Google Trends) sources. They pose significant challenges from the data management and storage and security and confidentiality points of view. This chapter discusses the advantages and the challenges that CBs face in using new sources of data to carry out their functions. In addition, it describes a few successful case studies in which new data sources have been incorporated by CBs to improve their economic and forecasting analyses.

1 Introduction

Over the past decade, the development of new technologies and social media has given rise to new data sources with specific characteristics in terms of their volume, level of detail, frequency, and structure (or lack of) (see [37]). In recent years, a large number of applications have emerged that exploit these new data sources in the areas of economics and finance, particularly in CBs.

In the specific area of economic analysis, the new data sources have significant potential for central banks (CBs), even taking into account that these institutions already make very intensive use of statistical data, both individual (microdata) and aggregate (macroeconomic), to perform their functions. In particular, these new sources allow for:

C. Ghirelli · S. Hurtado · J. J. Pérez (✉) · A. Urtasun
Banco de España, Madrid, Spain
e-mail: corinna.ghirelli@bde.es; samuel.hurtado@bde.es; javierperez@bde.es; aurtasun@bde.es

169

S. Consoli et al. (eds.), *Data Science for Economics and Finance*,
https://doi.org/10.1007/978-3-030-66891-4_8

1. Expanding the base data used to carry out financial stability and banking supervision functions (see, e.g., [14] and [32])
2. The use of new methodologies to improve economic analyses (see, e.g., [33])
3. A better understanding (due to more detailed data) and more agile monitoring (due to shorter time delays—almost real time) of economic activity (see [47] for an overview)
4. Improved measurement of agents' sentiments about the state of the economy and related concepts like uncertainty about key economic and policy variables (e.g., [4])
5. Improved measurement of agents' expectations regarding inflation or economic growth
6. Better assessment of economic policy and more possibilities for simulating alternative measures, owing chiefly to the availability of microdata that could be used to improve the characterization of agents' heterogeneity and, thus, to conduct a more in-depth and accurate analysis of their behavior (e.g., see [22] for application in education and [56] for application on social media)

According to Central Banking's annual survey, in 2019 over 60% of CBs used big data in their operations, and two-thirds of them used big data as a core or auxiliary input into the policy-making process. The most common uses for big data are nowcasting and forecasting, followed, among others, by stress-testing and fraud detection (see [20]). Some examples of projects carried out by CBs with new sources of data are: improving GDP forecasting exploiting newspaper articles [58] or electronic payments data (e.g., [3, 27]); machine learning algorithms to increase accuracy in predicting the future behavior of corporate loans (e.g., [55]); forecasting private consumption with credit card data (e.g., [18, 27]); exploiting Google Trends data to predict unemployment [24], private consumption [34, 19], or GDP [42]; web scraping from accommodation platforms to improve tourism statistics [48]; data from online portals of housing sales to improve housing market statistics [49]; sentiment analysis applied to financial market text-based data to study developments in the financial system [54]; and machine learning for outlier detection [31].

In this chapter, we delve into these ideas. First, in Sect. 2 we give a brief overview of some of the advantages and the challenges that CBs face when using these new data sources, while in Sect. 3 we describe a few successful case studies in which new data sources have been incorporated into a CBs' functioning. In particular, we focus on the use of newspaper data to measure uncertainty (two applications in Sect. 3.1), the link between the qualitative messages about the economic situation in the Bank of Spain's quarterly reports and quantitative forecasts (Sect. 3.2), and forecasting applications by means of machine learning methods and the use of non-standard data sources such as Google Trends (Sect. 3.3). Finally, in Sect. 4, we present some general conclusions.

2 New Data Sources for Central Banks

Central banks make intensive use of structured databases to carry out their functions, whether in the banking supervision, financial stability, or monetary policy domains, to mention the core ones.[1] Some examples of individual data are firms' balance sheets (see, e.g., [51] or [6]), information relating to the volume of credit granted by financial institutions to individuals and firms, or the data relating to agents' financial decisions (see, e.g., [5]). In the area of macroeconomics, the main source of information tends to be the national accounts or the respective central bank sources, although a great deal of other information on the economic and financial situation is also published by other bodies: e.g., social security data, payroll employment data (Bureau of Labor Statistics), stock prices (Bloomberg), and house prices (real estate advertising web platforms).

Thanks to technological developments, sources of information are being expanded significantly, in particular as regards their granularity and frequency. For instance, in many cases one can obtain information in almost real time about single actions taken by individuals or firms, and most of the time at higher frequencies than with traditional sources of data. For example, credit card transaction data, which can be used to approximate household consumption decisions, are potentially available in real time at a very reduced cost in terms of use, particularly when compared with the cost of conducting country-wide household surveys. By way of illustration, Chart 1 shows how credit card transactions performed very similarly to household consumption in Spain (for statistical studies exploiting this feature, see [40] and [13]).

The availability of vast quantities of information poses significant challenges in terms of the management, storage capacity and costs, and security and confidentiality of the infrastructure required. In addition, the optimal management of huge structured and unstructured datasets requires the integration of new professional profiles (data scientists and data engineers) at CBs and conveys the need for fully fledged digital transformations of these institutions. Moreover, the diverse nature of the new information sources requires the assimilation and development of techniques that transform and synthesize data, in formats that can be incorporated into economic analyses. For example, textual analysis techniques enable the information contained in the text to be processed and converted into structured data, as in Google Trends, online media databases, social media (e.g., Facebook and Twitter), web search portals (e.g., portals created for housing or job searches), mobile phone data, or satellite data, among others. From the point of view of the statistical treatment of the data, one concern often quoted (see [28]) is the statistical representativeness of the samples used based on the new data, which are developed without the strict requisites of traditional statistical theory (mainly in the field of surveys).

---

[1]The discussion in this section relies on our analysis in [37].

New data sources are expanding the frontier of statistics, in particular (but not exclusively) in the field of non-financial statistics. Examples are the initiatives to acquire better price measures in the economy using web-scraping techniques or certain external trade items, such as the estimation of tourist movements by tracking mobile networks (see [44]). Developing countries, which face greater difficulties in setting up solid statistics infrastructures, are starting to use the new data sources, even to conduct estimates of some national accounts aggregates (see [43]). The boom in new data sources has also spurred the development of technical tools able to deal with a vast amount of information. For instance, Apache Spark and Apache Hive are two very popular and successful products for processing large-scale datasets.[2] These new tools are routinely applied along with appropriate techniques (which include artificial intelligence, machine learning, and data analytics algorithms),[3] not only to process new data sources but also when dealing with traditional problems in a more efficient way. For example, in the field of official statistics, they can be applied to process structured microdata, especially to enhance their quality (e.g., to detect and remove outliers) or to reconcile information received from different sources with different frequency (e.g., see [60] and the references therein).

Finally, it should be pointed out that, somehow, the public monopoly over information that official statistical agencies enjoy is being challenged, for two main reasons. First, vast amounts of information are held by large, private companies that operate worldwide and are in a position to efficiently process them and generate, for example, indicators of economic and financial developments that "compete" with the "official" ones. Second, and related to the previous point, new techniques and abundant public-domain data can also be used by individuals to generate their own measures of economic and social phenomena and to publish this information. This is not a problem, per se, but one has to take into account that official statistics are based on internationally consolidated and comparable methodologies that serve as the basis for objectively assessing the economic, social, and financial situation and the response of economic policy. In this context, thus, the quality and transparency framework of official statistics needs to be strengthened, including by statistical authorities disclosing the methods used to compile official statistics so that other actors can more easily approach sound standards and methodologies. In addition, the availability of new data generated by private companies could be used to enrich official statistics. This may be particularly useful in nowcasting, where official

---

[2]Hive is a data warehouse system built for querying and analyzing big data. It allows applying structure to large amounts of unstructured data and integrates with traditional data center technologies. Spark is a big-data framework that helps extract and process large volumes of data.

[3]Data analytics refers to automated algorithms that analyze raw big data in order to reveal trends and metrics that would otherwise be lost in the mass of information. These techniques are typically used by large companies to optimize processes.

statistics are lagging: e.g., data on credit card transactions are an extremely useful indicator of private consumption.[4]

3 Successful Case Studies

3.1 Newspaper Data: Measuring Uncertainty

Applications involving text analysis (from text mining to natural language processing)[5] have gained special significance in the area of economic analysis. With these techniques, relevant information can be obtained from texts and then synthesized and codified in the form of quantitative indicators. First, the text is prepared (preprocessing), specifically by removing the part of the text that does not inform analysis (articles, non-relevant words, numbers, odd characters) and word endings, leaving only the root.[6] Second, the information contained in the words is synthesized using quantitative indicators obtained mainly by calculating the frequency of words or word groups. Intuitively, the relative frequency of word groups relating to a particular topic allows for the relative significance of this topic in the text to be assessed.

The rest of this section presents two examples of studies that use text-based indicators to assess the impact of economic policy uncertainty on the economy in Spain and the main Latin American countries: Argentina, Brazil, Chile, Colombia, Mexico, Perú, and Venezuela. These indicators have been constructed by the authors of this chapter based on the Spanish press and are currently used in the regular economic monitoring and forecasting tasks of the Bank of Spain.

3.1.1 Economic Policy Uncertainty in Spain

A recent branch of the literature relies on newspaper articles to compute indicators of economic uncertainty. Text data are indeed a valuable new source of information

---

[4]Data on credit card transactions are owned by credit card companies and, in principle, are available daily and with no lag. An application on this topic is described in Sect. 3.3.3.

[5]Text mining refers to processes to extract valuable information from the text, e.g., text clustering, concept extraction, production of granular taxonomies, and sentiment analysis. Natural language processing (NLP) is a branch of artificial intelligence that focuses on how to program computers to process and analyze large amounts of text data by means of machine learning techniques. Examples of applications of NLP include automated translation, named entity recognition, and question answering.

[6]The newest NLP models (e.g., transformer machine learning models) do not necessarily require preprocessing. For instance, in the case of BERT, developed by Google [25], the model already carries out a basic cleaning of the text by means of the tokenization process, so that the direct input for the pre-training of the model should be the actual sentences of the text.

since they reflect major current events that affect the decisions of economic agents and are available with no time lag.

In their leading paper, Baker et al. (see [4]) constructed an index of economic policy uncertainty (Economic Policy Uncertainty (EPU) index) for the United States, based on the volume of newspaper articles that contain words relating to the concepts of uncertainty, economy, and policy. Since this seminal paper, many researchers and economic analysts have used text-based uncertainty indicators in their analyses, providing empirical evidence of the negative effects on activity in many countries (e.g., see [50] for Germany, France, Italy, and Spain, [35] for China, or [23] for the Euro area). The authors of this chapter constructed an EPU index for Spain based on two leading Spanish newspapers: (*El País* and *El Mundo*). [38] recently developed a new Economic Policy Uncertainty index for Spain, which is based on the methodology of [4] but expands the press coverage from 2 to 7 newspapers, widens the time coverage starting from 1997 rather than from 2001, and fine-tunes the richness of the keywords used in the search expressions.[7]

The indicator shows significant increases or decreases relating to events associated, ex ante, with an increase or decrease in economic uncertainty, such as the terrorist attacks of September 11, 2001, in the United States, the collapse of Lehman Brothers in September 2008, the request for financial assistance by Greece in April 2010, the request for financial assistance to restructure the banking sector and savings banks in Spain in June 2012, the Brexit referendum in June 2016, or the episodes of political tension in the Spanish region of Catalonia in October 2017.

[38] found a significant dynamic relationship between this indicator and the main macroeconomic variables, such that unexpected increases in the economic policy uncertainty indicator have adverse macroeconomic effects. Specifically, an unexpected rise in uncertainty leads to a significant reduction of GDP, consumption, and investment. This result is in line with the findings in the empirical literature on economic uncertainty.

In addition, the authors of this chapter provide evidence on the relative role of enriching the keywords used in search expressions and widening both press and time coverage when constructing the index. Results are shown in Fig. 1, which compares macroeconomic responses to unexpected shocks in alternative EPU versions in which they vary in one of the aforementioned dimensions at a time, moving from the EPU index constructed by [4] to the new index. All of these dimensions are important since they all contribute to obtaining the expected negative sign in the responses. Expanding the time coverage is key to improving the precision of the estimates and to yielding significant results. The press coverage is also relevant.

---

[7]The new index is based on the four most widely read general newspapers in Spain and its three leading business newspapers: *El País*, *El Mundo*, *El Economista*, *Cinco Días*, *Expansión*, *ABC*, and *La Vanguardia*.

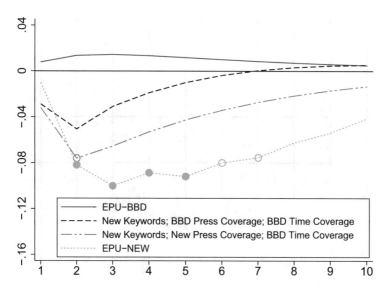

Fig. 1 The graph shows the impulse response function of the Spanish GDP growth rate up to 10 quarters after a positive shock of one standard deviation in the EPU for Spain. The x-axis represents quarters since the shock. The y-axis measures the Spanish GDP growth rate (in percentage points). Full (empty) circles indicate statistical significance at the 5 (10)% level; the solid line indicates no statistical significance. EPU-BBD: EPU index for Spain provided by [4]. EPU-NEW: EPU index for Spain constructed by [38]. Vector autoregression (VAR) models include the EPU index, spread, GDP growth rate, and consumer price index (CPI) growth rate; global EPU is included as an exogenous variable

3.1.2 Economic Policy Uncertainty in Latin America

By documenting the spillover effects of rising uncertainty across countries, the literature also demonstrates that rising economic uncertainty in one country can have global ramifications (e.g., [8, 9, 23, 59]). In this respect, [39] develop Economic Policy Uncertainty indexes for the main Latin American (LA) countries: Argentina, Brazil, Chile, Colombia, Mexico, Peru, and Venezuela. The objective of constructing these indexes is twofold: first, to measure economic policy uncertainty in LA countries in order to get a narrative of "uncertainty shocks" and their potential effects on economic activity in LA countries, and second, to explore the extent to which those LA shocks have the potential to spillover to Spain. This latter country provides an interesting case study for this type of "international spillover" given its significant economic links with the Latin American region.

The uncertainty indicators are constructed following the same methodology used for the EPU index for Spain [38], i.e., counting articles in the seven most important national Spanish newspapers that contain words related to the concepts of *economy*, *policy*, and *uncertainty*. In addition, however, we customize the text searches for the

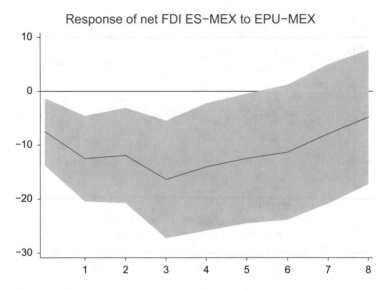

Fig. 2 The graph shows the impulse response function of Spanish net foreign direct investment (FDI) up to 10 quarters after a positive shock of one standard deviation in the Mexican EPU. The x-axis represents quarters since the shock. The y-axis measures the Spanish net FDI growth rate (in percentage points). Confidence intervals at the 5% level are reported

Latin American countries case by case.[8] Note that these indicators are also based on the Spanish press and thereby purely reflect variation in uncertainty in LA countries that is relevant to the Spanish economy, given the importance of the region to the latter. The premise is that the Spanish press accurately reflects the political, social, and economic situation in the LA region, given the existing close economic and cultural ties—including a common language for a majority of these countries. In this respect, one may claim that the indexes provide sensible and relevant measures of policy uncertainty for those countries. This is also in line with a branch of the literature that uses the international press to compute text-based indicators for broad sets of countries (see, e.g., [2] or [53]).

To explore the extent to which LA EPU shocks have the potential to spillover to Spain, the empirical analysis relies on two exercises. A first exercise studies the impact of LA EPU shocks on the performance of Spanish companies operating in the LA region. The underlying assumption is that higher uncertainty in one LA country would affect the investment decisions of Spanish companies that have subsidiaries in this Latin American country: i.e., investment in the LA country may be postponed due to the "wait-and-see effect" and/or the local uncertainty

---

[8]In particular, (1) we require that each article also contains the name of the LA country of interest; (2) among the set of keywords related to *policy*, we include the name of the central bank and the name of the government's place of work in the country of interest. For more details, see [39].

may foster investment decisions toward other foreign countries or within Spain. To carry out this exercise, the authors consider the stock market quotations of the most important Spanish companies that are also highly exposed to LA countries, controlling for the Spanish macroeconomic cycle. Results show that an unexpected positive shock in the EPU index of an LA country generates a significant drop in the companies' quotation growth rate in the first 2 months. This holds for all LA countries considered in the study and is confirmed by placebo tests, which consider Spanish companies that are listed in the Spanish stock market but do not have economic interests in the Latin American region. This suggests that, as expected, economic policy uncertainty in LA countries affects the quotations of Spanish companies that have economic interests in that region.

The second exercise studies the impact of Latin American EPU shocks on the following Spanish macroeconomic variables: the EPU index for Spain, exports and foreign direct investment (FDI) from Spain to Latin America, and the Spanish GDP. In this case as well, one would expect the spillover from one LA country's EPU to the Spanish EPU to be related to commercial relationships between both countries. The higher the exposure of Spanish businesses to a given country, the higher the spillover. To the extent that the EPU reflects uncertainty about the expected future economic policy situation in the country, unexpected shocks in the EPU of one LA country may affect the export and FDI decisions of Spanish companies. Finally, the relation between Latin American EPUs and the Spanish GDP is expected to be driven by the reduction in exports (indirect effect) and by the business decisions of multinational companies that have economic interests in the region. In particular, multinational companies take into account the economic performance of their subsidiaries when deciding upon investment and hiring in Spain. This, in turn, may affect the Spanish GDP. This second exercise is carried out at the quarterly level by means of VAR models, which document the spillover effects from Latin American EPU indexes to the Spanish EPU. Unexpected shocks in Latin American EPUs significantly dampen the commercial relationship between Spain and the Latin American countries in question. In particular, Spanish firms decrease their exports and FDI toward the countries that experience negative shocks in their EPU index. As an example, Fig. 2 shows the impulse response functions of Spanish net FDI to unexpected shocks in the Mexican EPU index.

3.2 The Narrative About the Economy as a Shadow Forecast: An Analysis Using the Bank of Spain Quarterly Reports

One text mining technique consists in the use of dictionary methods for sentiment analysis. To put it simply, a dictionary is a list of words associated with positive and negative sentiments. These lists can be constructed in several ways, ranging from purely manual to machine learning techniques.[9] Sentiment analysis is based on text database searches and requires the researcher to have access to the texts. In its simplest version, the searches allow calculating the frequency of positive and negative terms in a text. The sentiment index is defined as the difference (with some weights) between the two frequencies, that is, a text has a positive (negative) sentiment when the frequency of positive terms is higher (lower) than that of the negative terms. The newest applications of sentiment analysis are more sophisticated than this and rely on neural network architectures and transformer models, which are trained on huge datasets scraped from the web (e.g., all texts in Wikipedia), with the objective of predicting words based on their context. Many of these models take into account negations and intensifiers when computing the sentiment of the text, i.e., improving the results of dictionary-based sentiment exercises. As an example, the paper by [57] sets up a tool to extract opinions from a text by also taking into account the structure of sentences and the semantic relations between words.

In this section, we provide an example of sentiment analysis to show the usefulness of text data (following [26]). We rely on the most basic sentiment analysis technique, i.e., the simple counting of words contained in our own dictionary. Our application is based on the *Quarterly Economic Bulletin* on the Spanish economy by the Bank of Spain, published online since the first quarter of 1999. We consider the Overview section of the reports. The aim of the exercise is to construct an indicator (from Q1 1999) that reflects the sentiment of the Bank of Spain economic outlook reports, and the analysis shows that it mimics very closely the series of Bank of Spain GDP forecasts. This means that the (qualitative) narrative embedded in the text contains similar information to that conveyed by quantitative forecasts.[10]

To carry out the analysis, we create a dictionary of positive and negative terms in Spanish (90, among which some are roots, i.e., we have removed word endings) that are typically used in the economic language to describe the economy, e.g., words like *crecimiento* (growth) or *aumento* (increase) among positive terms, and *disminución* (decrease) or *reducción* (reduction) among negative ones. In order to control for wrong signs, we ignore these terms when they appear around (within nine words before or after) the words "unemployment" or "deficit." We assign a weight of $+1$ (-1) to the resulting counts of positive (negative) terms. Then, for each bulletin, we sum up all of the weighted counts of terms in the dictionary and divide the resulting number by the length of the bulletin. Then, we compare the resulting text-based index with the GDP growth projections conducted each quarter by the Bank of Spain, which in most of the samples under consideration were recorded internally but not published.

---

[9]Examples for English include the Bing Liu Opinion Lexicon [46] or SentiWordNet [30]. [52] created a Spanish dictionary based on the Bing Liu Opinion Lexicon: this list was automatically translated using the Reverso translator and subsequently corrected manually.

[10]Researchers at the Bank of Canada carried out a similar exercise: they applied sentiment analysis by means of machine learning methods on the monetary policy reports of the Bank of Canada. See [10].

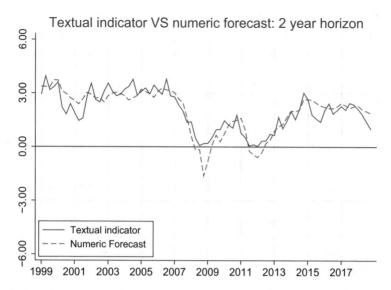

Fig. 3 The graph shows the textual indicator (solid blue line) against the numerical forecasts of the Bank of Spain (dashed red line). The y-axis measures the GDP growth rate (in percentage points). The black dotted line represents the observed GDP growth rate (the target variable of the forecast exercise)

We find a significant dynamic relationship between both series: the narrative text-based indicator follows the Spanish cycle and increases or decreases with the quantitative projections. In addition, the comparison shows that the economic bulletins are informative not only at the short-term forecast horizon but even more so at the 1-to-2-year forecast horizon. The textual indicator shows the highest correlation with the projections for a 2-year horizon. Figure 3 reports the textual indicator (solid blue line) against the GDP growth projection carried out by the Bank of Spain for the 2-year horizon (dashed red line). This evidence suggests that the narrative reflected in the text of the economic bulletins by the Bank of Spain follows very closely the underlying story told by the institution's GDP growth projections. This means that a "sophisticated" reader could infer GDP growth projections based on the text of the reports.

3.3   Forecasting with New Data Sources

Typically, central banks' forecasting exercises are carried out by combining soft indicators with the set of information provided by hard indicators (e.g., data from government statistical agencies such as the main macroeconomic variables: GDP, private consumption, and private investment, for instance).[11] The main limitation posed by hard data is that they are typically published with some lag and at a low frequency (e.g., quarterly). Soft indicators include, for instance, business and consumer confidence surveys. As such, these data provide qualitative information (hence, of a lower quality than hard data) typically available at a higher frequency than hard data. Thus, they provide additional and new information especially at the beginning of the quarter, when macroeconomic information is lacking, and their usefulness decreases as soon as hard data are released [34]. Text indicators are another type of soft indicator. Compared to the traditional survey-based soft indicators, text-based indicators show the following features:

1. They are cheaper from an economic point of view, in that they do not rely on monthly surveys but rather on subscriptions to press repository services.
2. They provide more flexibility since one can select the keywords depending on specific needs and get the entire time series (spanning backward), whereas in a survey, the inclusion of a new question would be reflected in the time series from that moment onward.

The rest of this section presents three applications aimed at improving forecasting. The first is based on sentiment analysis. The second application shows how machine learning can improve the accuracy of available forecasting techniques. Finally, the second application assesses the relative performance of alternative indicators based on new sources of data (Google Trends and credit card transactions/expenses).

3.3.1 A Supervised Method

As an empirical exercise, we construct a text-based indicator that helps track economic activity, as described in [1]. It is based on a similar procedure that is used to elaborate the economic policy uncertainty indicator, i.e., it relies on counting the number of articles in the Spanish press that contain specific keywords. In this case, we carry out a dictionary analysis as in the previous section, i.e., we set up a dictionary of positive and negative words that are typically used in portions of texts related to the GDP growth rate, the target variable of interest, so as to also capture the tone of the articles and, in particular, to what extent they describe upturns or downturns. For instance, words like "increase," "grow," or "raise are listed among the positive terms, while "decrease" and "fall" appear in the negative list. As with the EPU indicators, this one is also based on the Factiva Dow Jones repository of Spanish press and relies on seven relevant Spanish national newspapers: *El País*, *El Mundo*, *Cinco Días*, *Expansión*, *El Economista*, *La Vanguardia*, and *ABC*.

---

[11]Recently, [16] set up a model to efficiently exploit—jointly and in an efficient manner—a rich set of economic and financial hard and soft indicators available at different frequencies to forecast economic downturns in real time.

We place the following restrictions on all queries: (1) the articles are in Spanish; (2) the content of the article is related to Spain, based on Factiva's indexation; and (3) the article is about corporate or industrial news, economic news, or news about commodities or financial markets, according to Factiva's indexation. We then perform three types of queries for each newspaper:[12]

1. We count the number of articles that satisfy the aforementioned requirements. This will serve as the denominator for our indicator.
2. We count the number of articles that, in addition to satisfying the aforementioned conditions, contain upswing-related keywords. That is, the articles must contain the word *recuperacion** (recovery) or one of the following words, provided that they are preceded or followed by either *economic** (economic) or *economia* (economy) within a distance of five words: *aceler** (acceleration), *crec** (increase), *increment** (rise), *aument** (boost), *expansi** (growth), and *mejora** (improvement). In addition, in order to ensure that the news items are about the Spanish business cycle, we also require articles to contain the word *Españ** (Spain).
3. Similarly, we count the number of articles that, in addition to satisfying the aforementioned conditions, are about downswings. In particular, the articles have to contain the word *recession** (recession) or *crisis* (crisis), or one of the following words, provided that they are preceded or followed by either *economic** or *economia* within a distance of five words: *descen** (decrease), *ralentiz** (slowdown), *redu** (reduction), *disminu** (fall), *contraccion** (contraction), *decrec** (downturn), and *desaceler**(deceleration). The articles should also contain *Españ**.

Then, for each newspaper, we take the difference between the upturn- and downturn-related counts and scale the difference by the total number of economic articles in the same newspaper/month. Finally, we standardize the monthly series of scaled counts, average them across newspapers, rescale the resulting index to mean 0, and average it at the quarterly level.

The right panel in Fig. 4 shows the resulting textual indicator (solid blue line) against the GDP growth rate (red and dashed line).

Next, we test whether our textual indicator has some predictive power to nowcast the Spanish GDP growth rate. We perform a pseudo-real-time nowcasting exercise at the quarterly level as follows.[13] First, we estimate a baseline nowcasting model in which the GDP growth rate is nowcasted by means of an AR(1) process. Second, we estimate an alternative nowcasting model that adds our textual indicator and its lag to the GDP AR(1) process. Finally, we compare the forecast accuracy of both models. The alternative model provides smaller mean squared errors of predictions than the baseline one, which suggests that adding textual indicators to the AR(1)

---

[12]The search is carried out in Spanish. English translations are in parentheses.

[13]We use unrevised GDP data, so that our data should be a fair representation of the data available in real time.

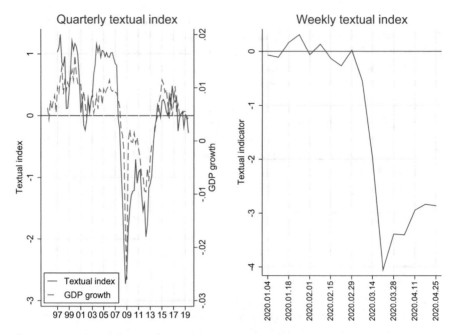

Fig. 4 The figure on the right shows the quarterly textual indicator of *economy* (blue and solid line) against the Spanish GDP growth rate (red and dashed line) until June 2019. The figure on the left shows the weekly textual indicator from January to March 2020

process improves the predictions of the baseline model. In addition, according to the Diebold–Mariano test, the forecast accuracy of the model improves significantly in the alternative model. The null hypothesis of this test is that both competing models provide the same forecast accuracy. By comparing the baseline with the alternative model, this hypothesis is rejected at the 10% level with a p-value of 0.063.[14]

A major advantage of newspaper-based indicators is that they can be updated in real time and are of high frequency. This has been extremely valuable since the Covid-19 outbreak, when traditional survey-based confidence indicators failed to provide timely signals about economic activity.[15] As an example, the right panel in Fig. 4 depicts the textual indicator at a weekly frequency around the Spanish lockdown (14 March 2020) and correctly captures the drastic reduction in Spanish economic activity around that time.

---

[14] A natural step forward would be to incorporate this text-based indicator into more structured nowcasting models that combine hard and soft indicators to nowcast GDP (e.g., [16]). The aim of the current exercise was to show the properties of our text-based indicator in the simplest framework possible.

[15] In [1], we compare this text-based indicator with the economic sentiment indicator (ESI) of the European Commission and show that, for Spain, the former significantly improves the GDP nowcast when compared with the ESI.

3.3.2 An Unsupervised Method

The latent Dirichlet allocation or LDA (see [11]) method can be used to estimate topics in text data. This is an unsupervised learning method, meaning that the data do not need to include a topic label and that the definition of the topics is not decided by the modeler but is a result of running the model over the data. It is appealing because, unlike other methods, it is grounded in a statistical framework: it assumes that the documents are generated according to a generative statistical process (the Dirichlet distribution) so that each document can be described by a distribution of topics and each topic can be described by a distribution of words. The topics are latent (unobserved), as opposed to the documents at hand and the words contained in each document.

The first step of the process is to construct a corpus with text data. In this instance, this is a large database of more than 780,000 observations containing all news pieces published by *El Mundo* (a leading Spanish newspaper) between 1997 and 2018, taken from the Dow Jones repository of Spanish press. Next, these text data have to be parsed and cleaned to end up with a version of the corpus that includes no punctuation, numbers, or special characters and is all lowercase and excludes the most common words (such as articles and conjunctions). This can then be fed to a language-specific stemmer, which eliminates variations of words (e.g., verb tenses) and reduces them to their basic stem (the simpler or, commonly, partial version of the word that captures its core meaning), and the result from this is used to create a bag-of-words representation of the corpus: a big table with one row for each piece of news and one column for each possible stemmed word, filled with numbers that represent how many times each word appears in each piece of news (note that this will be a very sparse matrix because most words from an extensive dictionary will not appear in most pieces of news).

This bag-of-words representation of the corpus is then fed to the LDA algorithm, which is used to identify 128 different topics that these texts discuss[16] and to assign to each piece of news an estimate of the probability that it belongs to each one of those topics. The algorithm analyzes the texts and determines which words tend to appear together and which do not, optimally assigning them to different topics so as to minimize the distance between texts assigned to any given topic and to maximize the distance between texts assigned to different topics.

The result is a database that contains, for each quarter from 1997 to 2018, the percentage of news pieces that fall within each of the 128 topics identified by the unsupervised learning model. A dictionary of positive and negative terms is also applied to each piece of news, and the results are aggregated into quarterly series that indicate how positive or negative are the news pieces relating to each topic.

---

[16]In LDA models, the number of topics to be extracted has to be chosen by the researcher. We run the model by varying the number of topics (we set this parameter equal to numbers that can be expressed as powers of two: 16, 32, 64, 128) and choose the model with 128 topics since it provides better results. Typically, the number of topics is chosen by minimizing the perplexity, which is a measure of the goodness-of-fit of the LDA.

We can now turn to a machine learning model using the data resulting from the analysis of Spanish newspapers to forecast Spanish GDP.[17] The term "machine learning" encompasses a very wide range of methods and algorithms used in different fields such as machine vision, recommender systems, or software that plays chess or go. In the context of economics, support vector machines, random forests, and neural networks can be used to analyze microdata about millions of consumers or firms and find correlations, patterns of behavior, and even causal relationships. CBs have incorporated machine learning techniques to enhance their operations, for instance, in the context of financial supervision, by training models to read banks' balance sheets and raise an alert when more scrutiny is required (e.g., see [21]). For time-series forecasting, ensemble techniques, including boosting and bagging, can be used to build strong forecasting models by optimally combining a large number of weaker models. In particular, ensemble modeling is a procedure that exploits different models to predict an outcome, either by using different modeling algorithms or using different training datasets. This allows reducing the generalization error of the prediction, as long as the models are independent. [7] provides an extensive evaluation of some of these techniques. In this subsection, we present one such ensemble model: a doubly adaptive aggregation model that uses the results from the LDA exercise in the previous subsection, coined DAAM-LDA. This model has the advantage that it can adapt to changes in the relationships in the data.

The ingredients for this ensemble forecasting model are a set of 128 very simple and weakly performing time-series models that are the result of regressing quarterly Spanish GDP growth on its first lag and the weight, positiveness, and negativeness of each topic in the current quarter. In the real-time exercise, the models are estimated every quarter and their first out-of-sample forecast is recorded. Since the share of each topic in the news and its positiveness or negativeness will tend to be indicators with a relatively low signal-to-noise ratio, and since most topics identified in the LDA exercise are not actually related to economics, most of these models will display a weak out-of-sample performance: only 4 out of the 128 outperform a simple random walk. Ensemble methods are designed specifically to build strong models out of such a set of weak models. One advantage is that one does not have to decide which topics are useful and which are not: the model automatically discards any topic that did not provide good forecasts in the recent periods.

One possible way to combine these forecasts would be to construct a nonlinear weight function that translates an indicator of the recent performance of each model at time t into its optimal weight for time $t + 1$. We constructed such a model, using as a weight function a neural network with just three neurons in its hidden layer, in order to keep the number of parameters and hyperparameters relatively low. We

---

[17] Basically, we rely on novel data to forecast an official statistic. An example of another application in which novel data replace official statistics is The Billion Prices Project, an academic initiative that computes worldwide real-time daily inflation indicators based on prices collected from online retailers (see http://www.thebillionpricesproject.com/). An alternative approach would be to enhance official statistics with novel data. This is not the target of this application.

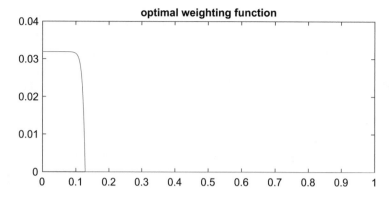

Fig. 5 This is the optimal function for transforming previous performance (horizontal axis) into the current weight of each weak model (vertical axis). It is generated by a neural network with three neurons in its hidden layer, so it could potentially have been highly nonlinear, but in practice (at least for this particular application), the optimal seems to be a simple step function

used a k-fold cross-validation procedure[18] to find the optimal memory parameter for the indicator of recent performance and the optimal regularization, which restricts the possibility that the neural network would overfit the data. The problem is that after all of this, even if the small neural network was able to generate all sorts of potentially very nonlinear shapes, the optimal weighting function would end up looking like a simple step function, as seen in Fig. 5.

To some extent, this was to be expected as it is already known in the forecasting literature that sophisticated weighting algorithms often have a hard time beating something less complex, like a simple average (see, e.g., [29]). In our case, though, since our weak models are really not well-performing, this would not be enough. So instead of spending the degrees of freedom on allowing for potentially highly nonlinear weights, the decision taken was to use a simple threshold function with just one parameter and then add complexity in other areas of the ensemble model, allowing the said threshold to vary over time.

This doubly adaptive aggregation model looks at the recent performance of each weak model in order to decide if it is used for $t + 1$ or not (i.e., weak models either enter into the average or they do not, and all models that enter have equal weight). The threshold is slowly adapted over time by looking at what would have been optimal in recent quarters, and both the memory coefficient (used for the indicator

---

[18]The k-fold cross-validation process works as follows: we randomly divide the data into k bins, train the model using $k - 1$ bins and different configurations of the metaparameters of the model, and evaluate the forecasting performance in the remaining bin (which was not used to train the model). This is done k times, leaving out one bin at a time for evaluation. The metaparameters that provide the best forecasting performance are selected for the final training, which uses all of the bins.

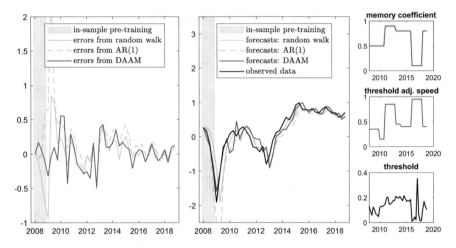

Fig. 6 Results from the real-time forecast exercise for Spanish quarterly GDP growth. DAAM-LDA is the doubly adaptive aggregation model with LDA data presented in this subsection

of recent performance of each weak model) and the allowed speed of adjustment of the threshold are re-optimized at the end of each year.

Importantly, the whole exercise is carried out in real time, using only past information in order to set up the parameters that are to be used for each quarter. Figure 6 summarizes the results from this experiment and also displays the threshold that is used at each moment in time, as well as the memory parameter and speed of adjustment of the threshold that are found to be optimal each year.

As seen in Table 1, the forecasts from DAAM-LDA can outperform a random walk, even if only 4 out of the 128 weak models that it uses as ingredients actually do so. If we restrict the comparison to just the last 4 years in the sample (2015–2018), we can include other state-of-the-art GDP nowcasting models currently in use at the Bank of Spain. In this restricted sample period, the DAAM-LDA model performs better than the random walk, the simple AR(1) model, and the Spain-STING model (see [17]). Still, the Bank of Spain official forecasts display unbeatable performance compared to the statistical methods considered in this section.

Table 1 Spanish GDP forecasting: root mean squared error in real-time out-of-sample exercise

	RW	AR(1)	BdE	DAAM-LDA	Spain-STING
2009–2018	0.290	0.476	0.082	0.240	–
2015–2018	0.110	0.155	0.076	0.097	0.121

Notes: Out-of-sample root mean squared error (RMSE) for different forecasts of Spanish quarterly GDP growth: random walk, simple AR(1) model, official Bank of Spain forecast, doubly adaptive aggregation model with LDA data, and Spain-STING

3.3.3 Google Forecast Trends of Private Consumption

The exercise presented in this section follows closely our paper [40]. In that paper, the question is whether new sources of information can help predict private household consumption. Typically, benchmark data to approximate private household spending decisions are provided by the national accounts and are available at a quarterly frequency ("hard data"). More timely data are usually available in the form of "soft" indicators, as discussed in the previous subsection of this chapter. In this case, the predictive power of new sources of data is ascertained in conjunction with the traditional, more proven, aforementioned "hard" and "soft" data.[19] In particular, the following sources of monthly data are considered: (1) data collected from automated teller machines (ATMs), encompassing cash withdrawals at ATM terminals, and point-of-sale (POS) payments with debit and credit cards; (2) Google Trends indicators, which provide proxies of consumption behavior based on Internet search patterns provided by Google; and (3) economic and policy uncertainty measures,[20] in line with another recent strand of the literature that has highlighted the relevance of the level of uncertainty prevailing in the economy for private agents' decision-making (e.g., see [12] and the references therein).

To exploit the data in an efficient and effective manner, [40] build models that relate data at quarterly and monthly frequencies. They follow the modeling approach of [45]. The forecasting exercise is based on pseudo-real-time data, and the target variable is private consumption measured by the national accounts. The sample for the empirical exercises starts by about 2000 and ends in 2017Q4.[21] As ATM/POS data are not seasonally adjusted, the seasonal component is removed by means of the TRAMO-SEATS software [41].

In order to test the relevant merits of each group of indicators, we consider several models that differ in the set of indicators included in each group. The estimated models include indicators from each group at a time, several groups at a time, and different combinations of individual models. As a mechanical benchmark, [40] use a random walk model whereby they repeat in future quarters the latest quarterly growth rate observed for private consumption. They focus on the forecast performance at the nowcasting horizon (current quarter) but also explore forecasts

---

[19]A growing literature uses new sources of data to improve forecasting. For instance, a number of papers use checks and credit and debit card transactions to nowcast private consumption (e.g., [36] for Canada, [27] for Portugal, [3] for Italy) or use Google Trends data (e.g., see [61], [19], and [34] for nowcasting private consumption in the United States, Chile, and France, respectively, or [15] for exchange rate forecasting).

[20]Measured alternatively by the IBEX stock market volatility index and the text-based EPU index provided by [4] for Spain

[21]The sample is restricted by the availability of some monthly indicators, i.e., Google Trends, the EPU index, and the Services Sector Activity Indicator are available from January 2004, January 2001, and January 2002, respectively.

at 1 to 4 quarters ahead of each of the current quarter forecast origins (first month of the quarter, second, and third).

The analysis yields the following findings. First, as regards models that use only indicators from each group, the ones that use quantitative indicators and payment cards (amounts) tend to perform better than the others in the nowcasting and, somewhat less so, in forecasting (1-quarter- and 4-quarters-ahead) horizons (see Panel A in Table 2). Relative root mean squared errors (RMSEs) are in almost all cases below one, even though from a statistical point of view, they are only different from quarterly random walk nowcasts and forecasts in a few instances. In general, the other models do not systematically best the quarterly random walk alternative. The two main exceptions are the model with qualitative indicators for the nowcasting horizons and the Google-Trends-based ones for the longer-horizon forecasts. The latter results might be consistent with the prior that Google-Trends-based indicators deliver information for today on steps to prepare purchases in the future.

Second, Panel B in Table 2 shows the results of the estimation of models that include quantitative indicators while adding, in turn, variables from the other groups (qualitative, payment cards (amounts), uncertainty, Google Trends). The improvement in nowcast accuracy is not generalized when adding more indicators, with the exception of the "soft" ones. Nonetheless, there is a significant improvement for longer forecast horizons when expanding the baseline model. In particular, for the 4-quarters-ahead one, uncertainty and Google-Trends-based indicators add significant value to the core "hard"-only-based model.

Finally, it seems clear that the combination (average) of models with individual groups of indicators improves the forecasting performance in all cases and at all horizons (see Panel C in Table 2). Most notably, the combination of the forecasts of models including quantitative indicators with those with payment cards (amounts) delivers, in general, the best nowcasting/forecasting performance for all horizons. At the same time, adding the "soft" forecasts seems to add value in the nowcasting phase. In turn, the combination of a broad set of models produces the lowest RMSE relative to the quarterly random walk in the 4-quarters-ahead forecast horizon.

So, to conclude, this study shows that even though traditional indicators do a good job nowcasting and forecasting private consumption in real time, novel data sources add value—most notably those based on payment cards but also, to a lesser extent, Google-Trends-based and uncertainty indicators—when combined with other sources.

Table 2 Relative RMSE statistics: ratio of each model to the quarterly random walk[a]

Panel A: models including indicators of only one group

	Nowcast			1-q-ahead			4-q-ahead		
	m1	m2	m3	m1	m2	m3	m1	m2	m3
Quantitative ("hard") indicators[b]	0.84	0.75 *	0.79	0.75 **	0.81	0.80	0.98	0.97	1.00
Qualitative ("soft") indicators[c]	1.01	0.85	0.85	1.11	1.05	1.05	1.09	1.10	1.29 *
Payment cards (amounts, am)[d]	0.79	0.82	0.88	0.65 ***	0.84	0.69**	0.74 **	0.84	0.83
Payment cards (numbers)[d]	1.05	1.15	1.13	0.90	1.10	0.98	0.75 **	0.81	0.79
Uncertainty indicators[e]	1.06	0.97	0.99	1.00	1.05	1.06	0.94	1.00	1.02
Google: aggregate of all indicators	1.04	1.06	1.06	0.85	1.03	1.03	0.71 **	0.79	0.79
Google: durable goods (lagged)	1.04	0.97	0.98	0.96	1.04	1.04	0.85 *	0.93	0.93

Panel B: models including indicators from different groups

	Nowcast			1-q-ahead			4-q-ahead		
	m1	m2	m3	m1	m2	m3	m1	m2	m3
Quantitative and qualitative	0.69 **	0.78	0.77	0.67 ***	0.76 *	0.72 *	0.79 *	0.82 *	0.80 *
Quantitative and payment cards (am)[d]	0.90	0.82	0.91	0.67 ***	0.79	0.78	0.86	0.89	0.91
Quantitative and uncertainty	0.88	0.86	0.75	0.74 **	0.91	0.93	0.69 **	0.76	0.76
Quantitative and Google (aggregate)	0.85	0.76	0.77	0.81 *	0.94	0.89	0.77 **	0.81 *	0.82
Quantitative and Google (durables)	0.91	0.95	0.87	0.69 **	0.83	0.88	0.72 **	0.76 *	0.77 *

(continued)

Table 2 (continued)

Panel C: combination of models

	Nowcast			1-q-ahead			4-q-ahead		
	m1	m2	m3	m1	m2	m3	m1	m2	m3
All models[f]	0.66 **	0.71 **	0.69 **	0.68 ***	0.77 *	0.68 **	0.73 **	0.78 *	0.78 *
Hard and payment cards (am)[d]	0.62 **	0.69 **	0.71 **	0.53 ***	0.69 **	0.52 ***	0.79 *	0.86	0.84
Hard, payment cards (am)[d] and soft	0.65 **	0.67 **	0.67 **	0.68 ***	0.74 **	0.59 ***	0.83 *	0.89	0.92
Hard and soft	0.68 **	0.66 **	0.66 **	0.77 **	0.75 **	0.69 **	0.91	0.94	1.02
Hard and Google (durables)	0.77 **	0.78 **	0.76 **	0.74 ***	0.83	0.78 *	0.85	0.91	0.90

Notes: The asterisks denote the Diebold–Mariano test results for the null hypothesis of equal forecast accuracy of two forecast methods. A squared loss function is used. The number in each cell represents the loss differential of the method in its horizontal line as compared to the quarterly random walk alternative. * (**) [***] denotes rejection of the null hypothesis at the 10% (5%) [1%] level. [a]Nowcast/forecast errors computed as the difference from the first released vintage of private consumption data. Forecasts are generated recursively over the moving window 2008Q1 (m1) to 2017Q4 (m3). [b]Social security registrations; Retail Trade Index; Activity Services Index. [c]PMI Services; Consumer Confidence Index. [d]Aggregate of payment cards via POS and ATMs. [e]Stock market volatility (IBEX); Economic Policy Uncertainty (EPU) index. [f]Combination of the results of 30 models, including models in which the indicators of each block are included separately, models that include the quantitative block and each other block, and versions of all the previous models but including lags of the variables

4 Conclusions

Central banks use structured data (micro and macro) to monitor and forecast economic activity. Recent technological developments have unveiled the potential of exploiting new sources of data to enhance the economic and statistical analyses of CBs. These sources are typically more granular and available at a higher frequency than traditional ones and cover structured (e.g., credit card transactions) and unstructured (e.g., newspaper articles, social media posts) sources. They pose significant challenges from the data management and storage and security and confidentiality points of view. In addition, new sources of data can provide timely information, which is extremely powerful in forecasting. However, they may entail econometric problems. For instance, in many cases they are not linked to the target variables by a causal relationship but rather reflect the same phenomena they aim to measure (for instance, credit card transactions are correlated with—and do not cause—consumption). Nevertheless, a causal relationship exists in specific cases, e.g., uncertainty shocks affect economic activity.

In this chapter, we first discussed the advantages and challenges that CBs face in using new sources of data to carry out their functions. In addition, we described a few successful case studies in which new data sources (mainly text data from newspapers, Google Trends data, and credit card data) have been incorporated into a CBs' functioning to improve its economic and forecasting analyses.

References

1. Aguilar, P., Ghirelli, C., Pacce, M., & Urtasun, A. (2020). *Can news help to measure economic sentiment? An application in Covid-19 times*. Working Papers 2027, Banco de España.
2. Ahir, H., Bloom, N., & Furceri, D. (2019). *The world uncertainty index*. Working Paper 19–027, Stanford Institute for Economic Policy Research.
3. Aprigliano, V., Ardizzi, G., & Monteforte, L. (2017). *Using the payment system data to forecast the Italian GDP*. Working paper No. 1098, Bank of Italy.
4. Baker, S. R., Bloom, N., & Davis, S. J. (2016). Measuring economic policy uncertainty. *The Quarterly Journal of Economics, 131*(4), 1593–1636.
5. Banco de España (2017). *Survey of household finances, 2014: Methods, results and changes since 2011*. Analytical Article No. 1/2017, Bank of Spain, January.
6. Banco de España (2018). *Central balance sheet data office*. Annual results of non-financial corporations 2017. https://www.bde.es/bde/en/areas/cenbal/
7. Barrow, D. K., & Crone, S. F. (2016). A comparison of AdaBoost algorithms for time series forecast combination. *International Journal of Forecasting, 32*(4), 1103–1119.
8. Bhattarai, S., Chatterjee, A., & Park, W. Y. (2019). Global spillover effects of US uncertainty. *Journal of Monetary Economics, 114*, 71–89. https://doi.org/10.1016/j.jmoneco.2019.05.008
9. Biljanovska, N., Grigoli, F., & Hengge, M. (2017). *Fear thy neighbor: Spillovers from economic policy uncertainty*. Working Paper No. 17/240, International Monetary Fund.
10. Binette, A., & Tchebotarev, D. (2019). *Canada's monetary policy report: If text could speak, what would it say?* Staff Analytical Note 2019–5, Bank of Canada.
11. Blei, D. M., Ng, A. Y. & Jordan, M. I. (2003). Latent dirichlet allocation. *Journal of Machine Learning Research, 3*, 993–1022.

12. Bloom, N. (2014). Fluctuations in uncertainty. *Journal of Economic Perspectives, 28*(2), 153–76.
13. Bodas, D., García, J., Murillo, J., Pacce, M., Rodrigo, T., Ruiz, P., et al. (2018). *Measuring retail trade using card transactional data*. Working Paper No. 18/03, BBVA Research.
14. Broeders, D., & J. Prenio (2018). *Innovative technology in financial supervision (Suptech): The experience of early users*. Financial Stability Institute Insights on Policy Implementation, Working paper No. 9, Bank for International Settlements, July.
15. Bulut, L. (2018). Google Trends and the forecasting performance of exchange rate models. *Journal of Forecasting, 37*(3), 303–315.
16. Cakmakli, C., Demircan, H., & Altug, S. (2018). *Modeling coincident and leading financial indicators for nowcasting and forecasting recessions: A unified approach*. Discussion Paper No. 13171, Center for Research in Economics and Statistics.
17. Camacho, M., & Perez-Quiros, G. (2011). Spain-sting: Spain short-term indicator of growth. *The Manchester School, 79*,594–616.
18. Carlsen, M., & Storgaard, P. E. (2010). *Dankort payments as a timely indicator of retail sales in Denmark*. Working paper No. 66, Bank of Denmark.
19. Carriére-Swallow Y., & Labbé, F. (2013). Nowcasting with google trends in an emerging market. *Journal of Forecasting, 32*(4), 289–298.
20. Hinge, D., & Šilytė, K. (2019). *Big data in central banks: 2019 survey results*. Central Banking, Article No. 4508326. https://www.centralbanking.com/central-banks/economics/data/4508326/big-data-in-central-banks-2019-survey-results
21. Chakraborty, C., & Joseph, A. (2017). *Machine learning at central banks*. Working paper No. 674, Bank of England.
22. Chetty, R., Friedman, J., & Rockoff, J. (2014). Measuring the impacts of teachers II: Teacher value-added and student outcomes in adulthood. *The American Economic Review, 104*(9), 2633–2679.
23. Colombo, V. (2013). Economic policy uncertainty in the us: Does it matter for the euro area? *Economics Letters, 121*(1), 39–42.
24. D'Amuri F., & Marcucci, J. (2017). The predictive power of Google searches in forecasting US unemployment, *International Journal of Forecasting, 33*(4), 801–816.
25. Devlin, J., Chang, M., Lee, K., & Toutanova, K. (2018). *BERT: Pre-training of Deep Bidirectional Transformers for Language Understanding*. arXiv:1810.04805v2.
26. Diaz-Sobrino, N., Ghirelli, C., Hurtado, S., Perez, J. J., & Urtasun, A. (2020), *The narrative about the economy as a shadow forecast: an analysis using bank of Spain quarterly reports*. Bank of Spain. Working Paper No. 2042. https://www.bde.es/f/webbde/SES/Secciones/Publicaciones/PublicacionesSeriadas/DocumentosTrabajo/20/Files/dt2042e.pdf
27. Duarte, C., Rodrigues, P. M., & Rua, A. (2017). A mixed frequency approach to the forecasting of private consumption with ATM/POS data. *International Journal of Forecasting, 33*(1), 61–75.
28. Einav, L., & Levin, J. (2014). The data revolution and economic analysis. *Innovation Policy and the Economy, 14*, 1–24.
29. Elliott, G., Granger, C. W. J., & Timmermann, A. (Eds.). (2006). *Handbook of economic forecasting*. Holland, Amsterdam: Elsevier.
30. Esuli, A., & Sebastiani, F. (2006). Sentiwordnet: A publicly available lexical resource for opinion mining. In *Proceedings of the 5th International Conference on Language Resources and Evaluation, LREC 2006* (pp. 417–422).
31. Farné, M., & Vouldis, A. T. (2018) *A methodology for automatised outlier detection in high-dimensional datasets: An application to Euro area banks' supervisory data*. Working Paper No. 2171, European Central Bank.
32. Fernández, A. (2019). *Artificial intelligence in financial services*. Analytical Articles, Economic Bulletin No. 2/2019, Bank of Spain.
33. Fernández-Villaverde, J., Hurtado, S., & Nuño, G. (2019). *Financial frictions and the wealth distribution*. Working Paper No. 26302, National Bureau of Economic Research.

34. Ferrara, L., & Simoni, A. (2019). *When are Google data useful to nowcast GDP? An approach via pre-selection and shrinkage.* Working paper No. 2019–04, Center for Research in Economics and Statistics.
35. Fontaine, I., Didier, L., & Razafindravaosolonirina, J. (2017). Foreign policy uncertainty shocks and US macroeconomic activity: Evidence from China. *Economics Letters, 155,* 121–125.
36. Galbraith, J. W., & Tkacz, G. (2015). *Nowcasting GDP with electronic payments data.* Working Paper No. 10, Statistics Paper Series, European Central Bank.
37. Ghirelli, C., Peñalosa, J., Pérez, J. J., & Urtasun, A. (2019a). *Some implications of new data sources for economic analysis and official statistics.* Economic Bulletin. Bank of Spain. May 2019.
38. Ghirelli, C., Pérez, J. J., & Urtasun, A. (2019). A new economic policy uncertainty index for Spain. *Economics Letters, 182,* 64–67.
39. Ghirelli, C., Pérez, J. J., & Urtasun, A. (2020). *Economic policy uncertainty in Latin America: measurement using Spanish newspapers and economic spillovers.* Working Papers 2024, Bank of Spain. https://ideas.repec.org/p/bde/wpaper/2024.html
40. Gil, M., Pérez, J. J., Sánchez, A. J., & Urtasun, A. (2018). *Nowcasting private consumption: Traditional indicators, uncertainty measures, credit cards and some internet data.* Working Paper No. 1842, Bank of Spain.
41. Gómez, V., & Maravall, A. (1996). *Programs TRAMO and SEATS: Instructions for the user,* Working paper No. 9628, Bank of Spain.
42. Götz, T. B., & Knetsch, T. A. (2019). Google data in bridge equation models for German GDP. *International Journal of Forecasting, 35*(1), 45–66.
43. Hammer, C. L., Kostroch, D. C., & Quirós, G. (2017). *Big data: Potential, challenges and statistical implications,* IMF Staff Discussion Note, 17/06, Washington, DC, USA: International Monetary Fund.
44. Hardy, A., Hyslop, S., Booth, K., Robards, B., Aryal, J., Gretzel, U., et al. (2017). Tracking tourists' travel with smartphone-based GPS technology: A methodological discussion. *Information Technology and Tourism, 17*(3), 255–274.
45. Harvey, A., & Chung, C. (2000). Estimating the underlying change in unemployment in the UK. *Journal of the Royal Statistical Society, Series A: Statistics in Society, 163*(3), 303–309.
46. Hu, M., & Liu, B. (2004). Mining and summarizing customer reviews. In *KDD-2004 - Proceedings of the Tenth ACM SIGKDD International Conference on Knowledge Discovery and Data Mining* (pp. 168–177). New York, NY, USA: ACM.
47. Kapetanios, G., & Papailias, F. (2018). *Big data & macroeconomic nowcasting: Methodological review.* ESCoE Discussion Paper 2018-12, Economic Statistics Centre of Excellence.
48. Lacroix R. (2019). The Bank of France datalake. In Bank for International Settlements (Ed.), IFC Bulletins chapters, *The use of big data analytics and artificial intelligence in central banking* (vol. 50). Basel: Bank for International Settlements.
49. Loberto, M., Luciani, A., & Pangallo, M. (2018). *The potential of big housing data: An application to the Italian real-estate market.* Working paper No. 1171, Bank of Italy.
50. Meinen, P., & Roehe, O. (2017). On measuring uncertainty and its impact on investment: Cross-country evidence from the euro area. *European Economic Review, 92,* 161–179.
51. Menéndez, Á., & Mulino, M. (2018). *Results of non-financial corporations in the first half of 2018.* Economic Bulletin No. 3/2018, Bank of Spain.
52. Molina-González, M. D., Martínez-Cámara, E., Martín-Valdivia, M.-T., & Perea-Ortega, J. M. (2013). Semantic orientation for polarity classification in Spanish reviews. *Expert Systems with Applications, 40*(18), 7250–7257.
53. Mueller, H., & Rauh, C. (2018). Reading between the lines: Prediction of political violence using newspaper text. *American Political Science Review, 112*(2), 358–375.
54. Nyman R., Kapadia, S., Tuckett, D., Gregory, D., Ormerod, P. & Smith, R. (2018). *News and narratives in financial systems: exploiting big data for systemic risk assessment.* Staff Working Paper No. 704, Bank of England.

55. Petropoulos, A., Siakoulis, V., Stavroulakis, E., & Klamargias, A. (2019). A robust machine learning approach for credit risk analysis of large loan level datasets using deep learning and extreme gradient boosting. In Bank for International Settlements (Ed.), *IFC Bulletins chapters, The use of big data analytics and artificial intelligence in central banking* (vol. 50). Basel: Bank for International Settlements.
56. Pew Research Center (2012). *Assessing the Representativeness of Public Opinion Surveys.* Mimeo. See: https://www.people-press.org/2012/05/15/assessing-the-representativeness-of-public-opinion-surveys/
57. Reforgiato Recupero, D., Presutti, V., Consoli, S., Gangemi, A., & Nuzzolese, A. G. (2015). Sentilo: Frame-based sentiment analysis. *Cognitive Computation, 7*(2), 211–225.
58. Thorsrud, L.A. (2020). Words are the new numbers: A newsy coincident index of business cycles. *Journal of Business and Economic Statistics. 38*(2), 393–409.
59. Trung, N. B. (2019). The spillover effect of the US uncertainty on emerging economies: A panel VAR approach. *Applied Economics Letters, 26*(3), 210–216.
60. Xu, C., Ilyas, I. F., Krishnan, S., & Wang, J. (2016). Data cleaning: Overview and emerging challenges. In *Proceedings of the ACM SIGMOD International Conference on Management of Data, 26-June-2016* (pp. 2201–2206).
61. Vosen, S., & Schmidt, T. (2011). Forecasting private consumption: Survey-based indicators vs. Google trends. *Journal of Forecasting, 30*(6), 565–578.

Sentiment Analysis of Financial News: Mechanics and Statistics

Argimiro Arratia, Gustavo Avalos, Alejandra Cabaña, Ariel Duarte-López, and Martí Renedo-Mirambell

Abstract This chapter describes the basic mechanics for building a forecasting model that uses as input sentiment indicators derived from textual data. In addition, as we focus our target of predictions on financial time series, we present a set of stylized empirical facts describing the statistical properties of lexicon-based sentiment indicators extracted from news on financial markets. Examples of these modeling methods and statistical hypothesis tests are provided on real data. The general goal is to provide guidelines for financial practitioners for the proper construction and interpretation of their own time-dependent numerical information representing public perception toward companies, stocks' prices, and financial markets in general.

1 Introduction

Nowadays several news technology companies offer sentiment data to assist the financial trading industry into the manufacturing of financial news sentiment indicators to feed as information to their automatic trading systems and for the making of investment decisions. Manufacturers of news sentiment-based trading models are faced with the problem of understanding and measuring the relationships among sentiment data and their financial goals, and further translating these into their forecasting models in a way that truly enhances their predictive power.

A. Arratia (✉) · G. Avalos · M. Renedo-Mirambell
Computer Science, Universitat Politècnica de Catalunya, Barcelona, Spain
e-mail: argimiro@cs.upc.edu; gavalos@cs.upc.edu; mrenedo@cs.upc.edu

A. Cabaña
Mathematics, Universitat Autónoma de Barcelona, Barcelona, Spain
e-mail: acabana@mat.uab.cat

A. Duarte-López
Acuity Trading Ltd., London, UK
e-mail: ariel.duarte@acuitytrading.com

Some issues that arise when dealing with sentiment data are: What are the sentiment data—based on news of a particular company or stock—saying about that company? How can this information be aggregated to a forecasting model or a trading strategy for the stock? Practitioners apply several ad hoc filters, as moving averages, exponential smoothers, and many other transformations to their sentiment data to concoct different indicators in order to exploit the possible dependence relation with the price or returns, or any other observable statistics. It is then of utmost importance to understand why a certain construct of a sentiment indicator might work or not, and for that matter it is crucial to understand the statistical nature of indicators based on sentiment data and analyze their insertion in econometric models. Therefore, we consider two main topics in sentiment analysis: the mechanics, or methodologies for constructing sentiment indicators, and the statistics, including stylized empirical facts about these variables and usage in price modeling.

The main purpose of this chapter is to give guidelines to users of sentiment data on the elements to consider in building sentiment indicators. The emphasis is on sentiment data extracted from financial news, with the aim of using the sentiment indicators for financial forecasting. Our general focus is on sentiment analysis for English texts. As a way of example, we apply this fundamental knowledge to construct six dictionary-based sentimental indicators and a ratio of stock's news volume. These are obtained by text mining streams of news articles from the *Dow Jones Newswires* (DJN), one of the most actively monitored source of financial news today. In the Empirical section (Sect. 4) we describe these sentimental and volume indicators, and further in the Statistics section (Sect. 3) analyze their statistical properties and predictive power for returns, volatility, and trading volume.

1.1 Brief Background on Sentiment Analysis in Finance

Extensive research literature in behavioral finance has shown evidence to the fact that investors do react to news. Usually, they show greater propensity for making an investment move based on bad news rather than on good news (e.g., as a general trait of human psychology [5, 39] or due to specific investors' trading attitudes [17]). Li [27] and Davis et al. [11] analyze the tone of qualitative information using term-specific word counts from corporate annual reports and earnings press releases, respectively. They go on to examine, from different perspectives, the contemporaneous relationships between future stock returns and the qualitative information extracted from texts of publicly available documents. Li finds that the two words "risk" and "uncertain" in firms' annual reports predict low annual earnings and stock returns, which the author interprets as under-reaction to "risk sentiment." Tetlock et al. [45] examine qualitative information in news stories at daily horizons and find that the fraction of negative words in firm-specific news stories forecasts low firm earnings. Loughran and McDonald [29] worked out particular lists of words specific to finance, extracted from 10-K filings, and tested whether these lists actually gauge tone. The authors found significant relations between their lists of words and returns, trading volume, subsequent return volatility, and unexpected earnings. These findings are corroborated by Jegadeesh and Wu [24] who designed a measure to quantify document tone and found significant relation between the tone of 10-Ks and market reaction for both negative and positive words. The important corollary of these works is that special attention should be taken to the nature and contents of the textual data used for sentiment analysis intended for financial applications. The selection of documents from where to build a basic lexicon has major influence on the accuracy of the final forecasting model, as sentiment varies according to context, and lists of words extracted from popular newspapers or social networks convey emotions differently than words from financial texts.

2 Mechanics of Textual Sentiment Analysis

We focus our exposition on sentiment analysis of text at the *aspect level*. This means that our concern is to determine whether a document, or a sentence within a document, expresses a positive, negative, or other sentiment emotions toward a target. For other levels and data corpuses, consult the textbook by Bing Liu [28].

In financial applications, the targets are companies, financial markets, commodities, or any other entity with financial value. We then use this sentiment information to feed forecasting models of variables quantifying the behavior of the financial entities of interest, e.g., price returns, volatility, and financial indicators.

A typical workflow for building forecasting models based on textual data goes through the following stages: (i) textual corpus creation and processing, (ii) sentiment computation, (iii) sentiment scores aggregation, and (iv) modeling.

(i) **Textual corpus management**. The first stage concerns the collecting of textual data and applying text mining techniques to clean and categorize terms within each document. We assume texts come in electronic format and that each document has a unique identifier (e.g., a filename) and a timestamp. Also, that through whatever categorization scheme used, we have identified within each document the targets of interest. Thus, documents can be grouped by common target and it is possible that a document appears in two different groups pertaining to two different targets.

Example 1 Targets (e.g., a company name or stock ticker) can be identified by keyword matching or name entity recognition techniques (check out the Stanford NER software.[1]) Alternatively, some news providers like *Dow Jones Newswires* include labels in their *xml* files indicating the company that the news is about.

---

[1] https://nlp.stanford.edu/software/CRF-NER.shtml.

(ii) **Computing sentiment scores.** Sentiment analysis is basically a text classification problem. Hence, one can tackle this algorithmic problem by
either 1) applying a supervised machine learning algorithm that is trained
on text already labeled as positive or negative (or any other emotion) or
2) using an unsupervised classification method based on the recognition
of some fixed syntactic patterns, or words, that are known to express a
sentiment (sentiment lexicon). The latter is frequently used by researchers
and practitioners in Finance, and it is the one applied on the data we have
at hand for sentiment analysis in this work. Hence, we will prioritize the
exposition of the lexicon-based unsupervised method and just give some
pointers to the literature in the machine learning approach to sentiment
classification.

($ii.A$) **Lexicon-based unsupervised sentiment classification.** The key component
of this text classification method is a dictionary of words, and more
general, syntactic patterns, that denote a specific sentiment. For example,
positiveness is conveyed by words such as *good, admirable, better*, etc. and
emoticons such as :-) or ; -] and alike, often used in short messages like
those in Twitter [8, 19]. These groups of words conform a *sentiment lexicon*
or *dictionary*.

Given a fixed sentiment $\mathscr{S}$ (e.g., positive, negative, ...), determined by some
lexicon $L(\mathscr{S})$, a basic algorithm to assign a $\mathscr{S}$-sentiment score to a document is to
count the number of appearances of terms from $L(\mathscr{S})$ in the document. This number
gives a measure of the strength of sentiment $\mathscr{S}$ in the document. In order to compare
the strengths of two different sentiments in a document, it would be advisable to rel-
ativize these numbers to the total number of terms in the document. There are many
enhancements of this basic sentiment scoring function, according to the different
values given to the terms in the lexicon (instead of each having an equal value of 1),
or if sign is considered to quantify direction of sentiment, and further considerations
on the context where, depending on neighboring words, the lexicon terms may
change their values or even shift from one sentiment to another. For example, *good*
is positive, but *not good* is negative. We shall revise some of these variants, but for
a detailed exposition, see the textbook by Liu [28] and references therein.

Let us now formalize a general scheme for a lexicon-based computation of a
time series of sentiment scores for documents with respect to a specific target (e.g.,
a company or a financial security). We have at hand $\lambda = 1, \ldots, \Lambda$ lexicons L_λ,
each defining a sentiment. We have K possible targets and we collect a stream of
documents at different times $t = 1, \ldots, T$. Let N_t be the total number of documents
with timestamp t. Let $D_{n,t,k}$ be the n-th document with timestamp t and make
mention of the k-th target, for $n = 1, \ldots, N_t$, $t = 1, \ldots, T$ and $k = 1, \ldots, K$.

Fix a lexicon L_λ and target G_k. A sentiment score based on lexicon L_λ for
document $D_{n,t,k}$ relative to target G_k can be defined as

$$S_{n,t}(\lambda, k) = \sum_{i=1}^{l_d} w_i S_{i,n,t}(\lambda, k) \tag{1}$$

where $S_{i,n,t}(\lambda, k)$ is the sentiment value given to unigram i appearing in the document and according to lexicon L_λ, being this value zero if the unigram is not in the lexicon. I_d is the total number of unigrams in the document $D_{n,t,k}$ and w_i is a weight, for each unigram that determines the way sentiment scores are aggregated in the document.

Example 2 If $S_{i,n,t} = 1$ (or 0 if unigram i is not in the lexicon), for all i, and $w_i = 1/I_d$, we have the basic sentiment density estimation used in [27, 29, 45] and several other works on text sentiment analysis, giving equal importance to all unigrams in the lexicon. A more refined weighting scheme, which reflects different levels of relevance of the unigram with respect to the target, is to consider $w_i = \texttt{dist}(i, k)^{-1}$, where $\texttt{dist}(i, k)$ is a word distance between unigram i and target k [16].

The sentiment score $S_{i,n,t}$ can take values in $\mathbb{R}$ and be decomposed into factors $v_i \cdot s_i$, where v_i is a value that accounts for a shift of sentiment (a *valence shifter*: a word that changes sentiments to the opposite direction) and s_i the sentiment value per se.

(*ii.A.*1) **On valence shifters.** Originally proposed and analyzed their contrarian effect on textual sentiment in [34], these are words that can alter a polarized word's meaning and belong to one of four basic categories: *negators, amplifiers, de-amplifiers*, and *adversative conjunction*. A negator reverses the sign of a polarized word, as in "that company is *not* good investment." An amplifier intensifies the polarity of a sentence, as, for example, the adverb *definitively* amplifies the negativity in the previous example: "that company is *definitively not* good investment." De-amplifiers (also known as downtoners), on the other hand, decrease the intensity of a polarized word (e.g., "the company is *barely* good as investment"). An adversative conjunction overrules the precedent clause's sentiment polarity, e.g., "I like the company *but* it is not worthy."

Shall we care about valence shifters? If valence shifters occur frequently in our textual datasets, then not considering them in the computation of sentiment scores in Eq. (1) will render an inaccurate sentiment valuation of the text. More so in the case of negators and adversative conjunctions which reverse or overrule the sentiment polarity of the sentence. For text from social networks such as Twitter or Facebook, the occurrence of valence shifters, particularly negators, has been observed to be considerably high (approximately 20% for several trending topics[2]), so certainly their presence should be considered in Eq. (1).

We have computed the appearance of valence shifters in a sample of 1.5 million documents from the *Dow Jones Newswires* set. The results of these calculations, which can be seen in Table 1, show low occurrence of downtoners and adversatives (around 3%), but negators in a number that may be worth some attention.

---

[2]https://cran.r-project.org/web/packages/sentimentr/readme/README.html.

Table 1 Occurrence % of valence shifters in 1.5 MM DJN documents

Text type	Negators	Amplifiers	Downtoners	Adversatives
DJN news articles	7.00%	14.13%	3.02%	3.02%

(*ii.A*.2) **Creating lexicons.** A starting point to compile a set of sentiment words is to use a structured dictionary (preferably online as WordNet) that lists synonyms and antonyms for each word. Then begin with a few selected words (keywords) carrying a specific sentiment and continue by adding some of the synonyms to the set, and to the complementary sentimental set add the antonyms. There are many clever ways of doing this sentiment keyword expansion using some supervised classification algorithms to find more words carrying similar emotion. An example is the work of Tsai and Tang [48] on financial keyword expansion using the continuous bag-of-words model on the 10-K mandated annual financial reports. Another clever supervised scheme based on network theory to construct lexicons is given in [35]. For a more extensive account of sentiment lexicon generation, see [28, Chap. 7] and the many references therein.

(*ii.B*) **Machine learning-based supervised sentiment classification.** Another way to classify texts is by using machine learning algorithms, which rely on a previously trained model to generate predictions. Unlike the lexicon-based method, these algorithms are not programmed to respond in a certain way according to the inputs received, but to extract behavior patterns from pre-labeled training datasets. The internal algorithms that shape the basis of this learning process have some strong statistical and mathematical components. Some of the most popular are Naïve Bayes, Support Vector Machines, and Deep Learning. The general stages of textual sentiment classification using machine learning models are the following:

Corpus development and preprocessing. The learning process starts from a manually classified corpus that after feature extraction will be used by the machine learning algorithm to find the best fitted parameters and asses the accuracy in a test stage. This is why the most important part for this process is the development of a good training corpus. It should be as large as possible and be representative of the set of data to be analyzed. After getting the corpus, techniques must be applied to reduce the noise generated by sentiment meaningless words, as well as to increase the frequency of each term through stemming or lemmatization. These techniques depend on the context to which it is applied. This means that a model trained to classify texts from a certain field could not be directly applied to another. It is then of key importance to have a manually classified corpus as good as possible.

Feature extraction. The general approach for extracting features consists of transforming the preprocessed text into a mathematical expression based on the detection of the co-occurrence of words or phrases. Intuitively, the text is broken down into a series of features, each one corresponding to an element of the input text.

Classification. During this stage, the trained model receives an unseen set of features in order to obtain an estimated class.

For further details, see [40, 28].

An example of sentiment analysis machine learning method is *Deep-MLSA* [13, 12]. This model consists of a multi-layer convolutional neural network classifier with three states corresponding to negative, neutral, and positive sentiments. Deep-MLSA copes very well with the short and informal character of social media tweets and has won the message polarity classification subtask of task 4 "Sentiment Analysis in Twitter" in the SemEval competition [33].

(*iii*) **Methods to aggregate sentiment scores to build indicators.** Fix a lexicon L_λ and target G_k. Once sentiment scores for each document related to target G_k are computed following the routine described in Eq. (1), proceed to aggregate these for each timestamp t to obtain the L_λ-based sentiment score for G_k at time t, denoted by $S_t(\lambda, k)$:

$$S_t(\lambda, k) = \sum_{n=1}^{N_t} \beta_n S_{n,t}(\lambda, k) \qquad (2)$$

As in Eq. (1), the weights β_n determine the way the sentiment scores for each document are aggregated. For example, considering $\beta_n = 1/\text{length}(D_{n,t,k})$ would give more relevance to short documents.

We obtain in this way a time series of sentiment scores, or sentiment indicator, $\{S_t : t = 1, \ldots T\}$, based on lexicon L_λ that defines a specific sentiment for target G_k. Variants of this L_λ-based sentiment indicator for G_k can be obtained by applying some filters F to S_t, thus $\{F(S_t) : t = 1, \ldots T\}$. For instance, apply a moving average to obtain a smoothed version of the raw sentiment scores series.

(*iv*) **Modeling.** Consider two basic approaches: either use the sentiment indicators as exogenous features for forecasting models, and test their relevance in forecasting price movements, returns of price, or other statistics of the price, or use them as external advisors for ranking the subjects (targets) of the news—which in our case are stocks—and create a portfolio. A few selected examples from the vast amount of published research on the subject of forecasting and portfolio management with sentiment data are [3, 4, 6, 21, 29, 44, 45, 49].

For a more extensive treatment of the building blocks for producing models based on textual data, see [1] and the tutorial for the **sentometrics** package in [2].

3 Statistics of Sentiment Indicators

In this second part of the chapter, we present some observed properties of the empirical data used in financial textual sentiment analysis, and statistical methods commonly used in empirical finance to help the researchers gain insight on the data for the purpose of building forecasting models or trading systems.

These empirical properties, or stylized facts, reported in different research papers, seem to be caused by and have an effect mostly on retail investors, according to a study by Kumar and Lee [26]. For it is accepted that institutional investors are informationally more rational in their trading behaviors (in great part due to a higher automatization of their trading processes and decision making), and consequently it is the retail investor who is more affected by sentiment tone in financial news and more prone to act on it, causing stock prices to drift away from their fundamental values. Therefore, it is important to keep in mind that financial text sentiment analysis and its applications would make more sense on markets with a high participation of retail investors (mostly from developed economies, such as the USA and Europe), as opposed to emerging markets. In these developed markets, institutional investors could still exploit the departures of stock prices from fundamental values because of the news-driven behavior of retail investors.

3.1 Stylized Facts

We list the most often observed properties of news sentiment data relative to market movements found in studies of different markets and financial instruments and at different time periods.

1. **Volume of news and volatility correlation.**   The longer the stock is on the news, the greater its volatility. This dependency among volume of news on a stock and its volatility has been observed for various stocks, markets, and for different text sources. For example, this relation has been observed with text data extracted from Twitter and stocks trading in S&P 500 in [3].
2. **Larger volume of news near earnings announcement dates.**   The volume of news about a company tends to increase significantly in the days surrounding the company's earnings announcement. This fact was observed by Tetlock, Saar-Tsechansky, and Macskassy in [45] for news appearing in *Wall Street Journal* and *Dow Jones Newswires* from 1980 to 2004, for companies trading in the S&P 500 index. The authors produced a histogram outlining the relationship between the number of company-specific news and the number of days since (respectively, until) the company's last (respectively, next) earnings announcement (which is the 0 in the plot). The authors did this for all companies collectively; we will update this histogram and show particular cases for individual companies.
   This fact suggests a possible statistical dependency relation among company-specific news and company's fundamentals.

3. **Negative sentiments are more related to market movements than positive ones.** This is observed in, e.g., [3, 29, 45] and [10], although the latter for data in the pre-Internet era, and the phenomenon is most prominent for mid and small-cap stocks.
4. **Stronger effects observed for mid and small-capitalization stocks.** This is suggested and analyzed in [10]. It is related to the fact that retail investors are those who mostly trade based on news sentiment, and this type of investors does not move big-cap stocks.

3.2   Statistical Tests and Models

In order to make some inference and modeling, and not remain confined to descriptive statistics, several tests on the indices, the targets, and their relationships can be performed. Also, models and model selection can be attempted.

3.2.1   Independence

Previous to using any indicator as a predictor, it is important to determine whether there is some dependency, in a statistical sense, among the target Y and the predictor X. We propose the use of an independence test based on the notion of *distance correlation*, introduced by Szekely et al. [43].

Given random variables X and Y (possibly multivariate), from a sample (X_1, Y_1), ..., (X_n, Y_n), the distance correlation is computed through the following steps:

1. Compute all Euclidean distances among pairs of observations of each vector $\|X_i - X_j\|$ and $\|Y_i - Y_j\|$ to get 2 $n \times n$ distance matrices, one for each vector.
2. Double-center each element: to each element, subtract the mean of its row and the mean of its column, and add the matrix mean.
3. Finally, compute the covariance of the n^2 centered distances.

Distance correlation is obtained by normalizing in such a way that, when computed with $X = Y$, the result is 1. It can be shown that, as $n \to \infty$, the distance covariance converges to a value that vanishes if and only if the vectors X and Y are independent. In fact, the limit is a certain distance between the characteristic function $\varphi_{(X,Y)}$ of the joint vector (X, Y) and the product of the characteristic functions of X and Y, $\varphi_X \varphi_Y$. From this description, some of the advantages of the distance correlation are clear: it can be computed for vectors, not only for scalars; it characterizes independence; since it is based on distances, X and Y can have different dimensions—we can detect dependencies between two groups, one formed by p variables and the other by q; and it is rotation-invariant.

The test of independence consists of testing the null hypothesis of zero distance correlation. The p-values are obtained by bootstrap techniques. The R package energy [38] includes the functions dcor and dcor.test for computing the distance correlation and the test of independence.

3.2.2 Stationarity

In the context of economic and/or social variables, we typically only observe one realization of the underlying stochastic process defining the different variables. It is not possible to obtain successive samples or independent realizations of it. In order to be able to estimate the "transversal" characteristics of the process, such as mean and variance, from its "longitudinal" evolution, we must assume that the transversal properties (distribution of the variables at each instant in time) are stable over time. This leads to the concept of stationarity.

A stochastic process (time series) is stationary (or strictly stationary) if the marginal distributions of all the variables are identical and the finite-dimensional distributions of any arbitrary set of variables depend only on the lags that separate them. In particular, the mean and the variance of all the variables are the same. Moreover, the joint distribution of any set of variables is translation-invariant (in time). Since in most cases of time series the joint distributions are very complicated (unless the data come from a very simple mechanism, such as i.i.d. observations), a usual procedure is to specify only the first- and second-order moments of the joint distributions, that is, $E\ X_t$ and $EX_{t+h}X_t$ for $t = 1, 2, \ldots, h = 0, 1, \ldots$, focusing on properties that depend only on these. A time series is weakly stationary if EX_t is constant and $EX_{t+h}X_t$ only depends on h (but not on t). This form of stationarity is the one that we shall be concerned with.

Stationarity of a time series can sometimes be assessed through Dickey–Fuller test [14], which is not exactly a test of the null hypothesis of stationarity, but rather a test for the existence of a unit root in autoregressive processes. The alternative hypothesis can either be that the process is stationary or that it is trend-stationary (i.e., stationary after the removal of a trend).

3.2.3 Causality

It is also important to assess the possibility of causation (and not just dependency) of a random process X_t toward another random process Y_t. In our case X_t being a sentiment index time series and Y_t being the stock's price return, or any other functional form of the price that we aim to forecast. The basic idea of causality is that due to Granger [20] which states that X_t causes Y_t, if Y_{t+k}, for some $k > 0$ can be better predicted using the past of both X_t and Y_t than it can by using the past of Y_t alone. This can be formally tested by considering a bivariate linear autoregressive model on X_t and Y_t, making Y_t dependent on both the histories of X_t and Y_t, together with a linear autoregressive model on Y_t, and then test for the null hypothesis of "X does not cause Y," which amounts to a test that all coefficients accompanying the lagged observations of X in the bivariate linear autoregressive model are zero. Then, assuming a normal distribution for the data, we can evaluate the null hypothesis through an F-test. This augmented vector autoregressive model for testing Granger causality is due to Toda and Yamamoto [47] and has the advantage of performing well with possibly non-stationary series.

There are several recent approaches to testing causality based on nonparametric methods, kernel methods, and information theory, among others, that cope with nonlinearity and non-stationarity, but disregarding the presence of side information (conditional causality); see, for example, [15, 30, 50]. For a test of conditional causality, see [41].

3.2.4 Variable Selection

The causality analysis reveals any cause–effect relationship between the sentiment indicators and any of the securities' price function as target. A next step is to analyze these sentiment indicators, individually or in an ensemble, as features in a regression model for any of the financial targets. A rationale for putting variables together could be at the very least what they might have in common semantically. For example, joint together in a model, all variables express a *bearish* (e.g., negativity) or *bullish* (e.g., positivity) sentiment. Nonetheless, at any one period of time, not all features in one of these groups might cause the target as well as their companions, and their addition in the model might add noise instead of value information. Hence, a regression model which discriminates the importance of variables is in order.

Here is where we propose to do a LASSO regression with all variables under consideration that explain the target. The LASSO, due to Tibshirani [46], optimizes the mean square error of the target and linear combination of the regressors, subject to a L_1 penalty on the coefficients of the regressors, which amounts to eliminating those which are significantly small, hence removing those variables that contribute little to the model. The LASSO does not take into account possible linear dependencies among the predictors that can lead to numerical unstabilities, so we recommend the previous verification that no highly correlated predictors are considered together. Alternatively, adding a L_2 penalty on the coefficients of the regressors can be attempted, leading to an elastic net.

4 Empirical Analysis

Now we put into practice the lessons learned so far.

A Set of Dictionary-Based Sentiment Indicators Combining the lexicons defined by Loughran and McDonald for [29] with extra keywords manually selected, we build six lexicons. For each one of these lexicons, and each company trading in the New York Stock Exchange market, we apply Eq. (1) to compute a sentiment score for each document extracted from a dataset of *Dow Jones Newswires* in the range of 1/1/2012 to 31/12/2019. We aggregate these sentiment scores on an hourly and a daily basis using Eq. (2) and end up with 2×6 hourly and daily period time series of news sentiment values for each NYSE stock. These hourly and daily sentiment indicators are meant to convey the following emotions: positive, Financial Up (finup), Financial Hype (finhype), negative, Financial Down (findown), and fear. Additionally, we created daily time series of the rate of volume of news referring to a stock with respect to all news collected within the same time frame. We termed this volume of news indicator as Relative Volume of Talk (RVT).

We use historic price data of a collection of stocks and their corresponding sentiment and news volume indicators (positive, finup, finhype, negative, fear, findown, and RVT) to verify the stylized facts of sentiment on financial securities and check the statistical properties and predictive power of the sentiment indicators to returns (ret), squared returns (ret2, as a proxy of volatility), and rate of change of trading volume (rVol). We sample price data with daily frequency from 2012 to 2019 and with hourly frequency (for high frequency tests) from 11/2015 to 11/2019. For each year we select the 50 stocks from the New York Stock Exchange market that have the largest volume of news to guarantee sufficient news data for the sentiment indicators. Due to space limitations, in the exhibits we present the results for 6 stocks from our dataset representatives of different industries: Walmart (WMT), Royal Bank of Scotland (RBS), Google (GOOG), General Motors (GM), General Electric (GE), and Apple Inc. (AAPL).

Stylized fact 1. We have observed, through independence tests based on distance correlation, that during relatively long periods of time, ret2, a proxy of volatility, and our RVT index are dependent variables, in particular, for "mediatic" companies, such as Amazon, Google, Apple, and blue chips in general. This is illustrated in Fig. 2.

Stylized fact 2. We take on the graphical idea of Tetlock et al. [45] representing the relation of news volume to earnings announcement date. However, instead of the bar plot of accumulated volumes for each date, we propose a more informative graphical representation of the distribution of the daily accumulated volumes of news (Fig. 1). This is constructed by drawing the boxplots [31] of volumes of news rather than its simple aggregation on the earning's day and previous/successive days. Moreover, since the effect of news on financial market behavior (and its reciprocal) around earnings announcement is noticeable only for short periods, we reduce the scope of analysis to 25 days after and before earnings announcement days (which are all placed as the 0 of the plot) and thus consider each news once (either as preceding or succeeding an earnings announcement).

In all the periods, the distribution of the number of news is highly asymmetric (all means are larger than medians), and their right tails are heavy, except on earning's day itself, where it looks more symmetric. From this new plot, we can see that, not only on earnings day but 1 and 2 weeks before and after earnings day, there is a rise in the volume of news. The most prominent increase in volume of news is seen the exact day of earnings announcement, and the day immediately after earnings

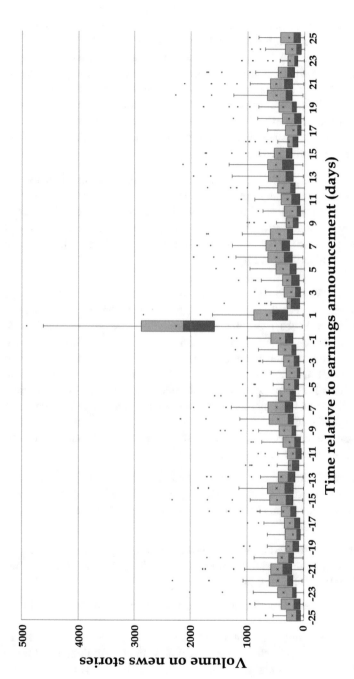

Fig. 1 Volume of news around earnings announcements. As in [45] we consider all firm-specific news stories about S&P 500 firms that appear in *Dow Jones Newswires* from 2012 to 2019, but consider a shorter range of 25 days before and 25 days after earnings announcements (time 0). For each news story, we compute the number of days until the corresponding firm's next earnings announcement or the number of days after the firm's last earnings announcement. Each story contributes only once to the volume of news $t \in [-25, 25]$ days away to or from the earnings announcement, and we plot a boxplot of this volume variable at each period t of days

announcement has also an abnormal increase with respect to the rest of the series of volumes, indicating a flourish of after-the-facts news. The number of extreme observations of each day is small: at most five companies exceed the standard limit (1.5 times the inter-quartile range) for declaring the value an "outlier".

We cannot then conclude from our representation of the media coverage of earnings announcements that the sentiments in the news may forecast fundamental indicators of the health of a company (e.g., price-to-earnings, price-to-book value, etc.) as it is done in [45], except perhaps for the few most talk-about companies, the outliers in our plot. However, we do speculate that the sentiment in the news following earnings announcements is the type of information useful for trading short sellers, as such has been considered in [17].

Stylized fact 3.   Again by testing independence among sentiment indices and market indicators (specifically, returns and squared returns), we have observed in our experiments that most of the time, sentiment indices related to negative emotions show dependency with ret and ret2 (mostly Financial Down and less intensive negative) more often than sentiment indices carrying positive emotions. This is illustrated in Fig. 2.

Independence and Variable Selection The distance correlation independence tests are exhibited in Fig. 2 and the results from LASSO regressions in Fig. 3. From these we observed the consistency of LASSO selection with dependence/independence among features and targets. The most sustained dependencies through time, and for the majority of stocks analyzed, are observed between RVT and ret2, RVT and rVol, findown and ret2, and finup and ret. LASSO selects RVT consistently with dependence results in the same long periods as a predictor of both targets ret2 and rVol, and it selects findown often as a predictor of ret2, and finup as a predictor of ret. On the other hand, positive is seldom selected by LASSO, just as this sentiment indicator results independent most of the time to all targets.

Stationarity Most of the indices we have studied follow some short-memory stationary processes. Most of them are Moving Averages, indicating dependency on the noise component, not on the value of the index, and always at small lags, at most 2.

Causality We have performed Granger causality tests on sentiment data with the corresponding stock's returns, squared returns, and trading volumes as the targets. We have considered the following cases:

- Data with daily frequency, performing the tests on monthly windows within the 2012–2019 period.
- Data with hourly frequency ranging from November 2015 to November 2019. In this case, we evaluated on both daily and weekly windows. Additionally, for the weekly windows, an additional test was run with overlapping windows starting every day.

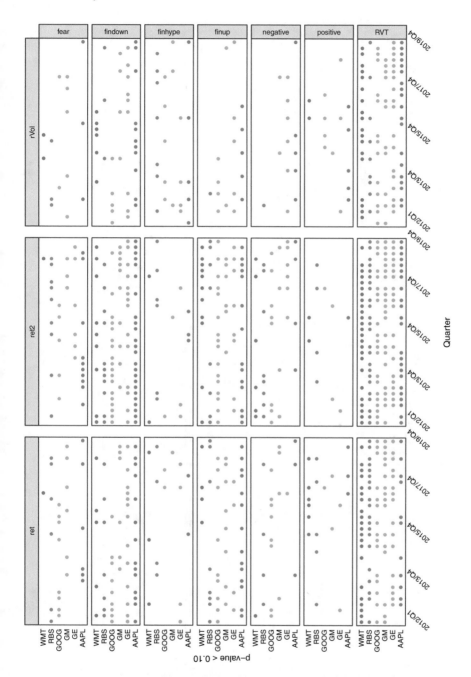

Fig. 2 Dependency through distance correlation tests (significance level at 0.1) performed on quarterly windows of daily data from 2012 to 2019

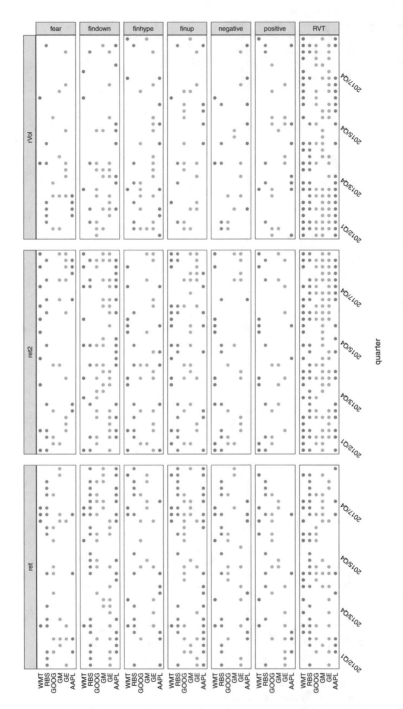

Fig. 3 Selected variables by LASSO tests performed on quarterly windows of daily data from 2012 to 2019

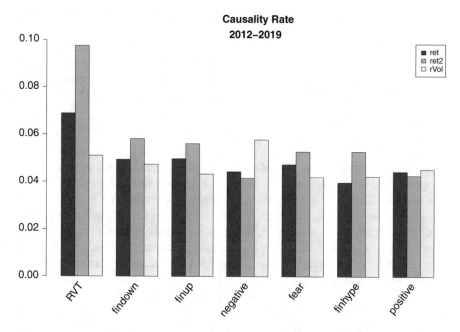

Fig. 4 Total success rate of the causality tests (significance level at 0.05) performed on monthly windows of daily data of the 2012–2019 period, across all stocks considered

In both cases, we find that for almost all variables, the tests only find causality in roughly 5% of the observations, which corresponds to the p-value (0.05) of the test. This means that the number of instances where causality is observed corresponds to the expected number of false positives, which would suggest that there is no actual causality between the sentiment indicators and the targets. The only pair of sentiment variable and target that consistently surpasses this value is RVT and ret2, for which causality is found in around 10% of the observations of daily frequency data (see Fig. 4).

Nonetheless, the lack of causality does not imply the lack of predictive power of the different features for the targets, only that the models will not have a causal interpretation in economic terms. Bear in mind that causality (being deterministic) is a stronger form of dependency and subsumes predictability (a random phenomenon).

5 Software

R

There has been a recent upsurge in R packages specific for topic modeling and sentiment analysis. The user has nowadays at hand several built-in functions in R to gauge sentiment in texts and construct his own sentiment indicators. We make a brief review below of the available R tools exclusively tailored for textual sentiment analysis. This list is by no means exhaustive, as new updates are quickly created due to the growing interest in the field, and that other sentiment analysis tools are already implicitly included in more general text mining packages as **tm** [32], **openNLP** [22], and **qdap** [37]. In fact, most of the current packages specific for sentiment analysis have strong dependencies on the aforementioned text mining infrastructures, as well as others from the CRAN Task View on Natural Language Processing[3]

SentimentAnalysis (2019-03): Performs a sentiment analysis of textual contents in R. Incorporates various existing dictionaries (e.g., Harvard IV or finance-specific dictionaries such as Loughran-McDonald), and it can also create customized dictionaries. The latter uses LASSO regularization as a statistical approach to select relevant terms based on an exogenous response variable [18].

RSentiment (2018-07): Analyzes the sentiment of a sentence in English and assigns score to it. It can classify the sentences to the following categories of sentiments: positive, negative, very positive, very negative, and neutral. For a vector of sentences, it counts the number of sentences in each category of sentiment. In calculating the score, negation and various degrees of adjectives are taken into consideration [9].

sentimentr (2019-03): Calculates text polarity sentiment [36].

sentometrics (2019-11): An integrated framework for textual sentiment time series aggregation and prediction. It contains all of the functions necessary to implement each one of the stages in the workflow described in Sect. 2 for building news sentiment-based forecasting models [2].

quanteda (2019-11): Quantitative analysis of textual data [7].

syuzhet (2017): Extracts sentiment and sentiment-derived plot arcs from the text [25].

Python

For Python's programmers there are also a large number of options for sentiment analysis. In fact, a quick search for "Sentiment Analysis" on The Python Package Index (PyPI)[4] returns about 6000 items. Here we include a reduced list of the most relevant modules.

Vader: Valence Aware Dictionary for sEntiment Reasoning is a rule-based model [23], mainly trained on the analysis of social texts (e.g., social media texts, movie reviews, etc.). Vader classifies the sentences in three categories: positive, negative, and neutral representing the ratios of proportions of text that fall into each category (the summation is 1 or close). It also provides a *compound* score which is computed by summing the valence scores of each word in the lexicon; this value is normalized between -1 and 1.[5] An implementation of Vader can also be found in the general-purpose library for Natural Language Processing *nltk*.

---

[3]https://cran.r-project.org/web/views/NaturalLanguageProcessing.html.

[4]https://pypi.org/.

[5]https://github.com/cjhutto/vaderSentiment#about-the-scoring.

TextBlob: From a given input text, the library[6] computes the sentiment in terms of polarity and subjectivity scores lying on the ranges $[-1.0, 1.0]$ and $[0.0, 1.0]$, respectively. For the subjectivity scores 0 means very objective and 1 is very subjective.

Pattern: It is a multipurpose package for web mining, NLP tasks, machine learning, and network analysis. The sentiment is outputed in the form of polarity and subjectivity, and these can be retrieved at document level or at word level [42].

pycorenlp: Provides an interface to the Stanford CoreNLP Java package from where several functionalities are inherited.[7] It provides sentiment annotations for each sentence included in a given text. The full list of CoreNLP wrappers can be found in its website.[8]

The survey in [51] introduces 24 utilities for sentiment analysis—9 of these tools have an API for common programming languages. However, several of these utilities are paid, but most of them provide free licenses for a limited period.

Acknowledgments The research of A. Arratia, G. Avalos, and M. Renedo-Mirambell is supported by grant TIN2017-89244-R from MINECO (Ministerio de Economía, Industria y Competitividad) and the recognition 2017SGR-856 (MACDA) from AGAUR (Generalitat de Catalunya). The research of A. Cabaña is partially supported by grant RTI2018-096072-B-I00 (Ministerio de Ciencia e Innovación, Spain).

The authors are grateful to the news technology company, Acuity Trading Ltd.[9] for providing the data for this research.

References

1. Algaba, A., Ardia, D., Bluteau, K., Borms, S., & Boudt, K. (2020). Econometrics meets sentiment: An overview of methodology and applications. *Journal of Economic Surveys, 34*(3), 512–547.
2. Ardia, D., Bluteau, K., Borms, S., & Boudt, K. (2020, forthcoming). The R package sentometrics to compute, aggregate and predict with textual sentiment. *Journal of Statistical Software.* https://doi.org/10.2139/ssrn.3067734
3. Arias, M., Arratia, A., & Xuriguera, R. (2013). Forecasting with twitter data. *ACM Transactions on Intelligent Systems and Technology (TIST), 5*(1), 8.
4. Baker, M., & Wurgler, J. (2007). Investor sentiment in the stock market. *Journal of Economic Perspectives, 21*(2), 129–152.
5. Baumeister, R. F., Bratslavsky, E., Finkenauer, C., & Vohs, K. D. (2001). Bad is stronger than good. *Review of General Psychology, 5*(4), 323–370.

---

[6]https://textblob.readthedocs.io/en/dev/quickstart.html#sentiment-analysis.

[7]https://pypi.org/project/pycorenlp/.

[8]https://stanfordnlp.github.io/CoreNLP/other-languages.html.

[9]http://acuitytrading.com/.

6. Beckers, B., Kholodilin, K. A., & Ulbricht, D. (2017). *Reading between the lines: Using media to improve German inflation forecasts*. Technical Report, DIW Berlin Discussion Paper. https://doi.org/10.2139/ssrn.2970466.
7. Benoit, K., Watanabe, K., Wang, H., Nulty, P., Obeng, A., Müller, S., et al. (2019). Quanteda: Quantitative Analysis of Textual Data. Version 1.5.2. https://cran.r-project.org/web/packages/quanteda/index.html
8. Bifet, A., & Frank, E. (2010). Sentiment knowledge discovery in Twitter streaming data. In *International Conference on Discovery Science. Lecture Notes in Computer Science* (vol. 6332, pp. 1–15).
9. Bose, S. (2018). *Rsentiment: Analyse Sentiment of English Sentences*. Version 2.2.2. https://CRAN.R-project.org/package=RSentiment
10. Chan, W.S. (2003). Stock price reaction to news and no-news: Drift and reversal after headlines. *Journal of Financial Economics, 70*(2), 223–260.
11. Davis, A. K., Piger, J. M., & Sedor, L. M. (2012). Beyond the numbers: Measuring the information content of earnings press release language. *Contemporary Accounting Research, 29*(3), 845–868.
12. Deriu, J., Lucchi, A., De Luca, V., Severyn, A., Muller, S., Cieliebak, M., et al. (2017). Leveraging large amounts of weakly supervised data for multi-language sentiment classification. In *26th International World Wide Web Conference, WWW 2017*, Art. no. 3052611 (pp. 1045–1052). https://arxiv.org/pdf/1703.02504.pdf
13. Deriu, J., Lucchi, A., Gonzenbach, M., Luca, V. D., Uzdilli, F., & Jaggi, M. (2016). Swiss-Cheese at SemEval-2016 task 4: Sentiment classification using an ensemble of convolutional neural networks with distant supervision. In *Proceedings of the 10th International Workshop on Semantic Evaluation (SemEval-2016)* (pp. 1124–1128)
14. Dickey, D. A., & Fuller, W. A. (1979). Distribution of the estimators for autoregressive time series with a unit root. *Journal of the American Statistical Association, 74*(366a), 427–431.
15. Diks, C., & Wolski, M. (2016). Nonlinear granger causality: Guidelines for multivariate analysis. *Journal of Applied Econometrics, 31*(7), 1333–1351.
16. Ding, X., Liu, B., & Yu, P. S. (2008). A holistic lexicon-based approach to opinion mining. In *WSDM'08 - Proceedings of the 2008 International Conference on Web Search and Data Mining* (pp. 231–240). New York, NY, USA: ACM.
17. Engelberg, J. E., Reed, A. V., & Ringgenberg, M. C. (2012). How are shorts informed?: Short sellers, news, and information processing. *Journal of Financial Economics, 105*(2), 260–278.
18. Feuerriegel, S., & Proellochs, N. (2019). *SentimentAnalysis: Dictionary-Based Sentiment Analysis* (2019). Version 1.3-3. https://CRAN.R-project.org/package=SentimentAnalysis
19. Go, A., Bhayani, R., & Huang, L. (2009). Twitter sentiment classification using distant supervision. *CS224N Project Report, Stanford, 1*(12), 2009.
20. Granger, C. (1969). Investigating causal relations by econometric models and cross-spectral methods. *Econometrica, 37*, 424–438.
21. Heston, S. L., & Sinha, N. R. (2017). News vs. sentiment: Predicting stock returns from news stories. *Financial Analysts Journal, 73*(3), 67–83.
22. Hornik, K. (2019). *openNLP: Apache OpenNLP Tools Interface*. R Package Version 0.2.7. https://cran.r-project.org/web/packages/openNLP/index.html
23. Hutto, C. J., & Gilbert, E. (2014). Vader: A parsimonious rule-based model for sentiment analysis of social media text. In *Proceedings of the 8th International Conference on Weblogs and Social Media, ICWSM 2014* (pp. 216–225).
24. Jegadeesh, N., & Wu, D. (2013). Word power: A new approach for content analysis. *Journal of Financial Economics, 110*(3), 712–729.
25. Jockers, M. L. (2017). *Syuzhet: Extract Sentiment and Plot Arcs from Text*. Version 1.0.4. https://CRAN.R-project.org/package=syuzhet
26. Kumar, A., & Lee, C. M. (2006). Retail investor sentiment and return comovements. *The Journal of Finance, 61*(5), 2451–2486.
27. Li, F. (2006). *Do stock market investors understand the risk sentiment of corporate annual reports?* Available at SSRN 898181 . http://www.greyfoxinvestors.com/wp-content/uploads/2015/06/ssrn-id898181.pdf

28. Liu, B. (2015). *Sentiment analysis: Mining opinions, sentiments, and emotions.* Cambridge: Cambridge University Press.
29. Loughran, T., & McDonald, B. (2011). When is a liability not a liability? Textual analysis, dictionaries, and 10-Ks. *The Journal of Finance, 66*(1), 35–65.
30. Marinazzo, D., Pellicoro, M., & Stramaglia, S. (2008). Kernel method for nonlinear granger causality. *Physical Review Letters, 100*(14), 144103.
31. McGill, R., Tukey, J. W., & Larsen, W. A. (1978). Variations of box plots. *The American Statistician, 32,* 12–16.
32. Meyer, D., Hornik, K., & Feinerer, I. (2008). Text mining infrastructure in R. *Journal of Statistical Software, 25*(5), 1–54.
33. Nakov, P., Ritter, A., Rosenthal, S., Sebastiani, F., & Stoyanov, V. (2016). Semeval-2016 task 4: Sentiment analysis in twitter. In *Proceedings of the 10th International Workshop on Semantic Evaluation (SemEval-2016)* (pp. 1–18).
34. Polanyi, L., & Zaenen, A. (2006). Contextual valence shifters. In *Computing attitude and affect in text: Theory and applications* (pp. 1–10). Berlin: Springer.
35. Rao, D., & Ravichandran, D. (2009). Semi-supervised polarity lexicon induction. In *EACL '09: Proceedings of the 12th Conference of the European Chapter of the Association for Computational Linguistics* (pp. 675–682). Stroudsburg, PA, USA: Association for Computational Linguistics.
36. Rinker, T. W. (2019). *Sentimentr: Calculate Text Polarity Sentiment.* Version 2.7.1. http://github.com/trinker/sentimentr
37. Rinker, T. W. (2020). *Qdap: Quantitative Discourse Analysis.* Buffalo, New York. Version 2.3.6 https://cran.r-project.org/web/packages/qdap/index.html
38. Rizzo, M. L., & Szekely, G. J. (2018) *Energy: E-Statistics: Multivariate Inference via the Energy of Data.* R package version 1.7-4. https://CRAN.R-project.org/package=energy.
39. Rozin, P., & Royzman, E. B. (2001). Negativity bias, negativity dominance, and contagion. *Personality and Social Psychology Review, 5*(4), 296–320.
40. Sebastiani, F. (2002). Machine learning in automated text categorization. *ACM Computing Surveys, 34*(1), 1–47.
41. Serès, A., Cabaña, A., & Arratia, A. (2016). Towards a sharp estimation of transfer entropy for identifying causality in financial time series. In *ECML-PKDD. Proceedings of the 1st Workshop MIDAS* (vol. 1774, pp. 31–42).
42. Smedt, T. D., & Daelemans, W. (2012). Pattern for python. *Journal of Machine Learning Research, 13*(Jun), 2063–2067.
43. Székely, G. J., Rizzo, M. L., & Bakirov, N. K. (2007). Measuring and testing dependence by correlation of distances. *The Annals of Statistics, 35*(6), 2769–2794.
44. Tetlock, P. C. (2007). Giving content to investor sentiment: The role of media in the stock market. *The Journal of Finance, 62,* 1139–1168.
45. Tetlock, P. C., Saar-Tsechansky, M., & Macskassy, S. (2008). More than words: Quantifying language to measure firm's fundamentals. *The Journal of Finance, 63*(3), 1437–1467.
46. Tibshirani, R. (1996). Regression shrinkage and selection via the lasso. *Journal of the Royal Statistical Society: Series B (Methodological), 58*(1), 267–288.
47. Toda, H. Y., & Yamamoto, T. (1995). Statistical inference in vector autoregressions with possibly integrated processes. *Journal of Econometrics, 66*(1–2), 225–250.
48. Tsai, M. F., & Wang, C. J. (2014). Financial keyword expansion via continuous word vector representations. In *Proceedings of the 2014 Conference on Empirical Methods in Natural Language Processing (EMNLP)* (pp. 1453–1458).
49. Uhl, M. W., Pedersen, M., Malitius, O. (2015). What's in the news? using news sentiment momentum for tactical asset allocation. *The Journal of Portfolio Management, 41*(2), 100–112.
50. Wibral, M., Pampu, N., Priesemann, V., Siebenhühner, F., Seiwert, H., Linder, M., et al. (2013). Measuring information-transfer delays. *PLoS ONE, 8*(2), Art. no. e55809.
51. Zucco, C., Calabrese, B., Agapito, G., Guzzi, P. H., & Cannataro, M. (2020). Sentiment analysis for mining texts and social networks data: Methods and tools. *Wiley Interdisciplinary Reviews: Data Mining and Knowledge Discovery, 10*(1), Art. no. e1333.

Semi-supervised Text Mining for Monitoring the News About the ESG Performance of Companies

Samuel Borms, Kris Boudt, Frederiek Van Holle, and Joeri Willems

Abstract We present a general monitoring methodology to summarize news about predefined entities and topics into tractable time-varying indices. The approach embeds text mining techniques to transform news data into numerical data, which entails the querying and selection of relevant news articles and the construction of frequency- and sentiment-based indicators. Word embeddings are used to achieve maximally informative news selection and scoring. We apply the methodology from the viewpoint of a sustainable asset manager wanting to actively follow news covering environmental, social, and governance (ESG) aspects. In an empirical analysis, using a Dutch-written news corpus, we create news-based ESG signals for a large list of companies and compare these to scores from an external data provider. We find preliminary evidence of abnormal news dynamics leading up to downward score adjustments and of efficient portfolio screening.

1 Introduction

Automated analysis of textual data such as press articles can help investors to better screen the investable universe. News coverage, how often news discusses a certain topic, and textual sentiment analysis, if news is perceived as positive or negative, serve as good proxies to detect important events and their surrounding perception.

S. Borms (✉)
Université de Neuchâtel, Neuchâtel, Switzerland

Vrije Universiteit, Brussels, Belgium
e-mail: samuel.borms@unine.ch

K. Boudt
Universiteit Gent, Ghent, Belgium

Vrije Universiteit, Brussels, Belgium
e-mail: kris.boudt@ugent.be

F. Van Holle · J. Willems
Degroof Petercam Asset Management, Brussels, Belgium
e-mail: f.vanholle@degroofpetercam.com; j.willems@degroofpetercam.com

© The Author(s) 2021
S. Consoli et al. (eds.), *Data Science for Economics and Finance*,
https://doi.org/10.1007/978-3-030-66891-4_10

Text-based signals have at least the advantage of timeliness and often also that of complementary information value. The challenge is to transform the textual data into useful numerical signals through the application of proper text mining techniques.

Key research in finance employing text mining includes [13, 14, 24, 3]. These studies point out the impact of textual sentiment on stock returns and trading volume. Lately, the focus has shifted to using text corpora for more specific goals. For instance, Engle et al. [11] form portfolios hedged against climate change news based on news indicators.

This chapter takes the use of textual data science in sustainable investment as a running example. Investors with a goal of socially responsible investing (SRI) consider alternative measures to assess investment risk and return opportunities. They evaluate portfolios by how well the underlying assets align with a corporate social responsibility (CSR) policy—for instance, if they commit to environmental-friendly production methods. A corporation's level of CSR is often measured along the environmental, social and corporate governance (ESG) dimensions.

Investors typically obtain an investable universe of ESG-compliant assets by comparing companies to their peers, using a best-in-class approach (e.g., including the top 40% companies) or a worst-in-class approach (e.g., excluding the bottom 40% companies). To do so, investors rely on in-house research and third-party agency reports and ratings. Berg et al. [6], Amel-Zadeh and Serafeim [2], and Escrig-Olmedo et al. [12], among others, find that these ESG ratings are diverse, not transparent, and lack standardization. Moreover, most agencies only provide at best monthly updates. Furthermore, ratings are often reporting-driven and not signal-driven. This implies that a company can be ESG-compliant "by the book" when it is transparent (akin to greenwashing), but that the ratings are not an accurate reflection of the true current underlying sustainability profile.

In the remainder of the chapter, we introduce a methodology to create and validate news-based indicators allowing to follow entities and topics of interest. We then empirically demonstrate the methodology in a sustainable portfolio monitoring context, extracting automatically from news an objective measurement of the ESG dimensions. Moniz [19] is an exception in trying to infer CSR-related signals from media news using text mining in this otherwise largely unexplored territory.

2 Methodology to Create Text-Based Indicators

We propose a methodology to extract meaningful time series indicators from a large collection of texts. The indicators should represent the dimensions and entities one is interested in, and their time variation should connect to real-life events and news stories. The goal is to turn the indicators into a useful decision-making signal. This is a hard problem, as there is no underlying objective function to optimize, text data are not easy to explore, and it is computationally cumbersome to iterate frequently. Our methodology is therefore semi-supervised, altering between rounds of algorithmic estimation and human expert validation.

2.1 From Text to Numerical Data

A key challenge is to transform the stream of qualitative textual data into quantitative indicators. This involves first the selection of the relevant news and the generation of useful metadata, such as the degree to which news discusses an entity or an ESG dimension, or the sentiment of the news message. We tackle this by using domain-specific keywords to query a database of news articles and create the metadata. The queried articles need to undergo a second round of selection, to filter out the irrelevant news. Lastly, the kept corpus is aggregated into one or more time series.

To classify news as relevant to sustainability, we rely on keywords generated from a word embedding space. Moniz [19] uses a latent topic model, which is a probabilistic algorithm that clusters a corpus into a variety of themes. Some of these themes can then be manually annotated as belonging to ESG. We decide to go with word embeddings as it gives more control over the inclusion of keywords and the resulting text selection. Another approach is to train a named entity recognition (NER) model, to extract specific categories of concepts. A NER model tailored to ESG concepts is hard to build from scratch, as it needs fine-grained labeled data.

The methodology laid out below assumes that the corpus is in a single language. However, it can be extended to a multi-language corpus in various ways. The go-to approach, in terms of accuracy, is to consider each language separately by doing the indicators construction independently for every language involved. After that, an additional step is to merge the various language-specific indicators into an indicator that captures the evolution across all languages. One could, for simplicity, generate keywords in one language and then employ translation. Another common way to deal with multiple languages is to translate all incoming texts into a target language and then proceed with the pipeline for that language.

2.1.1 Keywords Generation

Three types of keywords are required. The **query lexicon** is a list of keywords per dimension of interest (*in casu*, the three ESG dimensions). Its use is twofold: first, to identify the articles from a large database with at least one of these keywords, and second, to measure the relevance of the queried articles (i.e., more keywords present in an article means it is more relevant). The **sentiment lexicon** is a list of words with an associated sentiment polarity, used to calculate document-level textual sentiment. The polarity defines the average connotation a word has, for example, -1 for "violence" or 1 for "happy." **Valence shifters** are words that change the meaning of other words in their neighborhood. There are several categories of valence shifters, but we focus on amplifiers and deamplifiers. An amplifier strengthens a neighboring word, for instance, the word "very" amplifies the word "strong" in the case of "very strong." Deamplifiers do the opposite, for example, "hardly" weakens the impact of "good" when "hardly good." The reason to integrate valence shifters in the sentiment calculation is to better account for context in a text. The unweighted sentiment score of a document i with Q_i words under this approach is $s_i = \sum_{j=1}^{Q_i} v_{j,i} s_{j,i}$. The score $s_{j,i}$ is the polarity value attached in the sentiment lexicon to word j and is zero when the word is not in the lexicon. If word $j - 1$ is a valence shifter, its impact is measured by $v_{j,i} = 1.8$ for amplifiers or $v_{j,i} = 0.2$ for deamplifiers. By default, $v_{j,i} = 1$.

To generate the keywords, we rely on expansion through a word embedding space. Word embeddings are vector representations optimized so that words closer to each other in terms of linguistic context have a more similar quantitative representation. Word embeddings are usually a means to an end. In our case, based on an initial set of seed keywords, analogous words can be obtained by analyzing the words closest to them in the embedding space. Many word embeddings computed on large-scale datasets (e.g., on Wikipedia) are freely available in numerous languages.[1] The availability of pretrained word embeddings makes it possible to skip the step of estimating a new word embedding space; however, in this chapter, we describe a straightforward approach to do the estimation oneself.

Word2Vec [18] and GloVe [21] are two of the most well-known techniques to construct a word embedding space. More recent and advanced methods include fastText [7] and the BERT family [9]. Word2Vec is structured as a continuous bag-of-words (CBOW) or as a skip-gram architecture, both relying only on local word information. A CBOW model tries to predict a given word based on its neighboring words. A skip-gram model tries to use a given word to predict the neighboring words. GloVe [21] is a factorization method applied to the corpus word-word co-occurrence matrix. A co-occurrence matrix stores the number of times a column word appears in the context of a row word. As such, GloVe integrates both global (patterns across the entire corpus) and local (patterns specific to a small context window) statistics. The intuition is that words which co-occur frequently are assumed to share a related semantic meaning. This is apparent in the co-occurrence matrix, where these words as a row-column combination will have higher values.

GloVe's optimization outputs two v-dimensional vectors per word (the word vector and a separate context word vector), that is, $\boldsymbol{w}_1, \boldsymbol{w}_2 \in \mathbb{R}^v$. The final word vector to use is defined as $\boldsymbol{w} \equiv \boldsymbol{w}_1 + \boldsymbol{w}_2$. To measure the similarity between word vectors, say $\boldsymbol{w}_i$ and $\boldsymbol{w}_j$, the cosine similarity metric is commonly used. We define $cs_{ij} \equiv \boldsymbol{w}_i \boldsymbol{w}_j / \|\boldsymbol{w}_i\| \|\boldsymbol{w}_j\|$, where $\|.\|$ is the ℓ_2-norm. The measure $cs_{ij} \in [-1, 1]$, and the higher the more similar words i and j are in the embedding space.

Figure 1 displays the high-level process of expanding an initial set of seed words into the final three types of keywords needed. The seed words are the backbone of the analysis. They are defined manually and should relate strongly to the study domain. Alternatively, they can be taken from an existing lexicon, as done in [25] who start from the uncertainty terms in the Loughran and McDonald lexicon [17]. The seed words include both query seed words and sentiment seed words (often a

---

[1]For example, pretrained word embeddings by Facebook are available for download at https://fasttext.cc/docs/en/crawl-vectors.html.

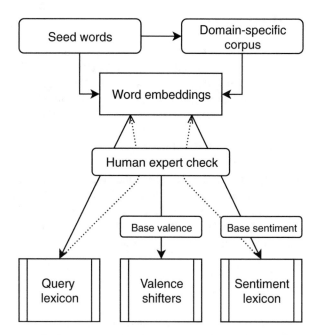

Fig. 1 Representation of the flow from seed words to the keywords of interest

subset of the former). The base valence and base sentiment word lists are existing dictionaries in need for a domain-specific twist to the application of interest.

All seed words are first used to query a more confined corpus from which the word embeddings will be estimated. The seed words are then expanded into the final query keywords by adding words that are similar, based on a ranking using the cs_{ij} metric and a human check. The human expert chooses between keeping the word, discarding the word, and assigning the word as a valence shifter. The same step is done for the sentiment seed words. As sentiment lexicons are typically larger, the words from a base sentiment lexicon not too far from the obtained query lexicon are added as well. The words coming from the word embeddings might be considered more important and thus weighted differently. The valence shifters are a combination of a base valence shifters list with the words assigned as a valence shifter. Section 3.2.1 further explains the implementation for the ESG use case.

This keywords generation framework has as limitation that it only considers unigrams, i.e., single words. Maintaining a valence shifters list adds a contextual layer in the textual sentiment calculation, and the number of keywords present in an article is a good overall indicator of the ESG relevance of news.

2.1.2 Database Querying

The database of texts is the large corpus that contains the subset of news relevant for the analysis. The task is to extract that subset as accurately as possible. The trade-off at play is that a large subset may guarantee full relevance, but it also adds more noise so it requires to think more carefully about the filtering step. In the process described in Fig. 1, a first query is needed to obtain a decent domain-specific corpus to estimate the embeddings.

Once the final query lexicon is composed, the batch of articles including the words in this lexicon as well as the entities to analyze needs to be retrieved and stored. To avoid a very time-consuming query, the querying is best approached as a loop over pairs of a given entity and the query lexicon keywords. A **list of entities** with the exact names to extract needs to be curated, possibly dynamic over time to account for name changes. Only the articles in which at least one entity name and at least one of the keywords is present are returned.

2.1.3 News Filtering

Keywords-based extraction does not guarantee that all articles retrieved are pertinent. It must be expected that a considerable degree of noise still remains. For example, press articles about a thief driving a BMW is not ESG-worthy news about the company BMW. Therefore, we recommend the following negative filters:

- Removal of texts that have no connection with the topic to study, for example, articles dealing with sports or lifestyle.
- Removal of articles that are too long (e.g., lengthy interviews) or too short (being more prone to a biased measurement of relevance and sentiment). Instead of removing the longer-than-usual articles, one could proceed with the leading paragraph(s) or a summary.
- Removal of exact duplicated entries or highly related (near-duplicated) entries.
- Removal of texts that are subject to database-specific issues, such as articles with a wrong language tag.

The level of filtering is a choice of the researcher. For instance, one can argue to leave (near-)duplicates in the corpus if one wants to represent the total news coverage, irrespective of whether the news rehashes an already published story or not. In this sense, it is also an option to reweight an article based on its popularity, proxied by the number of duplicates within a chosen interval of publication or by the number of distinct sources expressing related news.

2.1.4 Indicators Construction

A corpus with N documents between daily time points $t = 1, \ldots, T$ has a $N \times p$ matrix Z associated to it. This matrix maps the filtered corpus for a given entity to p numerical metadata variables. It stores the values used for optional additional filtering and ultimately for the aggregation into the time series indicators. Every row corresponds to a news article with its time stamp. The number of articles at time t is equal to N_t, such that $N \equiv N_1 + \ldots + N_T$.

The ultimate indices are obtained applying a function $f : Z \mapsto I$, where I is a $U \times P$ time series matrix that represents the "suite" of P final text-based indices, with $U \leq T$. The (linear or nonlinear) aggregation function depends on the use case.

Specific computation of the metadata and the aggregation into indices are elaborated upon in the application described in Sect. 3.

2.2 Validation and Decision Making

Not all ESG information is so-called material. The created indicators only become useful when explicitly mapped into practical and validated decision-making signals.

Qualitative validation involves surveying the news to assess the remaining irrelevance of the articles. It also includes a graphical check in terms of peaks around the appearance of important events. Quantitative validation statistically measures the leading properties in regard to a certain target variable (e.g., existing sustainability scores) and the effectiveness of an investment strategy augmented with text-based information (in terms of out-of-sample risk and return and the stability and interpretation of formed portfolios).

In a real-life setting, when wanting to know which companies face a changing sustainability profile ("positives") and which not ("negatives"), false positives are acceptable but false negatives are typically not; in the same vein doctors do not want to tell sick patients they are healthy. It is more important to bring up all cases subject to a potentially changed underlying ESG profile (capturing all the actual positives at the cost of more false positives), rather than missing out on some (the false negatives) but bringing only the certain cases to the surface (merely a subset of the true positives). In machine learning classification lingo, this would mean aiming for excellent recall performance. An analyst will always proceed to investigation based on the signals received before recommending a portfolio action. Still, only an amount of signals that can reasonably be coped with should get through.

3 Monitoring the News About Company ESG Performance

In this section, we further motivate the integration of news-based ESG indices in sustainable investment practices. Secondly, we implement the described methodology and validate its applicability.

3.1　Motivation and Applications

We believe there is a high added value of news-implied time-varying ESG indicators for asset managers and financial analysts active in both risk management and investment. These two main types of applications in the context of sustainable investment are motivated below.

3.1.1　Text-Based ESG Scoring as a Risk Management Tool

According to [22], social preferences are the driving factor behind why investors are willing to forgo financial performance when investing in SRI-compliant funds. This class of investors might be particularly interested in enhanced ESG risk management. An active sustainable portfolio manager should react appropriately when adverse news comes out, to avoid investors becoming worried, as the danger of reputational damage lurks.

　The degree to which a company is sustainable does not change much at a high frequency, but unexpected events such as scandals may immediately cause a corporation to lose its ESG-compliant stamp. An investor relying on low-frequency rating updates may be invested wrongly for an extended time period. Thus, it seems there is the need for a timelier filter, mainly to exclude corporations that suddenly cease to be ESG-compliant. News-based indicators can improve this type of negative screening. In fact, both negative and positive ESG screenings are considered among the most important future investment practices [2]. A universe of stocks can be split into a sustainable and a non-sustainable subuniverse. The question is whether news-based indicators can anticipate a change in the composition of the subuniverses.

　Portfolio managers need to be proactive by choosing the right response among the various ESG signals they receive, arriving from different sources and at different times. In essence, this makes them an "ESG signals aggregator." The more signals, the more flexibility in the ESG risk management approach. An important choice in the aggregation of the signals is which value to put on the most timely signal, usually derived from news analysis.

　Overall, the integration of textual data can lead to a more timely and a more conservative investment screening process, forcing asset managers as well as companies to continuously do well at the level of ESG transparency and ESG news presence.

3.1.2　Text-Based ESG Scoring as an Investment Tool

Increased investment performance may occur while employing suitable sustainable portfolio strategies or strategies relying on textual information. These phenomena are not new, but doing both at the same time has been less frequently investigated. A global survey by Amel-Zadeh and Serafeim [2] shows that the main reason for senior investment professionals to follow ESG information is investment performance. Their survey does not discuss the use of news-based ESG data. Investors can achieve improved best-in-class stock selection or do smarter sector rotation. Targeted news-based indices can also be exploited as a means to tilt portfolios toward certain sustainability dimensions, in the spirit of Engle et al. [11]. All of this can generate extra risk-adjusted returns.

3.2 Pipeline Tailored to the Creation of News-Based ESG Indices

To display the methodology, we create text-based indices from press articles written in Dutch, for an assortment of European companies. We obtain the news data from the combined archive of the Belga News Agency and Gopress, covering all press sources in Belgium, as well as the major press outlets from the Netherlands. The data are not freely available.

The pipeline is incremental with respect to the companies and dimensions monitored. One can add an additional company or an extra sustainability (sub)dimension by coming up with new keywords and applying it to the corpus, which will result in a new specified time series output. This is important for investors that keep an eye on a large and changing portfolio, who therefore might benefit from the possibility of building the necessary corpus and indicators incrementally. The keywords and indicators can be built first with a small corpus and then improved based on a growing corpus. Given the historical availability of the news data, it is always easy to generate updated indicators for backtesting purposes. If one is not interested in defining keywords, one can use the keywords used in this work, available upon request.

3.2.1 Word Embeddings and Keywords Definition

We manually define the seed words drawing inspiration from factors deemed of importance by Vigeo Eiris and Sustainalytics, leading global providers of ESG research, ratings, and data. Environmental factors are for instance climate change and biodiversity, social factors are elements such as employee relations and human rights, and governance factors are, for example, anti-bribery and gender diversity. We define a total of 16, 18, and 15 seed words for the environmental, social, and governance dimensions, respectively. Out of those, we take 12 negative sentiment seed words. There are no duplicates across categories. Table 1 shows the seed words.

The time horizon for querying (and thus training the word embeddings) spans from January 1996 to November 2019. The corpus is queried separately for each dimension using each set of seed words. We then combine into a large corpus, consisting of 4,290,370 unique news articles. This initial selection assures a degree

Table 1 Dutch E, S, G, and negative sentiment seed words

E	S	G	Sentiment[a]
milieu (*environment*), energie (*energy*), mobiliteit (*mobility*), nucleair (*nuclear*), klimaat (*climate*), biodiversiteit (*biodiversity*), koolstof (*carbon*), vervuiling (*pollution*), water, verspilling (*waste*), ecologie (*ecology*), duurzaamheid (*sustainability*), uitstoot (*emissions*), hernieuwbaar (*renewable*), olie (*oil*), olielek (*oil leak*)	samenleving (*society*), gezondheid (*health*), mensenrechten (*human rights*), sociaal (*social*), discriminatie (*discrimination*), inclusie (*inclusion*), donatie (*donation*), staking (*strike*), slavernij (*slavery*), stakeholder, werknemer (*employee*), werkgever (*employer*), massaontslag (*mass fire*), arbeid (*labor*), community, vakbond (*trade union*), depressie (*depression*), diversiteit (*diversity*)	gerecht (*court*), budget, justitie (*justice*), bestuur (*governance*), directie (*management*), omkoping (*bribery*), corruptie (*corruption*), ethiek (*ethics*), audit, patentbreuk (*patent infringement*), genderneutraal (*gender neutral*), witwaspraktijken (*money laundering*), dierproeven (*animal testing*), lobbyen (*lobbyism*), toploon (*top wage*)	vervuiling, verspilling, olielek, discriminatie, staking, slavernij, massaontslag, depressie, omkoping, corruptie, patentbreuk, witwaspraktijken

[a] These are a subset of the words in E, S, and G

of domain specificity in the obtained word vectors, as taking the entire archive would result in a too general embedding.

We tokenize the corpus into unigrams and take as vocabulary the 100,000 most frequent tokens. A preceding cleaning step drops Dutch stop words, all words with less than 4 characters, and words that do not appear in at least 10 articles or in more than 10% of the corpus. We top the vocabulary with the 49 ESG seed words.

To estimate the GloVe word embeddings, we rely on the R package **text2vec** [23]. We choose a symmetric context window of 7 words and set the vector size to 200. Word analogy experiments in [21] show that a larger window or a larger vector size does not result in significantly better accuracy. Hence, this hyperparameters choice offers a good balance between expected accuracy and estimation time. In general, small context windows pick up substitutable words (e.g., due to enumerations), while large windows tend to better pick up topical connections. Creating the word embeddings is the most time-consuming part of the analysis, which might take from start to finish around half a day on a regular laptop. Figure 2 shows the fitted embedding space, shrunk down to two dimensions, focused on the seed words "duurzaamheid" and "corruptie."

To expand the seed words, for every seed word in each dimension, we start off with the 25 closest words based on cs_{ij}, i.e., those with the highest cosine similarity. By hand, we discard irrelevant words or tag words as an amplifying or as a deamplifying valence shifter. An example in the first valence shifter category is "chronische" (*chronic*), and an example in the second category is "afgewend" (*averted*). We reposition duplicates to the most representative category. This leads to

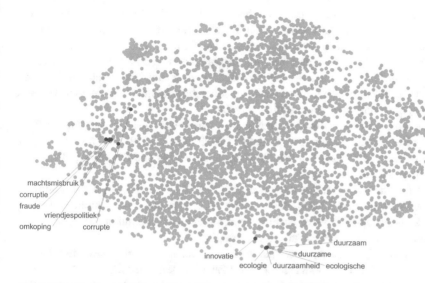

Fig. 2 Visualization of the embedding for a 5% fraction of the 100,049 vocabulary words. The t-distributed stochastic neighbor embedding (t-SNE) algorithm implemented in the R package **Rtsne** [15] is used with the default settings to reduce the 200-dimensional space to a two-dimensional space. In red, focal seed words "duurzaamheid" and "corruptie," and in green the respective five closest words according to the cosine similarity metric given the original high-dimensional word embeddings

197, 226, and 166 words, respectively, for the environmental, social, and governance dimensions.

To expand the sentiment words, we take the same approach. The obtained words (151 in total) receive a polarity score of -2 in the lexicon. From the base lexicon entries that also appear in the vocabulary, we discard the words for which none of its closest 200 words is an ESG query keyword. If at least one of these top 200 words is a sentiment seed word, the polarity is set to -1 if not already. In total, the sentiment lexicon amounts to 6163 words, and we consider 84 valence shifters.

3.2.2 Company Selection and Corpus Creation

To query the news related to companies, we use a reasonable trade-off between their commonplace name and their legal name.[2] Counting the total entity occurrences

---

[2]Suffixes (e.g., N.V. or Ltd.) and too generic name parts (e.g., International) are excluded. We also omit companies with names that could be a noun or a place (for instance, Man, METRO, Partners, Restaurant, or Vesuvius). Our querying system is case-insensitive, but case sensitivity would solve the majority of this problem. We only consider fully merged companies, such as Unibail-Rodamco-Westfield and not Unibail-Rodamco.

(measured by $n_{i,t}$; see Sect. 3.2.3) happens less strict by also accounting for company subnames. Our assumption is that often the full company name is mentioned once, and further references are made in an abbreviated form. As an example, to query news about the company Intercontinental Hotels, we require the presence of "Intercontinental" and "Hotels," as querying "Intercontinental" alone would result in a lot of unrelated news. To count the total matches, we consider both "Intercontinental" and "Intercontinental Hotels."

We look at the 403 European companies that are included in both the Sustainalytics ESG dataset (ranging from August 2009 to July 2019) and (historically) in the S&P Europe 350 stock index between January 1999 and September 2018. The matching is done based on the tickers.

We run through all filters enumerated in Sect. 2.1.3. Articles without minimum 450 or with more than 12,000 characters are deleted. To detect near-duplicated news, we use the locality-sensitive hashing approximate nearest neighbor algorithm [16] as implemented in the R package **textreuse** [20].

In total, 1,453,349 company-specific and sustainability-linked news articles are queried, of which 1,022,898 are kept after the aforementioned filtering. On average 33.4% of the articles are removed. Most come from the removal of irrelevant articles (20.5 p.p.); only a minor part is the result of filtering out too short and too long articles (6.4 p.p.). Pre-filtering, 42.2%, 71%, and 64.3% are marked belonging to the E, S, or G dimension, respectively. Post-filtering, the distribution is similar (38.1%, 70.2%, and 65.9%). Additionally, we drop the articles which have only one entity mention. The total corpus size falls to 365319. The strictness of this choice is to avoid the inclusion of news in which companies are only mentioned in passing [19]. Furthermore, companies without at least 10 articles are dropped. We end up with 291 of the companies after the main filtering procedure and move forward to the index construction with for each company a corpus.

3.2.3 Aggregation into Indices

As discussed in Sect. 2.1.4, we define a matrix $\mathbf{Z}_e$ for every entity e (i.e., a company) as follows:

$$
\mathbf{Z}_e =
\begin{bmatrix}
n_{1,1} & n_{1,1}^E & n_{1,1}^S & n_{1,1}^G & a_{1,1}^E & a_{1,1}^S & a_{1,1}^G & s_{1,1} \\
n_{2,1} & n_{2,1}^E & n_{2,1}^S & n_{2,1}^G & a_{2,1}^E & a_{2,1}^S & a_{2,1}^G & s_{2,1} \\
\vdots & \vdots & \vdots & \vdots & \vdots & \vdots & \vdots & \vdots \\
n_{i,t} & n_{i,t}^E & n_{i,t}^S & n_{i,t}^G & a_{i,t}^E & a_{i,t}^S & a_{i,t}^G & s_{i,t} \\
\vdots & \vdots & \vdots & \vdots & \vdots & \vdots & \vdots & \vdots \\
n_{N^e-1,T} & n_{N^e-1,T}^E & n_{N^e-1,T}^S & n_{N^e-1,T}^G & a_{N^e-1,T}^E & a_{N^e-1,T}^S & a_{N^e-1,T}^G & s_{N^e-1,T} \\
n_{N^e,T} & n_{N^e,T}^E & n_{N^e,T}^S & n_{N^e,T}^G & a_{N^e,T}^E & a_{N^e,T}^S & a_{N^e,T}^G & s_{N^e,T}
\end{bmatrix}.
\tag{1}
$$

The computed metadata for each news article are the number of times the company is mentioned (column 1); the total number of detected keywords for the E, S, and G dimensions (columns 2 to 4); the proportions of the E, S, and G keywords w.r.t. one another (columns 5 to 7); and the textual sentiment score (column 8). More specifically, n counts the number of entity mentions; n^E, n^S, and n^G count the number of dimension-specific keywords; and s is the textual sentiment score. The proportion $a_{i,t}^d$ is equal to $n_{i,t}^d / (n_{i,t}^E + n_{i,t}^S + n_{i,t}^G)$, for d one of the sustainability dimensions. It measures something distinct from keywords occurrence—for example, two documents can have the same number of keywords of a certain dimension yet one can be about one dimension only and the other about all three.

The sentiment score is calculated as $s_{i,t} = \sum_{j=1}^{Q_{i,t}} \omega_{j,i,t} v_{j,i,t} s_{j,i,t}$, where $Q_{i,t}$ is the number of words in article i at time t, $s_{j,i,t}$ is the polarity score for word j, $v_{j,i,t}$ is the valence shifting value applied to word j, and $\omega_{j,i,t}$ is a weight that evolves as a U-shape across the document.[3] To do the sentiment computation, we use the R package **sentometrics** [4].[4]

The metadata variables can also be used for further filtering, requiring, for instance, a majority proportion of one dimension in an article to include it. We divide Z_e into $Z_{e,E}$, $Z_{e,S}$, and $Z_{e,G}$. In those subsets, we decide to keep only the news entries for which $n_{i,t}^d \geq 3$ and $a_{i,t}^d > 0.5$, such that each sustainability dimension d is represented by articles maximally related to it. This trims down the total corpus size to 166020 articles.[5]

For a given dimension d, the time series matrix that represents the suite of final text-based indices is a combination of 11 frequency-based and 8 sentiment-adjusted indicators. We do the full-time series aggregation in two steps. This allows separating out the first simple from the subsequent (possibly time) weighted daily aggregation. We are also not interested in relative weighting within a single day; rather we will utilize absolute weights that are equally informative across the entire time series period.

We first create daily $T \times 1$ frequency vectors f, p, d, and n and a $T \times 1$ vector s of a daily sentiment indicator. For instance, $f = (f_1, \ldots, f_t, \ldots, f_T)'$ and $f_{[k,u]} = (f_k, \ldots, f_t, \ldots, f_u)'$. The elements of these vectors are computed starting from the

---

[3]Notably, $\omega_{j,i,t} = c \left(j - (Q_{i,t} + 1)/2 \right)^2$ with c a normalization constant. Words earlier and later in the document receive a higher weight than words in the middle of the document.

[4]See the accompanying package website at https://sentometricsresearch.github.io/sentometrics for code examples, and the survey paper by Algaba et al. [1] about the broader sentometrics research field concerned with the construction of sentiment indicators from alternative data such as texts.

[5]For some companies the previous lower bound of 10 news articles is breached, but we keep them aboard. The average number of documents per company over the embedding time horizon is 571.

submatrix $\boldsymbol{Z}_{e,d}$, with at any time $N_t^{e,d}$ articles, as follows:

$$f_t = N_t^{e,d}, \quad p_t = 1/N_t^{e,d} \sum_{i=1}^{N_t^{e,d}} a_{i,t}^d, \quad d_t = \sum_{i=1}^{N_t^{e,d}} n_{i,t}^d, \quad n_t = \sum_{i=1}^{N_t^{e,d}} n_{i,t}. \tag{2}$$

For sentiment, $s_t = 1/N_t^{e,d} \sum_{i=1}^{N_t^{e,d}} s_{i,t}$. Missing days in $t = 1, \ldots, T$ are added with a zero value. Hence, we have that f is the time series of the number of selected articles, p is the time series of the average proportion of dimension-specific keyword mentions, d is the time series of the number of dimension-specific keyword mentions, and n is the time series of the number of entity mentions. Again, these are all specific to the dimension d.

The second step aggregates the daily time series over multiple days. The weighted frequency indicators are computed as $\boldsymbol{f}'_{[k,u]} \boldsymbol{B}_{[k,u]} \boldsymbol{W}_{[k,u]}$, with $\boldsymbol{B}_{[k,u]}$ a $(u - k + 1) \times (u - k + 1)$ diagonal matrix with the time weights $\boldsymbol{b}_{[k,u]} = (b_k, \ldots, b_t, \ldots, b_u)'$ on the diagonal, and $\boldsymbol{W}_{[k,u]}$ a $(u - k + 1) \times 7$ metadata weights matrix defined as:

$$\boldsymbol{W}_{[k,u]} = \begin{bmatrix} p_k & g(d_k) & h(n_k) & p_k g(d_k) & p_k h(n_k) & g(d_k)h(n_k) & p_k g(d_k)h(n_k) \\ \vdots & \vdots & \vdots & \vdots & \vdots & \vdots & \vdots \\ p_t & g(d_t) & h(n_t) & p_t g(d_t) & p_t h(n_t) & g(d_t)h(n_t) & p_t g(d_t)h(n_t) \\ \vdots & \vdots & \vdots & \vdots & \vdots & \vdots & \vdots \\ p_u & g(d_u) & h(n_u) & p_u g(d_u) & p_u h(n_u) & g(d_u)h(n_u) & p_u g(d_u)h(n_u) \end{bmatrix},$$

$$\tag{3}$$

where $g(x) = \ln(1 + x)$ and $h(x) = x$. In our application, we choose to multiplicatively emphasize the number of keywords and entity mentions but alleviate the effect of the first, as in rare cases disproportionately many keywords pop up. The value p_t is a proportion between 0 and 1 and requires no transformation. The aggregate for the last column is $\sum_{t=k}^{u} f_t b_t p_t \ln(1 + d_t) n_t$, for instance.

The aggregations repeated for $u = \tau, \ldots, T$, where τ pinpoints the size of the first aggregation window, give the time series. They are assembled in a $U \times 7$ matrix of column vectors. Every vector represents a different weighting of the obtained information in the text mining step.

We opt for a daily moving fixed aggregation window $[k, u]$ with $k \equiv u - \tau + 1$. As a time weighting parameter, we take $b_t = \alpha_t / \sum_{t=k}^{u} \alpha_t$, with $\alpha_t = \exp\left(0.3 \left(\frac{t}{\tau} - 1\right)\right)$. We set τ to 30 days. The chosen exponential time weighting scheme distributes half of the weight to the last 7 days in the 30-day period, therefore ensuring that peaks are not averaged away. To omit any time dynamic, it is sufficient to set $b_t = 1$.

The non-weighted frequency measures for time u are computed as $b'_{[k,u]}A_{[k,u]}$, where $A_{[k,u]}$ is a $(u - k + 1) \times 4$ weights matrix defined as:

$$A_{[k,u]} = \left[f_{[k,u]} \; p_{[k,u]} \; d_{[k,u]} \; n_{[k,u]} \right]. \qquad (4)$$

The frequency-based time series indicators are all stored into a $U \times 11$ matrix.

The computation of the (weighted) sentiment values follows the same logic as described and results in a $U \times 8$ matrix. The final indices combined are in a $U \times 19$ matrix $I_{e,d}$. We do this for the 3 ESG dimensions, for a total of 57 unique text-based sustainability indicators, for each of the 291 companies.

3.2.4 Validation

We first present a couple of sustainability crisis cases and how they are reflected in our indicators relative to the scores from Sustainalytics. Figure 3 shows the evolution of the indicators for the selected cases.

Figure 3a displays Lonmin, a British producer of metals active in South Africa, whose mine workers and security were at the center of strikes mid-August 2012 leading to unfortunate killings. This is a clear example of a news-driven sustainability downgrade. It was picked up by our constructed news indicators, in that news coverage went up and news sentiment went down, and later reflected in a severe downgrade by Sustainalytics in their social score. Similar patterns are visible for the Volkswagen Dieselgate case (Fig. 3b), for the Libor manipulation scandal (Fig. 3c, which besides Barclays, also other financial institutions are impacted), and for a corruption lawsuit at Finmeccanica (Fig. 3d).

The main conclusions are the following. First, not all Sustainalytics downgrades (or sustainability changes in general) are covered in the press. Second, our indicators pick up severe cases faster, avoiding the lag of a few weeks or longer before adjustments in Sustainalytics scores are observed. The fact that media analysis does not pick up all events, but when it does, it does so fast(er), is a clear argument in favor of combining news-based ESG data with traditional ESG data.

In these illustrations, the general pattern is that the peak starts to wear out before the change in Sustainalytics score is published. Smoother time scaling would result in peaks occurring later, sometimes after the Sustainalytics reporting date, as well as phasing out slower (i.e., more persistence). This is because the news reporting is often clustered and spread out over several days. Likewise, an analysis run without the strict relevance filtering revealed less obvious peaks. Therefore, for (abnormal) peak detection, we recommend short-term focused time weighting and strict filtering.

In addition to the qualitative validation of the indicators, we present one possible way to quantitatively measure their ability to send early warnings for further investigation. We perform an ex-post analysis. Early warnings coming from the news-based indicators are defined as follows. We first split the period prior to a downward re-evaluation by Sustainalytics (a drop larger than 5) into two blocks

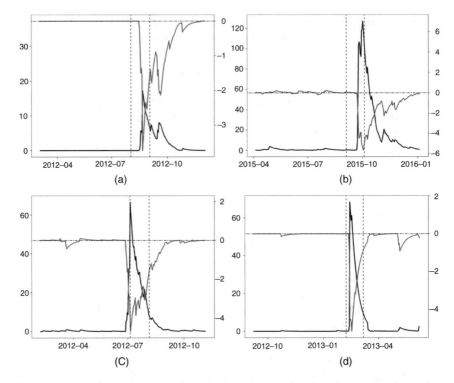

Fig. 3 News-based indicators around a selection of severe Sustainalytics downgrades (a drop larger than 5 on their 0–100 scale). The vertical bars indicate the release date of the downgraded score and 1 month before. The time frame shown is 6 months prior and 3 months after the release date. In black the average of the 11 frequency-based indicators (left axis) and in red of the 8 sentiment-based measures (right axis, with a horizontal line through zero). (**a**) Lonmin (Social). (**b**) Volkswagen (ESG). (**c**) Barclays (Governance). (**d**) Finmeccanica (Governance)

of 3 months. The first 3-month block is the reference period. The indicator values in the second 3-month block are continuously benchmarked against an extreme outcome of the previous block. For the frequency-based indicators, a hypothetical early warning signal is sent when the indicator surpasses the 99% quantile of the daily values in the reference block. For the sentiment-based indicators, a signal is sent if the indicator dips below the 1% reference quantile. Less signals will be passed on if the cut-offs are more extreme, but they will more likely be relevant.

Table 2 displays the results of the analysis for the averaged frequency-based and sentiment-based indicators. Between 11% and 34% of downgrades correspond with more abnormal news dynamics as defined. When so, on average about 50 days ahead of a realized downgrade, an initial news-based early warning is sent. Note that these early warnings should be interpreted as reasonable *first* signals, not necessarily the optimal ones, nor the only ones. There is ample room to fine-tune these metrics, and especially the amplitude of the signals generated in line with investment needs, as hinted to in Sect. 2.2.

Table 2 Ex-post early warning ability of news-based indicators

	Events	Detected		Time gain (days)	
		f	s	f	s
E	53%	19%	11%	48	48
S	53%	34%	24%	52	52
G	63%	25%	19%	51	46
ESG	24%	28%	18%	52	47

This table shows ex-post early warning performance statistics. The "events" column is the proportion of the 291 companies analyzed that faced at least one substantial Sustainalytics downgrade released at a day t_D. The "detected" column is the proportion of downgrades for which minimum one early warning was generated within 3 months before t_D. The "time gain (days)" column is the average number of days the first early warning precedes t_D. The analysis is done for the average of the 11 frequency-based indicators (f) and of the 8 sentiment-based measures (s)

3.3 Stock and Sector Screening

Another test of the usefulness of the created indices is to input them in a sustainable portfolio construction strategy. This allows studying the information content of the indices in general, of the different types of indices (mainly frequency-based against sentiment-based), and of the three ESG dimensions. The analysis should be conceived as a way to gauge the value of using textual data science to complement standard ESG data, not as a case in favor of ESG investing in itself.

We run a small horse race between three straightforward monthly screening strategies. The investable universe consists of the 291 analyzed companies. The strategies employed are the following:

- Invest in the 100 top-performing companies. [S1]
- Invest in the companies excluding the 100 worst-performing ones. [S2]
- Invest in the companies in the 10 top-performing sectors. [S3]

All strategies equally weight the monthly rebalanced selection of companies. We include 24 sectors formed by combining the over 40 peer groups defined in the Sustainalytics dataset. The notion of top-performing companies (resp. worst-performing) means having, at rebalancing date, the lowest (resp. the highest) news coverage or the most positive (resp. the most negative) news sentiment. The strategies are run with the indicators individually for each ESG dimension. To benchmark, we run the strategies using the scores from Sustainalytics and also compare with a portfolio equally invested in the total universe.

We take the screening one step further by imposing for all three strategies that companies should perform among the best both according to the news-based indicators and according to the ratings from Sustainalytics. We slightly modify the strategies per approach to avoid retaining a too limited group of companies; strategy S1 looks at the 150 top-performing companies, strategy S2 excludes the 50 worst-performing companies, and strategy S3 picks the 15 top-performing sectors. The total investment portfolio consists of the intersection of the selected companies by the two approaches.

We split the screening exercise in two out-of-sample time periods. The first period covers February 1999 to December 2009 (131 months), and the second period covers January 2010 to August 2018 (104 months). The rebalancing dates are at every end of the month and range from January 1999 to July 2018.[6] To screen based on our news-based indicators, we take the daily value at rebalancing date. For the Sustainalytics strategy, we take the most recently available monthly score, typically dating from 2 to 3 weeks earlier.

An important remark is that to estimate the word embeddings, we use a dataset whose range (i.e., January 1996–November 2019) is greater than that of the portfolio analysis. This poses a threat of lookahead bias—meaning, at a given point in time, we will have effectively already considered news data beyond that time point. This would be no problem if news reporting style is fixed over time, yet word use in news and thus its relationships in a high-dimensional vector space are subject to change.[7]It would be more correct (but also more compute intensive) to update the word embeddings rolling forward through time, for example, once a year. The advantage of a large dataset is an improved overall grasp of the word-to-word semantic relationships. Assuming the style changes are minor, and given the wide scope of our dataset, the impact on the outcome of the analysis is expected to be small.

3.3.1   Aggregate Portfolio Performance Analysis

We analyze the strategies through aggregate comparisons.[8] The results are summarized in Table 3. We draw several conclusions.

First, in both subsamples, we notice a comparable or better performance for the S2 and S3 investment strategies versus the equally weighted portfolio. The sector screening procedure seems especially effective. Similarly, we find that our news indicators, both the news coverage and the sentiment ones, are a more valuable screening tool, in terms of annualized Sharpe ratio, than using Sustainalytics scores. The approach of combining the news-based signals with the Sustainalytics ratings leads for strategies S1 and S2 to better outcomes compared to relying on the Sustainalytics ratings only. Most of the Sharpe ratios across ESG dimensions for the combination approach are close to the unscreened portfolio Sharpe ratio. The worst-

---

[6]Within this first period, the effective corpus size is 87611 articles. Within the second period, it is 60,977 articles. The two periods have a similar monthly average number of articles.

[7]An interesting example is *The Guardian* who declared in May 2019 to start using more often "climate emergency" or "climate crisis" instead of "climate change."

[8]As a general remark, due to the uncertainty in the expected return estimation, the impact of any sustainability filter on the portfolio performance (e.g., the slope of the linear function; Boudt et al. [8] derive to characterize the relationship between a sustainability constraint and the return of mean-tracking error efficient portfolios) is hard to evaluate accurately.

Table 3 Sustainable portfolio screening (across strategies)

		(a) News engine							
		E		S		G		ESG	
		f	s	f	s	f	s	f	s
P1	S1	0.50	0.46	**0.59**	0.40	**0.57**	0.50	**0.55**	0.46
	S2	**0.53**	0.49	**0.61**	0.47	**0.60**	**0.54**	**0.58**	0.50
	S3	**0.65**	0.46	**0.76**	0.44	**0.62**	**0.59**	**0.69**	0.50
P2	S1	0.88	0.81	0.93	0.85	0.91	0.86	0.91	0.84
	S2	**1.03**	0.99	0.99	**1.02**	**1.04**	**1.03**	**1.02**	**1.01**
	S3	**1.11**	**1.02**	**1.02**	0.98	0.99	**1.17**	**1.06**	**1.08**
All	S1	0.64	0.59	**0.72**	0.56	**0.70**	0.63	0.69	0.60
	S2	**0.71**	0.68	**0.76**	0.67	**0.77**	**0.73**	**0.75**	0.69
	S3	**0.82**	0.66	**0.86**	0.64	**0.76**	**0.81**	**0.82**	**0.71**

		(b) Sustainalytics			
		E	S	G	ESG
P2	S1	0.81	0.82	0.98	0.88
	S2	0.92	0.91	0.98	0.94
	S3	**1.07**	0.89	0.98	**1.00**

		(c) News engine + Sustainalytics							
		E		S		G		ESG	
		f	s	f	s	f	s	f	s
P2	S1	0.93	0.86	0.84	0.79	**1.09**	**1.08**	0.96	0.92
	S2	0.97	0.94	0.99	0.98	0.99	0.98	0.99	0.97
	S3	0.41	0.93	0.33	**1.03**	0.46	0.85	0.41	0.98

Table 3a shows the annualized Sharpe ratios for all strategies (S1–S3), averaged across the strategies on the 11 frequency-based indicators (f) and on the 8 sentiment-based indicators (s). The ESG column invests equally in the related E, S, and G portfolios. Table 3b shows the Sharpe ratios for all strategies using Sustainalytics scores. Table 3c refers to the strategies based on the combination of both signals. P1 designates the first out-of-sample period (February 1999 to December 2009), P2 the second out-of-sample period (January 2010 to August 2018), and All the entire out-of-sample period. An equally weighted benchmark portfolio consisting of all 291 assets obtains a Sharpe ratio of 0.52 (annualized return of 8.4%), of 1.00 (annualized return of 12.4%), and of 0.70 (annualized return of 10.1%) over P1, P2, and All, respectively. The screening approaches performing at least as good as the unscreened portfolio are indicated in bold

in-class exclusion screening (strategy S2) performs better than the best-in-class inclusion screening (strategy S1), of which only a part is explained by diversification benefits.

There seems to be no performance loss when applying news-based sustainability screening. It is encouraging to find that the portfolios based on simple universe screening procedures contingent on news analysis are competitive with an unscreened portfolio and with screenings based on ratings from a reputed data provider.

Second, the indicators adjusted for sentiment are not particularly more informative than the frequency-based indicators. On the contrary, in the first subsample, the news coverage indicators result in higher Sharpe ratios. Not being covered (extensively) in the news is thus a valid screening criterion. In general, however, there is little variability in the composed portfolios across the news-based indicators, as many included companies simply do not appear in the news, and thus the differently weighted indices are the same.

Third, news has in both time periods satisfactory relative value. The Sharpe ratios are low in the first subsample due to the presence of the global financial crisis. The good performance in the second subperiod confirms the universally growing importance and value of sustainability screening. It is also consistent with the study of Drei et al. [10], who find that, between 2014 and 2019, ESG investing in Europe led to outperformance.

Fourth, the utility of each dimension is not uniform across time or screening approach. In the first subperiod, the social dimension is best. In the second period, the governance dimension seems most investment worthy, but closely followed by the other dimensions. Drei et al. [10] observe an increased relevance of the environmental and social dimensions since 2016, whereas the governance dimension has been the most rewarding driver overall [5]. An average across the three dimension-specific portfolios also performs well, but not better.

The conclusions stay intact when looking at the entire out-of-sample period, which covers almost 20 years.

3.3.2 Additional Analysis

We also assess the value of the different weighting schemes. Table 4 shows the results for strategy S3 across the 8 sentiment indices, in the second period. It illustrates that the performance discrepancy between various weighting schemes for the sentiment indicators is not clear-cut. More complex weighting schemes, in this application, do not clearly beat the simpler weighting schemes.

Table 4 Sustainable portfolio screening (across sentiment indicators)

	s1	s2	s3	s4	s5	s6	s7	s8
E	1.01	1.04	0.99	0.99	1.04	1.02	1.04	1.02
S	0.98	0.99	0.98	1.00	0.98	0.95	0.95	1.01
G	1.14	1.17	1.20	1.20	1.16	1.19	1.09	1.17
ESG	1.06	1.09	1.08	1.08	1.08	1.08	1.05	1.08

This table shows the annualized Sharpe ratios in P2 for the screening strategy S3, built on the sentiment-based indicators, being $s1$ and $s2$–$s8$ as defined through the weighting matrix in (3)

An alternative approach for the strategies on the frequency-based indicators is to invert the ranking logic, so that companies with a high news coverage benefit and low or no news coverage are penalized. We run this analysis but find that the results worsen markedly, indicating that attention in the news around sustainability topics is not a good screening metric.

To test the sensitivity to the strict filtering choice of leaving out articles not having at least three keywords and more than half of all keywords related to one dimension, we rerun the analysis keeping those articles in. Surprisingly, some strategies improve slightly, but not all. We did not examine other filtering choices.

We also tested a long/short strategy but the results were poor. The long leg performed better than the short leg, as expected, but there was no reversal effect for the worst-performing stocks.

Other time lag structures (different values for τ or different functions in B) are not tested, given this would make the analysis more a concern of market timing than of assessing the lag structure. A short-term indicator catches changes earlier, but they may have already worn out by the rebalancing date, whereas long-term indicators might still be around peak level or not yet. We believe fine-tuning the time lag structure is more crucial for peak detection and visualization.

4   Conclusion

This chapter presents a methodology to create frequency-based and sentiment-based indicators to monitor news about the given topics and entities. We apply the methodology to extract company-specific news indicators relevant to environmental, social, and governance matters. These indicators can be used to timely detect abnormal dynamics in the ESG performance of companies, as an input in risk management and investment screening processes. They are not calibrated to automatically make investment decisions. Rather, the indicators should be seen as an additional source of information to the asset manager or other decision makers.

We find that the indicators often anticipate substantial negative changes in the scores of the external ESG research provider Sustainalytics. Moreover, we also find that the news indices can be used as a sole input to screen a universe of stocks and construct simple but well-performing investment portfolios. In light of the active sustainable investment manager being an "ESG ratings aggregator," we show that combining the news signals with the scores from Sustainalytics leads to a portfolio selection that performs equally well as the entire universe.

Given the limited reach of our data (we use Flemish and Dutch news to cover a wide number of European stocks), better results are expected with geographically more representative news data as well as a larger universe of stocks. Hence, the information potential is promising. It would be useful to investigate the benefits local news data bring for monitoring companies with strong local ties.

Additional value to explore lies in more meaningful text selection and index weighting. Furthermore, it would be of interest to study the impact of more fine-grained sentiment calculation methods. Summarization techniques and topic modeling are interesting text mining tools to obtain a drill down of sustainability subjects or for automatic peak labeling.

Acknowledgments We are grateful to the book editors (Sergio Consoli, Diego Reforgiato Recupero, and Michaela Saisana) and three anonymous referees, seminar participants at the CFE (London, 2019) conference, Andres Algaba, David Ardia, Keven Bluteau, Maxime De Bruyn, Tim Kroencke, Marie Lambert, Steven Vanduffel, Jeroen Van Pelt, Tim Verdonck, and the Degroof Petercam Asset Management division for stimulating discussions and helpful feedback. Many thanks to Sustainalytics (https://www.sustainalytics.com) for providing us with their historical dataset, and to Belga for giving us access to their news archive. This project received financial support from Innoviris, swissuniversities (https://www.swissuniversities.ch), and the Swiss National Science Foundation (http://www.snf.ch, grant #179281).

References

1. Algaba, A., Ardia, D., Bluteau, K., Borms, S., & Boudt, K. (2020). Econometrics meets sentiment: An overview of methodology and applications. *Journal of Economic Surveys, 34*(3), 512–547. https://doi.org/10.1111/joes.12370
2. Amel-Zadeh, A., & Serafeim, G. (2018). Why and how investors use ESG information: Evidence from a global survey. *Financial Analysts Journal, 74*(3), 87–103. https://doi.org/10.2469/faj.v74.n3.2
3. Antweiler, W., & Frank, M. (2004). Is all that talk just noise? The information content of internet stock message boards. *Journal of Finance, 59*(3), 1259–1294. https://doi.org/10.1111/j.1540-6261.2004.00662.x
4. Ardia, D., Bluteau, K., Borms, S., & Boudt, K. (2020). The R package **sentometrics** to compute, aggregate and predict with textual sentiment. *Forthcoming in Journal of Statistical Software*. https://doi.org/10.2139/ssrn.3067734
5. Bennani, L., Le Guenedal, T., Lepetit, F., Ly, L., & Mortier, V. (2018). *The alpha and beta of ESG investing*. Amundi working paper 76. http://research-center.amundi.com
6. Berg, F., Koelbel, J., & Rigobon, R. (2019). *Aggregate confusion: The divergence of ESG ratings*. MIT Sloan School working paper 5822–19. https://doi.org/10.2139/ssrn.3438533
7. Bojanowski, P., Grave, E., Joulin, A., & Mikolov, T. (2017). Enriching word vectors with subword information. *Transactions of the Association for Computational Linguistics, 5*, 135–146. https://doi.org/10.1162/tacl_a_00051
8. Boudt, K., Cornelissen, J., & Croux, C. (2013). The impact of a sustainability constraint on the mean–tracking error efficient frontier. *Economics Letters, 119*, 255–260. https://doi.org/10.1016/j.econlet.2013.03.020
9. Devlin, J., Chang, M. W., Lee, K., & Toutanova, K. (2018). *BERT: Pre–training of deep bidirectional transformers for language understanding*. Working paper, arXiv:1810.04805, https://arxiv.org/abs/1810.04805v2
10. Drei, A., Le Guenedal, T., Lepetit, F., Mortier, V., Roncalli, T., & Sekine, T. (2019). *ESG investing in recent years: New insights from old challenges*. Amundi discussion paper 42. http://research-center.amundi.com
11. Engle, R., Giglio, S., Kelly, B., Lee, H., & Stroebel, J. (2020). Hedging climate change news. *Review of Financial Studies, 33*(3), 1184–1216. https://doi.org/10.1093/rfs/hhz072
12. Escrig-Olmedo, E., Muñoz-Torres, M. J., & Fernandez-Izquierdo, M. A. (2010). Socially responsible investing: Sustainability indices, ESG rating and information provider agencies. *International Journal of Sustainable Economy, 2*, 442–461.

13. Heston, S., & Sinha, N. (2017). News vs. sentiment: Predicting stock returns from news stories. *Financial Analysts Journal, 73*(3), 67–83. https://doi.org/10.2469/faj.v73.n3.3
14. Jegadeesh, N., & Wu, D. (2013). Word power: A new approach for content analysis. *Journal of Financial Economics, 110*, 712–729. https://doi.org/10.1016/j.jfineco.2013.08.018
15. Krijthe, J., van der Maaten, L. (2018). *Rtsne: T-distributed Stochastic Neighbor Embedding using a Barnes-Hut Implementation.* R Package Version 0.15. https://CRAN.R-project.org/package=Rtsne
16. Leskovec, J., Rajaraman, A., & Ullman, J. (2014). *Mining of massive datasets.* Chapter Finding Similar Items (pp. 72–134). Cambridge: Cambridge University Press. https://doi.org/10.1017/CBO9781139924801
17. Loughran, T., & McDonald, B. (2011). When is a liability not a liability? Textual analysis, dictionaries, and 10-Ks. *Journal of Finance, 66*, 35–65. https://doi.org/10.1111/j.1540-6261.2010.01625.x
18. Mikolov, T., Sutskever, I., Chen, K., Corrado, G., & Dean, J. (2013). Distributed representations of words and phrases and their compositionality. In *Proceedings of the 26th International Conference on Neural Information Processing Systems* (pp. 3111–3119). http://dl.acm.org/citation.cfm?id=2999792.2999959
19. Moniz, A. (2016). *Inferring the financial materiality of corporate social responsibility news.* Working paper, SSRN 2761905. https://doi.org/10.2139/ssrn.2761905
20. Mullen, L. (2016). *textreuse: Detect Text Reuse and Document Similarity.* R Package Version 0.1.4. https://CRAN.R-project.org/package=textreuse
21. Pennington, J., Socher, R., & Manning, C. (2014). GloVe: Global vectors for word representation. In *Proceedings of the 2014 Conference on Empirical Methods in Natural Language Processing* (pp. 1532–1543). New York, NY, USA: ACM. https://doi.org/10.3115/v1/D14-1162
22. Riedl, A., & Smeets, P. (2017). Why do investors hold socially responsible mutual funds? *Journal of Finance, 72*(6), 2505–2550. https://doi.org/10.1111/jofi.12547
23. Selivanov, D., & Wang, Q. (2018). *text2vec: Modern Text Mining Framework for R.* R Package Version 0.5.1. https://CRAN.R-project.org/package=text2vec
24. Tetlock, P. C., Saar-Tsechansky, M., & Macskassy, S. (2008). More than words: Quantifying language to measure firms' fundamentals. *Journal of Finance, 63*(3), 1437–1467. https://doi.org/10.1111/j.1540-6261.2008.01362.x
25. Theil, C. K., Štajner, S., & Stuckenschmidt, H. (2018). Word embeddings–based uncertainty detection in financial disclosures, In *Proceedings of the First Workshop on Economics and Natural Language Processing* (pp. 32–37). https://doi.org/10.18653/v1/W18-3104

Extraction and Representation of Financial Entities from Text

Tim Repke and Ralf Krestel

Abstract In our modern society, almost all events, processes, and decisions in a corporation are documented by internal written communication, legal filings, or business and financial news. The valuable knowledge in such collections is not directly accessible by computers as they mostly consist of unstructured text. This chapter provides an overview of corpora commonly used in research and highlights related work and state-of-the-art approaches to extract and represent financial entities and relations.

The second part of this chapter considers applications based on knowledge graphs of automatically extracted facts. Traditional information retrieval systems typically require the user to have prior knowledge of the data. Suitable visualization techniques can overcome this requirement and enable users to explore large sets of documents. Furthermore, data mining techniques can be used to enrich or filter knowledge graphs. This information can augment source documents and guide exploration processes. Systems for document exploration are tailored to specific tasks, such as investigative work in audits or legal discovery, monitoring compliance, or providing information in a retrieval system to support decisions.

1 Introduction

Data is frequently called the oil of the twenty-first century.[1] Substantial amounts of data are produced by our modern society each day and stored in big data centers. However, the actual value is only generated through statistical analyses and data mining. Computer algorithms require numerical and structured data,

---

[1]E.g., https://www.economist.com/leaders/2017/05/06/the-worlds-most-valuable-resource-is-no-longer-oil-but-data.

T. Repke (✉) · R. Krestel
Hasso Plattner Institute, University of Potsdam, Potsdam, Germany
e-mail: tim.repke@hpi.de; ralf.krestel@hpi.de

© The Author(s) 2021
S. Consoli et al. (eds.), *Data Science for Economics and Finance*,
https://doi.org/10.1007/978-3-030-66891-4_11

such as in relational databases. Texts and other unstructured data contain a lot of information that is not readily accessible in a machine-readable way. With the help of *text mining*, computers can process large corpora of text. Modern *natural language processing* (NLP) methods can be used to extract structured data from text, such as mentions of companies and their relationships. This chapter outlines the fundamental steps necessary to construct a *knowledge graph* (KG) with all the extracted information. Furthermore, we will highlight specific state-of-the-art techniques to further enrich and utilize such a knowledge graph. We will also present text mining techniques that provide numerical representations of text for structured semantic analysis.

Many applications greatly benefit from an integrated resource for information in exploratory use cases and analytical tasks. For example, journalists investigating the Panama Papers needed to untangle and sort through vast amounts of data, search entities, and visualize found patterns hidden in the large and very heterogeneous leaked set of documents and files [10]. Similar datasets are of interest for data journalists in general or in the context of computational forensics [19, 13]. Auditing firms and law enforcement need to sift through massive amounts of data to gather evidence of criminal activity, often involving communication networks and documents [28]. Current computer-aided exploration tools,[2] offer a wide range of features from data ingestion, exploration, analysis, to visualization. This way, users can quickly navigate the underlying data based on extracted attributes, which would otherwise be infeasible due to the often large amount of heterogeneous data.

There are many ways to represent unstructured text in a machine-readable format. In general, the goal is to reduce the amount of information to provide humans an overview and enable the generation of new insights. One such representation are *knowledge graphs*. They encode facts and information by having nodes and edges connecting these nodes forming a graph.[3] In our context, we will consider nodes in the graph as named entities, such as people or companies, and edges as their relationships. This representation allows humans to explore and query the data on an abstracted level and run complex analyses. In economics and finance, this offers access to additional data sources. Whereas internally stored transactions or balance sheets at a bank only provide a limited view of the market, information hidden in news, reports, or other textual data may offer a more global perspective.

For example, the context in which data was extracted can be a valuable additional source of information that can be stored alongside the data in the knowledge graph. Topic models [8] can be applied to identify distinct groups of words that best describe the key topics in a corpus. In recent years, embeddings significantly gained popularity for a wide range of applications [64]. Embeddings represent a piece of text as a high-dimensional vector. The distance between vectors in such a vector space can be interpreted as semantic distance and reveals interesting relationships.

---

[2]E.g., extraction and indexing engine (https://www.nuix.com/), network analysis and visualization (https://linkurio.us/), or patent landscapes (https://clarivate.com/derwent/).

[3]Knowledge graphs are knowledge bases whose knowledge is organized as a graph.

This chapter focuses on the construction and application of knowledge graphs, particularly on company networks. In the first part, we describe an NLP pipeline's key steps to construct (see Sect. 2) and refine (see Sect. 3) such a knowledge graph. In the second part, we focus on applications based on knowledge graphs. We differentiate them into syntactic and semantic exploration. The syntactic exploration (see Sect. 5) considers applications that directly operate on the knowledge graph's structure and meta-data. Typical use cases assume some prior knowledge of the data and support the user by retrieving and arranging the relevant extracted information. In Sect. 6 we extend this further to the analogy of semantic maps for interactive visual exploration. Whereas syntactic applications follow a localized bottom-up approach for the interactions, semantic exploration usually enables a top-down exploration, starting from a condensed global overview of all the data.

2 Extracting Knowledge Graphs from Text

Many business insights are hidden in unstructured text. Modern NLP methods can be used to extract that information as structured data. In this section, we mainly focus on named entities and their relationships. These could be mentions of companies in news articles, credit reports, emails, or official filings. The extracted entities can be categorized and linked to a knowledge graph. Several of those are publicly accessible and cover a significant amount of relations, namely, Wikidata [77], the successor of DBpedia [34], and Freebase [9], as well as YAGO [76]. However, they are far from complete and usually general-purpose, so that specific domains or details might not be covered. Thus, it is essential to extend them automatically using company-internal documents or domain-specific texts.

The extraction of named entities is called *names entity recognition* (NER) [23] and comprises two steps: first, detecting the boundaries of the mention within the string of characters and second, classifying it into types such as ORGANIZATION, PERSON, or LOCATION. Through *named entity linking* (NEL) [70],[4] a mention is matched to its corresponding entry in the knowledge graph (if already known). An unambiguous assignment is crucial for integrating newly found information into a knowledge graph. For the scope of this chapter, we consider a fact to be a relation between entities. The most naïve approach is to use entity co-occurrence in text. *Relationship extraction* (RELEX) identifies actual connections stated in the text, either with an *open* or *closed* approach. In a closed approach, the relationships are restricted to a predefined set of relations, whereas the goal with an open approach is to extract all connections without restrictions.

Figure 1 shows a simplified example of a company network extracted from a small text excerpt. Instead of using the official legal names, quite different colloquial names, acronyms, or aliases are typically used when reporting about companies.

---

[4]Also called *entity resolution*, *entity disambiguation*, *entity matching*, or *record linkage*.

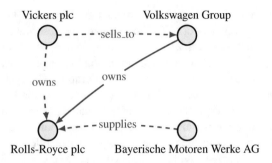

Fig. 1 Network of information extracted from the excerpt: *VW* purchased *Rolls-Royce & Bentley from Vickers on 28 July 1998. From July 1998 until December 2002, BMW continued to* supply *engines for the Rolls-Royce Silver Seraph (Excerpt from https://en.wikipedia.org/wiki/ Volkswagen_Group. Accessed on 22.02.2020).*

There are three main challenges in entity linking: 1) *name variations* as shown in the example with "VW" and "Volkswagen"; 2) *entity ambiguity*, where a mention can also refer to multiple different knowledge graph entries; and 3) *unlinkable entities* in the case that there is no corresponding entry in the knowledge graph yet. The resulting graph in Fig. 1 depicts a sample knowledge graph generated from facts extracted from the given text excerpt. Besides the explicitly mentioned entities and relations, the excerpt also contains many implied relationships; for example, a sold company is owned by someone else after the sell. Further, relationships can change over time, leading to edges that are only valid for a particular time. This information can be stored in the knowledge graph and, e.g., represented through different types of edges in the graph. Through *knowledge graph completion*, it is possible to estimate the probability whether a specific relationship between entities exists [74].

In the remainder of this section, we provide a survey of techniques and tools for each of the three steps mentioned above: NER (Sect. 2.2), NEL (Sect. 2.2), and RELEX (Sect. 2.3).

2.1 Named Entity Recognition (NER)

The first step of the pipeline for knowledge graph construction from text is to identify mentions of named entities. Named entity recognition includes several subtasks, namely, identifying proper nouns and the boundaries of named entities and classifying the entity type. The first work in this area was published in 1991 and proposed an algorithm to automatically extract company names from financial news to build a database for querying [54, 46]. The task gained interest with MUC-6, a shared task to distinguish not only types, such as person, location, organization, but also numerical mentions, such as time, currency, and percentages [23]. Traditionally, research in this area is founded in computational linguistics, where the goal is to parse and describe the natural language with statistical rule-based methods. The foundation for that is to correctly tokenize the unstructured text, assign part-of-speech tags (also known as POS tagging), and create a parse tree that describes the sentence's dependencies and overall structure. Using this information, linguists defined rules that describe typical patterns for named entities.

Handcrafted rules were soon replaced by machine learning approaches that use tags mentioned above and so-called surface features. These surface features describe syntactic characteristics, such as the number of characters, capitalization, and other derived information. The most popular supervised learning methods for the task are hidden Markov models and conditional random fields due to their ability to derive probabilistic rules from sequence data [6, 42]. However, supervised learning requires large amounts of annotated training data. Bootstrapping methods can automatically label text data using a set of entity names as a seed. These semi-supervised methods do so by marking occurrences of these seed entities in the text and using contextual information to annotate more data automatically. For an overview of related work in that area, we refer to Nadeau et al. [47]. In recent years, deep learning approaches gained popularity. They have the advantage that they do not require sophisticated pre-processing, feature engineering, or potentially error-prone POS tagging and dependency parsing. Especially recurrent neural networks are well suited since they take entire sequences of tokens or characters into account. For an overview of currently researched deep learning models for NER, we refer readers to the extensive survey by Yadav and Bethard [80].

Although the task of automatically identifying company names in text, there is still a lot of research dedicated to named entity recognition. Due to their structural heterogeneity, recognizing company names is particularly difficult compared to person or location names. Examples of actual German company names show the complexity of the task. Not only are some of the names very long ("Simon Kucher & Partner Strategy & Marketing Consultants GmbH"), they interleave abstract names with common nouns, person names, locations, and legal forms, for example: "Loni GmbH," "Klaus Traeger," and "Clean-Star GmbH & Co Autowaschanlage Leipzig KG." Whereas in English almost all capitalized proper nouns refer to named entities, it is significantly harder to find entity mentions in other languages, for example, in German, where all common nouns are capitalized [17]. Loster et al. [65, 36] dedicate a series of papers to the recognition of financial entities in text. In particular, they focus on correctly determining the full extent of a mention by using tries, which are tree structures, to improve dictionary-based approaches [39].

The wealth of publications and the availability of open-source libraries reflect the importance and popularity of NER. The following overview shows the most successful projects used in research and industry alike.

GATE ANNIE   The General Architecture for Text Engineering (GATE),[5] first released in 1995, is an extensive and mature open-source Java toolkit for many

---

[5]https://gate.ac.uk/.

aspects of natural language processing and information extraction tasks. ANNIE, A Nearly-New Information Extraction system, is the component for named entity extraction implementing a more traditional recognition model [15]. GATE provides all necessary tools to build a complete system for knowledge graph construction in combination with other components.

NLTK   The Natural Language Toolkit (NLTK),[6] first released in 2001, is one of the most popular Python libraries for natural language processing [7]. It provides a wealth of easy-to-use API for all traditional text processing tasks and named entity recognition capabilities.

OpenNLP   The Apache OpenNLP project[7] is a Java library for the most common text processing tasks and was first released in 2004 [27]. It provides implementations of a wide selection of machine learning-based NLP research designed to extend data pipelines built with Apache Flink or Spark.

CoreNLP   The Stanford NLP research group released their first version of CoreNLP[8] in 2006 as a Java library with Python bindings [52]. It is actively developed and provides NLP tools ranging from traditional rule-based approaches to models from recently published state-of-the-art deep learning research.

spaCy   Only recently in 2015, spaCy[9] was released as a Python/Cython library, focusing on providing high performance in terms of processing speed. It also includes language models for over 50 languages and a growing community of extensions. After an initial focus on processing speed, the library now includes high-quality language models based on recent advances in deep learning.

For a detailed comparison of the frameworks mentioned above, we refer readers to the recently published study by Schmitt et al. [68].

2.2   Named Entity Linking (NEL)

The problem of linking named entities is rooted in a wide range of research areas (Fig. 2). Through named entity linking, the strings discovered by NER are matched to entities in an existing knowledge graph or extend it. Wikidata is a prevalent knowledge graph for many use cases. Typically, there is no identical string match for an entity mention discovered in the text and the knowledge graph. Organizations are rarely referred to by their full legal name, but rather an acronym or colloquial variation of the full name. For example, VW could refer to Vorwerk, a manufacturer for household appliances, or Volkswagen, which is also known as Volkswagen Group

---

[6]http://nltk.org/.

[7]https://opennlp.apache.org.

[8]https://nlp.stanford.edu/software/.

[9]https://spacy.io/.

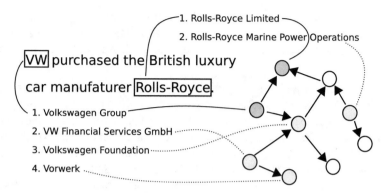

Fig. 2 Example for ranking and linking company mentions to the correct entity in a set of candidates from the knowledge graph

or Volkswagen AG. At the time of writing, there are close to 80 entries in Wikidata[10] when searching for "Volkswagen," excluding translations, car models, and other non-organization entries. Entity linking approaches use various features to match the correct real-world entity. These features are typically based on the entity mention itself or information about the context in which it appeared. Thereby, they face similar challenges and use comparable approaches as research in record linkage and duplicate detection. Shen et al. [70] provide a comprehensive overview of applications, challenges, and a survey of the main approaches. As mentioned earlier, there are three main challenges when linking named entities, namely, name variations, entity ambiguity, and unlinkable entities. In this subsection, we discuss these challenges using examples to illustrate them better. We also present common solutions to resolve them and close with an overview of entity linking systems.

Name Variations A real-world entity is referred to in many different ways, such as the full official name, abbreviations, colloquial names, various known aliases, or simply with typos. These variations increase the complexity of finding the correct match in the knowledge base. For example, Dr. Ing. h.c. F. Porsche GmbH, Ferdinand Porsche AG, and Porsche A.G. are some name variations for the German car manufacturer Porsche commonly found in business news. Entity linking approaches traditionally take two main steps [70]. The first step selects candidate entries for the currently processed mention from the knowledge base. The second step performs the actual linking by choosing the correct candidate. The candidate generation reduces the number of possible matches, as the disambiguation can become computationally expensive. The most common approach is to use fuzzy string comparisons, such as an edit distance like the Levenshtein distance or the Jaccard index for overlapping tokens. Additionally, a few rules for name expansion can generate possible abbreviations or extract potential acronyms from names.

---

[10]https://www.wikidata.org/w/index.php?search=volkswagen.

These rules should use domain-specific characteristics, for example, common legal forms (Ltd. ↦ Limited) as well as names (International Business Machines ↦ IBM). If an existing knowledge base is available, a dictionary of known aliases can be derived.

Entity Ambiguity A mentioned entity could refer to multiple entries in the knowledge graph. For example, Volkswagen could not only refer to the group of car manufacturers but also the financial services, international branches, or the local car dealership. Only the context, the company mention appears in, may help identify the correct entry, by taking keywords within the sentence (local context) or the document (global context) into account. The entity disambiguation, also called *entity ranking*, selects the correct entry among the previously generated set of candidates of possible matches from the knowledge base. This second linking step aims to estimate the likelihood of a knowledge base entry being the correct disambiguation for a given mention. These scores create a ranking of candidates. Typically, the one with the highest score is usually chosen to be the correct match. Generally, ranking models follow either a supervised or unsupervised approach. Supervised methods that use annotated data mentions are explicitly linked to entries in the knowledge base to train classifiers, ranking models, probabilistic models, or graph-based methods. When there is no annotated corpus available, data-driven unsupervised learning or information retrieval methods can be used. Shen et al. [70] further categorize both approaches into three paradigms. *Independent ranking methods* consider entity mentions individually without leveraging relations between other mentions in the same document and only focusing on the text directly surrounding it. On the other hand, *collective ranking methods* assume topical coherence for all entity mentions in one document and link all of them collectively. Lastly, *collaborative ranking methods* leverage the textual context of similar entity mentions across multiple documents to extend the available context information.

Unlinkable Entities Novel entities have no corresponding entries in the knowledge graph yet. It is important to note that NEL approaches should identify such cases and not just pick the best possible match. Unlinkable entities may be added as new entries to the knowledge graph. However, this depends on the context and its purpose. Suppose HBO_2 was found in a sentence and is supposed to be linked to a knowledge base of financial entities. If the sentence is about inorganic materials, this mention most likely refers to metaboric acid and should be dismissed, whereas in a pharmaceutical context, it might refer to the medical information systems firm HBO & Company. In that case, it should be added as a new entity and not linked to the already existing television network HBO. Entity linking systems deal with this in different ways. They commonly introduce a NIL entity, which represents a universal unlinkable entity, into the candidate set or a threshold for the likelihood score.

Other Challenges Growing size and heterogeneity of KGs are further challenges. Scalability and speed is a fundamental issue for almost all entity ranking systems. A key part to solve this challenge is a fast comparison function to generate candidates with a high recall to reduce the number of computations of similarity scores. State-of-the-art approaches that use vector representations have the advantage that nearest neighborhood searches within a vector space are almost constant [41]. However, training them requires large amounts of data, which might not be available in specific applications. Furthermore, targeted adaptations are not as trivial as with rule-based or feature-based systems. Another challenge for entity ranking systems are heterogeneous sources. Whereas multi-language requirements can be accounted for by separate models, evolving information over time imposes other difficulties. Business news or other sources continuously generate new facts that could enrich the knowledge graph further. However, with a growing knowledge graph, the characteristics of the data change. Models tuned on specific characteristics or trained on a previous state of the graph may need regular updates.

Approaches There are numerous approaches for named entity linking. Traditional approaches use textual fragments surrounding the entity mention to improve the linking quality over just using a fuzzy string match. Complex joint reasoning and ranking methods negatively influence the disambiguation performance in cases with large candidate sets. Zou et al. [83] use multiple bagged ranking classifiers to calculate a consensus decision. This way, they can operate on subsets of large candidate sets and exploit previous disambiguation decisions whenever possible. As mentioned before, not every entity mention can be linked to an entry in the knowledge graph. On the other hand, including the right entities in the candidate set is challenging due to name variations and ambiguities. Typically, there is a trade-off between the precision (also called linking correctness rate) of a system and its recall (also called linking coverage rate). For example, simply linking mentions of VW in news articles to the most popular entry in the knowledge graph is probably correct. All common aliases are well known and other companies with similar acronyms appear less frequently in the news, which leads to high precision and recall. In particular applications, this is more challenging. Financial filings often contain references to numerous subsidiaries with very similar names that need to be accurately linked. CohEEL is an efficient method that uses random walks to combine a precision-oriented and a recall-oriented classifier [25]. They achieve wide coverage while maintaining a high precision, which is of high importance for business analytics.

The research on entity linking shifted toward deep learning and embedding-based approaches in recent years. Generally, they learn high-dimensional vector representations of tokens in the text and knowledge graph entries. Zwicklbauer et al. [85] use such embeddings to calculate the similarity between an entity mention and its respective candidates from the knowledge graph. Given a set of training data in which the correct links are annotated in the text, they learn a robust similarity measure. Others use the annotated mentions in the training data as special tokens in the vocabulary and project words and entities into a common vector space [81, 21]. The core idea behind DeepType [53] is to support the linking process by providing type information about the entities from an existing knowledge graph to the disambiguation process, which they train in an end-to-end fashion. Such approaches require existing knowledge graphs and large sets of training data. Although this can be generated semi-automatically with open information extraction methods, maintaining a high quality can be challenging. Labeling high-quality training data manually is infeasible while maintaining high coverage. Active learning methods can significantly reduce the required amount of annotated data. DeepMatcher offers a ready-to-use implementation of a neural network that makes use of fully automatically learned attribute and word embeddings to train an entity similarity function with targeted human annotation [45].

2.3 Relationship Extraction (RELEX)

Relationship extraction identifies triples of two entities and their relation that appear in a text. Approaches follow one of two strategies: mining of *open-domain* triples or *fixed-domain* triples. In an open-domain setting, possible relations are not specified in advance and typically just use a keyword between two entities. Stanford's OpenIE [3] is a state-of-the-art information extraction system that splits sentences into sets of clauses. These are then shortened and segmented into triples. Figure 3 shows the relations extracted by OpenIE from the example used in Fig. 1. One such extracted triple would be (BMW, supply, Rolls-Royce).

Such a strategy is useful in cases where no training data or no ontology is available. An ontology is a schema (for a knowledge graph) that defines the types of possible relations and entities. In the following section, we provide more details on standardized ontologies and refinement. One disadvantage of open-domain extraction is that synonymous relationships lead to multiple edges in the knowledge graph. Algorithms can disambiguate the freely extracted relations after enriching the knowledge graph with data from all available text sources. In a fixed-domain setting, all possible relation types are known ahead of time. Defining a schema has the advantage that downstream applications can refer to predefined relation types. For example, in Fig. 1 we consider relations such as ORG owns ORG, which is implicitly matched by *"VW purchased Rolls-Royce."*

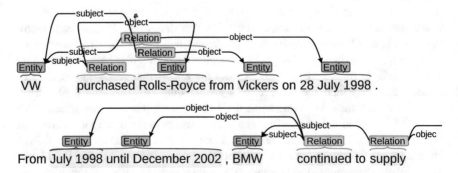

Fig. 3 Relations recognized by OpenIE in text from Fig. 1; output is visualized by CoreNLP (An online demo of CoreNLP is available at https://corenlp.run/.)

The naïve way to map relations mentioned in the text to a schema is to provide a dictionary for each relation type. An algorithm can automatically extend a dictionary from a few manually annotated sentences with relation triples or a seed dictionary. Agichtein and Gravano published the very famous *Snowball* algorithm, which follows this approach [1]. In multiple iterations, the algorithm grows the dictionary based on an initially small set of examples. This basic concept is applied in semi-supervised training to improve more advanced extraction models. The collection of seed examples can be expanded after every training iteration. This process is also called distant supervision. However, it can only detect relationship types already contained in the knowledge graph and cannot discover new relationship types. A comprehensive discussion of distant supervision techniques for relation extraction is provided by Smirnova [71]. Zuo et al. demonstrated the domain-specific challenges of extracting company relationships from text [84].

Recent approaches mostly focus on deep learning architectures to identify relations in a sequence of words. Wang et al. [78] use convolutional layers and attention mechanisms to identify the most relevant syntactic patterns for relation extraction. Others employ recurrent models to focus on text elements in sequences of variable length [33]. Early approaches commonly used conditional random fields (CRF) on parse trees, representing the grammatical structure and dependencies in a sentence. Nguyen et al. [48] combine modern neural BiLSTM architectures with CRFs for an end-to-end trained model to improve performance. Based on the assumption that if two entities are mentioned in the same text segment, Soares et al. [73] use BERT [16] to learn relationship embeddings. These embeddings are similar to dictionaries with the advantage that embedding vectors can be used to easily identify the matching relation type for ambiguous phrases in the text.

3 Refining the Knowledge Graph

In the previous section, we described the key steps in constructing a knowledge graph, namely, named entity extraction, entity linking, and relationship extraction. This process produces a set of triples from a given text corpus that forms a knowledge graph's nodes and edges. As we have shown in the previous section, compiling a duplicate-free knowledge graph is a complex and error-prone task. Thus, these triples need refinement and post-processing to ensure a high-quality knowledge graph. Any analysis based on this graph requires the contained information to be as accurate and complete as possible.

Manual refinement and standards are inevitable for high-quality results. For better interoperability, the Object Management Group, the standards consortium that defined UML and BPMN, among other things, specified the Financial Industry Business Ontology (FIBO).[11] This ontology contains standard identifiers for rela-

---

[11] https://www.omg.org/spec/EDMC-FIBO/BE/.

tionships and business entities. The Global Legal Identifier Foundation (GLEIF)[12] is an open resource that assigns unique identifiers to legal entities and contains statistics for around 1.5 million entries at the time of writing.

Using existing knowledge graphs as a reference together with standardized ontologies is a good foundation for the manual refinement process. However, the sheer size of these datasets requires support by automated mechanisms in an otherwise unattainable task. With CurEx, Loster et al. [37] demonstrate the entire pipeline of curating company networks extracted from text. They discuss the challenges of this system in the context of its application in a large financial institution [38]. Knowledge graphs about company relations are also handy beyond large-scale analyses of the general market situation. For example, changes in the network, as reported in SEC filings,[13] are of particular interest to analysts. Sorting through all mentioned relations is typically impractical. Thus, automatically identifying the most relevant reported business relationships in newly released filings can significantly support professionals in their work. Repke et al. [60] use the surrounding text, where a mentioned business relation appears, to create a ranking to enrich dynamic knowledge graphs. There are also other ways to supplement the available information about relations. For example, a company network with weighted edges can be constructed from stock market data [29]. The authors compare the correlation of normalized stock prices with relations extracted from business news in the same time frame and found that frequently co-mentioned companies oftentimes share similar patterns in the movements of their stock prices.

Another valuable resource for extending knowledge graphs are internal documents, as they contain specialized and proprietary domain knowledge. For example, the graph can also be extended beyond just company relations and include key personnel and semantic information. In the context of knowledge graph refinement, it is essential to provide high-quality and clean input data to the information extraction pipeline. The Enron Corpus [30], for example, has been the basis for a lot of research in many fields. This corpus contains over 500,000 emails from more than 150 Enron employees. The text's structure and characteristics in emails are typically significantly different from that of news, legal documents, or other reports. With Quagga,[14] we published a deep learning-based system to pre-process email text [55]. It identifies the parts of an email text that contains the actual content. It disregards additional elements, such as greetings, closing words, signatures, or automatically inserted meta-data when forwarding or replying to emails. This meta-data could extend the knowledge graph with information about who is talking to whom about what, which is relevant for internal investigations.

---

[12]https://search.gleif.org/.

[13]https://www.sec.gov/edgar.shtml.

[14]https://github.com/HPI-Information-Systems/QuaggaLib.

4 Analyzing the Knowledge Graph

Knowledge about the structure of the market is a highly valuable asset. This section focuses on specific applications in the domain of business intelligence for economics and finance. Especially financial institutions have to have a detailed overview of the entire financial market, particularly the network of organizations in which they invest. Therefore, Ronnqvist et al. [63] extracted bank networks from text to quantify interrelations, centrality, and determinants.

In Europe, banks are required by law to estimate their systemic risk. The network structure of the knowledge graph allows the investigation of many financial scenarios, such as the impact of corporate bankruptcy on other market participants within the network. In this particular scenario, the links between the individual market participants can be used to determine which companies are affected by bankruptcy and to what extent. Balance sheets and transactions alone would not suffice to calculate that risk globally, as it only provides an ego-network and thus a limited view of the market. Thus, financial institutions have to integrate their expertise in measuring the economic performance of their assets and a network of companies to simulate how the potential risk can propagate. Constantin et al. [14] use data from the financial network and market data covering daily stock prices of 171 listed European banks to predict bank distress.

News articles are a popular source of information for analyses of company relationships. Zheng and Schwenkler demonstrate that company networks extracted from news can be used to measure financial uncertainty and credit risk spreading from a distressed firm [82]. Others also found that the return of stocks reflects economic linkages derived from text [67]. We have shown that findings like this are controversial [29]. Due to the connectedness within industry sectors and the entire market, stock price correlation patterns are very common. Large companies and industry leaders heavily influence the market and appear more frequently in business news than their smaller competitors. Additionally, news are typically slower than movements on the stock market, as insiders receive information earlier through different channels. Thus, observation windows have to be in sync with the news cycle for analyses in this domain.

News and stock market data can then be used to show, for example, how the equity market volatility is influenced by newspapers [4]. Chahrour et al. [11] make similar observations and construct a model to show the relation between media coverage and demand-like fluctuations orthogonal to productivity within a sector. For models like this to work, company names have to be detected in the underlying texts and linked to the correct entity in a knowledge graph. Hoberg and Phillips extract an information network from product descriptions in 10-K statements filed with the SEC [26]. With this network, they examine how industry market structure and competitiveness change over time.

These examples show that knowledge graphs extracted from text can model existing hypotheses in economics. A well-curated knowledge graph that aggregates large amounts of data from a diverse set of sources would allow advanced analyses and market simulations.

5   Exploring the Knowledge Graph

Knowledge graphs of financial entities enable numerous downstream tasks. These include automated enterprise valuation, identifying the sentiment toward a particular company, or discovering political and company networks from textual data. However, knowledge graphs can also support the work of accountants, analysts, and investigators. They can query and explore relationships and structured knowledge quickly to gather the information they need. For example, visual analytics helps to monitor the financial stability in company networks [18]. Typically, such applications display small sub-graphs of the entire company network as a so-called node-link diagram. Circles or icons depict companies connected by straight lines. The most popular open-source tools for visualizing graphs are Cytoscape [49] and Gephi [5]. They mostly focus on visualization rather than capabilities to interactively explore the data. Commercial platforms, on the other hand, such as the NUIX Engine[15] or Linkurious,[16] offer more advanced capabilities. These include data pre-processing and analytics frequently used in forensic investigations, e.g., by journalists researching the Panama Papers leak.

There are different ways to visualize a network, most commonly by node-link diagrams as described above. However, already with a small number of nodes and edges, the readability is hard to maintain [51]. Edge bundling provides for better clarity of salient high-level structures [35]. The downside is that individual edges can become impossible to follow. Other methods for network visualization focus on the adjacency matrix of the graph. Each row and column corresponds to a node in the graph and cells are colored according to the edge weight or remain empty. The greatest challenge is to arrange the rows and columns in such a way that salient structures become visible. Sankey diagrams are useful to visualize hierarchically clustered networks to show the general flow of information or connections [12]. For more details have a look at the excellent survey of network visualization methods by Gibson et al. [22]. There is no one best type of visualization. It depends on the specific application to identify the ideal representation to explore the data.

Repke et al. developed "Beacon in the Dark" [59], a platform incorporating various modes of exploration. It includes a full system pipeline to process and integrate structured and unstructured data from email and document corpora. It also consists of an interface with coordinated multiple views to explore the data in a topic sphere (semantics), by tags that are automatically assigned, and the communication and entity network derived from meta-data and text. The system goes beyond traditional approaches by combining communication meta-data and integrating additional information using advanced text mining methods and social network analysis. The objectives are to provide a data-driven *overview* of the dataset to determine initial leads without knowing anything about the data. The system also

---

[15]NUIX Analytics extracts and indexes knowledge from unstructured data (https://www.nuix.com).

[16]Linkurious Enterprise is a graph visualization and analysis platform (https://linkurio.us/).

offers extensive filters and rankings of available information to focus on relevant aspects and finding necessary data. With each interaction, the interface components update to provide the appropriate context in which a particular information snippet appears.

6 Semantic Exploration Using Visualizations

Traditional interfaces for knowledge graphs typically only support node-to-node exploration with basic search and filtering capabilities. In the previous section, we have shown that some of them also integrate the underlying source data. However, they only utilize the meta-data, not the semantic context provided by the text in which entities appear. Furthermore, prior knowledge about the data is required to formulate the right queries as a starting point for exploration. In this section, we focus on methods to introduce semantics into the visualization of knowledge graphs. This integration enables users to explore the data more intuitively and provides better explanatory navigation.

Semantic information can be added based on the context in which entity mentions appear. The semantics could simply be represented by keywords from the text surrounding an entity. The benefit is that users not only interact with raw network data but with descriptive information. Text mining can be used to automatically enrich the knowledge about companies, for example, by assigning the respective industry or sentiment toward products or an organization itself. Such an approach is even possible without annotated data. Topic models assign distributions of topics to documents and learn which words belong to which topics. Early topic models, such as latent semantic indexing, do so by correlating semantically related terms from a collection of text documents [20]. These models iteratively update the distributions, which is computationally expensive for large sets of documents and long dictionaries. Latent semantic analysis, the foundation for most current topic models, uses co-occurrence of words [31]. Latent Dirichlet allocation jointly models topics as distributions over words and documents as distributions over topics [8]. This allows to summarize large document collections by means of topics.

Recently, text mining research has shifted toward embedding methods to project words or text segments into a high-dimensional vector space. The semantic similarity between words can be calculated based on distance measures between their vectors. Initial work in that area includes word2vec [44] and doc2vec [32]. More recent popular approaches such as BERT [16] better capture the essential parts of a text. Similar approaches can also embed graph structures. RDF2vec is an approach that generates sequences of connections in the graph leveraging local information about its sub-structures [62]. They show some applications that allow the calculation of similarities between nodes in the graph. There are also specific models to incorporate text and graph data to optimize the embedding space. Entity-centric models directly assign vectors to entities instead of just tokens. Traditionally, an embedding model assumes a fixed dictionary of known words or character ngrams that form these words. By annotating the text with named entity information before training the model, unique multi-word entries in the dictionary directly relate to known entities. Almasian et al. propose such a model for entity-annotated texts [2]. Other interesting approaches build networks of co-occurring words and entities. TopExNet uses temporal filtering to produce entity-centric networks for topic exploration in news streams [75]. For a survey of approaches and applications of knowledge graph embeddings, we refer the readers to [79].

Topic models, document embeddings, and entity embeddings are useful tools for systematic data analysis. However, on their own, they are not directly useable. In the context of book recommendations, embeddings have been used to find similar books using combinations of embeddings for time and place of the plot [61]. Similar approaches could be applied in the domain of financial entities, for example, to discover corresponding companies in a different country. In use cases without prior knowledge, it might be particularly helpful to get an overview of all the data. Also, for monitoring purposes, a bird's-eye view of the entire dataset can be beneficial. The most intuitive way is to organize the information in the form of an interactive map. Sarlin et al. [66] used self-organizing maps to arrange economic sectors and countries to create maps. Coloring the maps enables them to visually compare different financial stability metrics across multiple time frames around periods with high inflation rates or an economic crisis.

The idea of semantic landscapes is also popular in the area of patent research. The commercial software ThemeScape by Derwent[17] produces landscapes of patents that users can navigate similar to a geographical map. Along with other tools, they enable experts to find related patents or identify new opportunities quickly. Smith et al. built a system to transform token co-occurrence information in texts to semantic patterns. Using statistical algorithms, they generate maps of words that can be used for content analysis in knowledge discovery tasks [72]. Inspired by that, the New York Public Library made a map of parts of their catalog.[18] Therefore, they use a force-based network layout algorithm to position the information. It uses the analogy of forces that attract nodes to one another when connected through an edge or otherwise repel them. The network they use is derived from co-occurring subject headings and terms, which were manually assigned tags to organize their catalog. Sen et al. created a map of parts of the Wikipedia in their Cartograph project [69]. This map, as shown in Fig. 4, uses embedded pages about companies and dimensionality reduction to project the information on a two-dimensional canvas [40, 50]. Structured meta-data about pages is used to compute borders between "countries" representing different industry sectors. Maps like this provide an intuitive alternative interface for users to discover related companies. Most recently, the Open Syllabus Project[19] released their interactive explorer. Like Cartograph, this enables users to navigate through parts of the six million syllabi

---

[17]https://clarivate.com/derwent.

[18]https://www.nypl.org/blog/2014/07/31/networked-catalog.

[19]Open Syllabus Explorer visualization shows the 164,720 texts (http://galaxy.opensyllabus.org/).

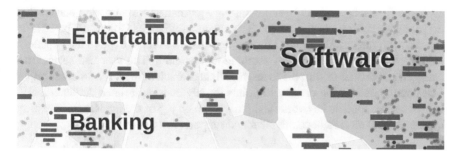

Fig. 4 Screenshot of part of the Cartograph map of organizations and their sectors

collected by the project. To do so, they first create a citation network of all publications contained in the visualization. Using this network, they learn a node embedding [24] and reduce the number of dimensions for rendering [43].

The approaches presented above offer promising applications in business analytics and exploring semantically infused company networks. However, even though the algorithms use networks to some extent, they effectively only visualize text and rely on manually tagged data. Wikipedia, library catalogs, and the syllabi corpus are datasets that are developed over many years by many contributors who organize the information into structured ontologies. In business applications, data might not always have this additional information available, and it is too labor-intensive to curate the data manually. Furthermore, when it comes to analyzing company networks extracted from text, the data is comprised of both the company network and data provenance information. The methods presented above only visualize either the content data or the graph structure. In data exploration scenarios, the goal of getting a full overview of the dataset at hand is insurmountable with current tools. We provide a solution that incorporates both, the text sources *and* the entity network, into exploratory landscapes [56]. We first embed the text data and then use multiple objectives to optimize for a good network layout and semantically correct the layout of source documents during the dimensionality reduction [58]. Figure 5 shows a small demonstration of the resulting semantic-infused network

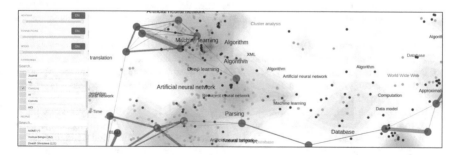

Fig. 5 Screenshot of the *MODiR* interface prototype showing an excerpt of a citation network

layout [57]. Users exploring such data, e.g., journalists investigating leaked data or young scientists starting research in an unfamiliar field, need to be able to interact with the visualization. Our prototype allows users to explore the generated landscape as a digital map with zooming and panning. The user can select from categories or entities to shift the focus, highlight characterizing keywords, and adjust a heatmap based on the density of points to only consider related documents. We extract region-specific keywords and place them on top of the landscape. This way, the meaning of an area becomes clear and supports fast navigation.

7 Conclusion

In this chapter, we provided an overview of methods to automatically construct a knowledge graph from text, particularly a network of financial entities. We described the pipeline starting from named entity recognition, over linking and matching those entities to real-world entities, to extracting the relationships between them from text. We emphasized the need to curate the extracted information, which typically contains errors that could negatively impact its usability in subsequent applications. There are numerous use cases that require knowledge graphs connecting economic, financial, and business-related information. We have shown how these knowledge graphs are constructed from heterogeneous textual documents and how they can be explored and visualized to support investigations, analyses, and decision making.

References

1. Agichtein, E., & Gravano, L. (2000). *Snowball*: Extracting relations from large plain-text collections. In *Proceedings of the Joint Conference on Digital Libraries (JCDL)* (pp. 85–94). New York, NY, USA: ACM Press.
2. Almasian, S., Spitz, A., & Gertz, M. (2019). Word embeddings for entity-annotated texts. In *Proceedings of the European Conference on Information Retrieval (ECIR)*. Lecture Notes in Computer Science (vol. 11437, pp. 307–322). Berlin: Springer.
3. Angeli, G., Premkumar, M. J. J., & Manning, C. D. (2015). Leveraging linguistic structure for open domain information extraction. In *Proceedings of the Annual Meeting of the Association for Computational Linguistics (ACL)* (pp. 344–354). Stroudsburg, PA, USA: Association for Computational Linguistics.
4. Baker, S. R., Bloom, N., Davis, S. J., & Kost, K. J. (2019). *Policy news and stock market volatility*. Working Paper 25720, National Bureau of Economic Research.
5. Bastian, M., Heymann, S., Jacomy, M. (2009). Gephi: An open source software for exploring and manipulating networks. In *Proceedings of the International Semantic Web Conference (ISWC)*. Palo Alto, CA, USA: The AAAI Press.
6. Bikel, D. M., Miller, S., Schwartz, R. M., & Weischedel, R. M. (1997). Nymble: A high-performance learning name-finder. In *Applied Natural Language Processing Conference (ANLP)* (pp. 194–201). Stroudsburg, PA, USA: Association for Computational Linguistics.
7. Bird, S., Klein, E., & Loper, E. (2009). *Natural language processing with Python*. Sebastopol, CA, USA: O'Reilly.

8. Blei, D. M., Ng, A. Y., & Jordan, M. I. (2003). Latent Dirichlet allocation. *Journal of Machine Learning Research 3*(Jan), 993–1022.
9. Bollacker, K. D., Evans, C., Paritosh, P., Sturge, T., & Taylor, J. (2008). Freebase: A collaboratively created graph database for structuring human knowledge. In *Proceedings of the ACM Conference on Management of Data (SIGMOD)* (pp. 1247–1250).
10. Chabin, M. A. (2017). Panama papers: A case study for records management? *Brazilian Journal of Information Science: Research Trends, 11*(4), 10–13.
11. Chahrour, R., Nimark, K., & Pitschner, S. (2019). *Sectoral media focus and aggregate fluctuations.* Swedish House of Finance Research Paper Series 19–12, SSRN.
12. Chang, C., Bach, B., Dwyer, T., & Marriott, K. (2017). Evaluating perceptually complementary views for network exploration tasks. In *Proceedings of the SIGCHI conference on Human Factors in Computing Systems (CHI)* (pp. 1397–1407). New York, Ny, USA: ACM Press.
13. Coddington, M. (2015). Clarifying journalism's quantitative turn: A typology for evaluating data journalism, computational journalism, and computer-assisted reporting. *Digital Journalism, 3*(3), 331–348.
14. Constantin, A., Peltonen, T. A., & Sarlin, P. (2018). Network linkages to predict bank distress. *Journal of Financial Stability, 35*, 226–241.
15. Cunningham, H., Maynard, D., Bontcheva, K., & Tablan, V. (2002). GATE: A framework and graphical development environment for robust NLP tools and applications. In *Proceedings of the Annual Meeting of the Association for Computational Linguistics (ACL)* (pp. 168–175). Stroudsburg, PA, USA: Association for Computational Linguistics.
16. Devlin, J., Chang, M., Lee, K., & Toutanova, K. (2019). BERT: Pre-training of deep bidirectional transformers for language understanding. In *Proceedings of the Conference of the North American Chapter of the Association for Computational Linguistics (NAACL-HLT)* (pp. 4171–4186). Stroudsburg, PA, USA: Association for Computational Linguistics.
17. Faruqui, M., & Padó, S. (2010). Training and evaluating a German named entity recognizer with semantic generalization. In *Proceedings of the Conference on Natural Language Processing (KONVENS)* (pp. 129–133).
18. Flood, M. D., Lemieux, V. L., Varga, M., & Wong, B. W. (2016). The application of visual analytics to financial stability monitoring. *Journal of Financial Stability, 27*, 180–197.
19. Franke, K., & Srihari, S.N. (2007). Computational forensics: Towards hybrid-intelligent crime investigation. In *Proceedings of the International Symposium on Information Assurance and Security (IAS)* (pp. 383–386). New York City, NY, USA: IEEE.
20. Furnas, G. W., Deerwester, S. C., Dumais, S. T., Landauer, T. K., Harshman, R. A., Streeter, L. A., et al. (1988). Information retrieval using a singular value decomposition model of latent semantic structure. In *Proceedings of the ACM Conference on Information Retrieval (SIGIR)* (pp. 465–480). New York City, NY, USA: ACM Press.
21. Ganea, O., & Hofmann, T. (2017). Deep joint entity disambiguation with local neural attention. In *Proceedings of the Conference on Empirical Methods in Natural Language Processing (EMNLP)* (pp. 2619–2629). Stroudsburg, PA, USA: Association for Computational Linguistics.
22. Gibson, H., Faith, J., & Vickers, P. (2013). A survey of two-dimensional graph layout techniques for information visualisation. *Information Visualization, 12*(3–4), 324–357.
23. Grishman, R., & Sundheim, B. (1996). Message understanding conference- 6: A brief history. In *Proceedings of the International Conference on Computational Linguistics (COLING)* (pp. 466–471).
24. Grover, A., & Leskovec, J. (2016). node2vec: Scalable feature learning for networks. In *Proceedings of the ACM SIGKDD Conference on Knowledge Discovery and Data Mining (KDD)* (pp. 855–864). New York City, NY, USA: ACM Press.
25. Grütze, T., Kasneci, G., Zuo, Z., & Naumann, F. (2016). CohEEL: Coherent and efficient named entity linking through random walks. *Journal of Web Semantics, 37–38*, 75–89.
26. Hoberg, G., & Phillips, G. (2016). Text-based network industries and endogenous product differentiation. *Journal of Political Economy, 124*(5), 1423–1465.

27. Ingersoll, G., Morton, T., & Farris, A. (2012). *Taming text*. Shelter Island, NY, USA: Manning Publications.
28. Karthik, M., Marikkannan, M., & Kannan, A. (2008). An intelligent system for semantic information retrieval information from textual web documents. In *International Workshop on Computational Forensics (IWCF)* (pp. 135–146). Heidelberg: Springer.
29. Kellermeier, T., Repke, T., & Krestel, R. (2019). Mining business relationships from stocks and news. In V. Bitetta, I. Bordino, A. Ferretti, F. Gullo, S. Pascolutti, & G. Ponti (Eds.), *Proceedings of MIDAS 2019*, Lecture Notes in Computer Science (vol. 11985, pp. 70–84). Heidelberg: Springer.
30. Klimt, B., & Yang, Y. (2004). The Enron corpus: A new dataset for email classification research. In *Proceedings of the European Conference on Machine Learning (ECML)* (pp. 217–226). Heidelberg: Springer.
31. Landauer, T. K., Foltz, P. W., & Laham, D. (1998). An introduction to latent semantic analysis. *Discourse Processes, 25*(2–3), 259–284.
32. Le, Q., & Mikolov, T. (2014). Distributed representations of sentences and documents. In *Proceedings of the International Conference on Machine Learning (ICML)* (pp. 1188–1196). Brookline, Ma, USA: JMLR Inc. and Microtome Publishing.
33. Lee, J., Seo, S., & Choi, Y. S. (2019). Semantic relation classification via bidirectional LSTM networks with entity-aware attention using latent entity typing. *Symmetry, 11*(6), 785.
34. Lehmann, J., Isele, R., Jakob, M., Jentzsch, A., Kontokostas, D., Mendes, P.N., et al. (2015). DBpedia - A large-scale, multilingual knowledge base extracted from wikipedia. *Semantic Web, 6*(2), 167–195.
35. Lhuillier, A., Hurter, C., & Telea, A. (2017). State of the art in edge and trail bundling techniques. *Computer Graphics Forum, 36*(3), 619–645.
36. Loster, M., Hegner, M., Naumann, F., & Leser, U. (2018). Dissecting company names using sequence labeling. In *Proceedings of the Conference "Lernen, Wissen, Daten, Analysen" (LWDA). CEUR Workshop Proceedings* (vol. 2191, pp. 227–238). CEUR-WS.org.
37. Loster, M., Naumann, F., Ehmueller, J., & Feldmann, B. (2018). Curex: A system for extracting, curating, and exploring domain-specific knowledge graphs from text. In *Proceedings of the International Conference on Information and Knowledge Management (CIKM)* (pp. 1883–1886). New York City, NY, USA: ACM Press.
38. Loster, M., Repke, T., Krestel, R., Naumann, F., Ehmueller, J., Feldmann, B., et al. (2018). The challenges of creating, maintaining and exploring graphs of financial entities. In *Proceedings of the International Workshop on Data Science for Macro-Modeling with Financial and Economic Datasets (DSMM@SIGMOD)* (pp. 6:1–6:2). New York City, NY, USA: ACM Press.
39. Loster, M., Zuo, Z., Naumann, F., Maspfuhl, O., & Thomas, D. (2017). Improving company recognition from unstructured text by using dictionaries. In *Proceedings of the International Conference on Extending Database Technology (EDBT)* (pp. 610–619). OpenProceedings.org.
40. Maaten, L. V. D., & Hinton, G. (2008). Visualizing data using t-SNE. *Journal of Machine Learning Research (JMLR), 9*, 2579–2605.
41. Malkov, Y. A., & Yashunin, D. A. (2020). Efficient and Robust Approximate Nearest Neighbor Search Using Hierarchical Navigable Small World Graphs. *IEEE Transactions on Pattern Analysis and Machine Intelligence, 42*(4), Art. No. 8594636, 824-836.
42. McCallum, A., & Li, W. (2003). Early results for named entity recognition with conditional random fields, feature induction and web-enhanced lexicons. In *Proceedings of the Conference on Computational Natural Language Learning (CoNLL)* (pp. 188–191). Stroudsburg, PA, USA: Association for Computational Linguistics.
43. McInnes, L., & Healy, J. (2018). UMAP: Uniform manifold approximation and projection for dimension reduction. *CoRR, abs/1802.03426*.
44. Mikolov, T., Sutskever, I., Chen, K., Corrado, G. S., & Dean, J. (2013). Distributed representations of words and phrases and their compositionality. In *Proceedings of the Conference on Neural Information Processing Systems (NIPS)* (pp. 3111–3119). San Diego, CA, USA: NIPS Foundation.

45. Mudgal, S., Li, H., Rekatsinas, T., Doan, A., Park, Y., Krishnan, G., et al. (2018). Deep learning for entity matching: A design space exploration. In *Proceedings of the ACM Conference on Management of Data (SIGMOD)* (pp. 19–34). New York City, NY, USA: ACM Press.
46. Nadeau, D., & Sekine, S. (2007). A survey of named entity recognition and classification. *Lingvisticae Investigationes, 30*(1), 3–26.
47. Nadeau, D., Turney, P. D., & Matwin, S. (2006). Unsupervised named-entity recognition: Generating gazetteers and resolving ambiguity. In L. Lamontagne & M. Marchand (Eds.), *Proceedings of the Conference of the Canadian Society for Computational Studies of Intelligence.* Lecture Notes in Computer Science (vol. 4013, pp. 266–277). Berlin: Springer.
48. Nguyen, D. Q., & Verspoor, K. (2019). End-to-end neural relation extraction using deep biaffine attention. In *Proceedings of the European Conference on Information Retrieval (ECIR).* Lecture Notes in Computer Science (vol. 11437, pp. 729–738). Berlin: Springer.
49. Otasek, D., Morris, J. H., Bouças, J., Pico, A. R., & Demchak, B. (2019). Cytoscape automation: Empowering workflow-based network analysis. *Genome Biology, 20*(1), 1–15.
50. Pezzotti, N., Lelieveldt, B. P., van der Maaten, L., Höllt, T., Eisemann, E., & Vilanova, A. (2017). Approximated and user steerable t-SNE for progressive visual analytics. *IEEE Transactions on Visualization and Computer Graphics (TVCG), 23*(7), 1739–1752.
51. Pohl, M., Schmitt, M., & Diehl, S. (2009). Comparing the readability of graph layouts using eyetracking and task-oriented analysis. In *Computational Aesthetics 2009: Eurographics Workshop on Computational Aesthetics,* Victoria, British Columbia, Canada, 2009 (pp. 49–56).
52. Qi, P., Dozat, T., Zhang, Y., & Manning, C. D. (2018). Universal dependency parsing from scratch. In *Proceedings of the CoNLL 2018 Shared Task: Multilingual Parsing from Raw Text to Universal Dependencies* (pp. 160–170). Stroudsburg, PA, USA: Association for Computational Linguistics.
53. Raiman, J., & Raiman, O. (2018). DeepType: Multilingual entity linking by neural type system evolution. In *Proceedings of the National Conference on Artificial Intelligence (AAAI)* (pp. 5406–5413). Palo Alto, CA, USA: AAAI Press.
54. Rau, L. F. (1991). Extracting company names from text. In *Proceedings of the IEEE Conference on Artificial Intelligence Application* (vol. 1, pp. 29–32). Piscataway: IEEE.
55. Repke, T., & Krestel, R. (2018). Bringing back structure to free text email conversations with recurrent neural networks. In *Proceedings of the European Conference on Information Retrieval (ECIR)* (pp. 114–126). Heidelberg: Springer.
56. Repke, T., & Krestel, R. (2018). Topic-aware network visualisation to explore large email corpora. In *International Workshop on Big Data Visual Exploration and Analytics (BigVis), Proceedings of the International Conference on Extending Database Technology (EDBT)* (pp. 104–107). CEUR-WS.org.
57. Repke, T., & Krestel, R. (2020). Exploration interface for jointly visualised text and graph data. In *Proceedings of the International Conference on Intelligent User Interfaces (IUI)* (pp. 73–74). Geneva: ACM Press.
58. Repke, T., & Krestel, R. (2020). Visualising large document collections by jointly modeling text and network structure. In *Proceedings of the Joint Conference on Digital Libraries (JCDL)* (pp. 279–288). Geneva: ACM Press.
59. Repke, T., Krestel, R., Edding, J., Hartmann, M., Hering, J., Kipping, D., et al. (2018). Beacon in the dark: A system for interactive exploration of large email corpora. In *Proceedings of the International Conference on Information and Knowledge Management (CIKM)* (pp. 1871–1874). New York, NY, USA: ACM Press.
60. Repke, T., Loster, M., & Krestel, R. (2017). Comparing features for ranking relationships between financial entities based on text. In *Proceedings of the International Workshop on Data Science for Macro-Modeling with Financial and Economic Datasets (DSMM@SIGMOD)* (pp. 12:1–12:2). New York, NY, USA: ACM Press.
61. Risch, J., Garda, S., & Krestel, R. (2018). Book recommendation beyond the usual suspects - embedding book plots together with place and time information. In *Proceedings of the International Conference on Asia-Pacific Digital Libraries (ICADL).* Lecture Notes in Computer Science (vol. 11279, pp. 227–239). Berlin: Springer.

62. Ristoski, P., Rosati, J., Noia, T. D., Leone, R. D., & Paulheim, H. (2019). RDF2Vec: RDF graph embeddings and their applications. *Semantic Web, 10*(4), 721–752.
63. Rönnqvist, S., & Sarlin, P. (2015). Bank networks from text: interrelations, centrality and determinants. *Quantitative Finance, 15*(10), 1619–1635.
64. Ruder, S., Vulic, I., & Søgaard, A. (2019). A survey of cross-lingual word embedding models. *Journal of Artificial Intelligence Research (JAIR), 65,* 569–631.
65. Samiei, A., Koumarelas, I., Loster, M., & Naumann, F. (2016). Combination of rule-based and textual similarity approaches to match financial entities. In *Proceedings of the International Workshop on Data Science for Macro-Modeling, (DSMM@SIGMOD)* (pp. 4:1–4:2). New York, NY, USA: ACM Press.
66. Sarlin, P. (2013). Exploiting the self-organizing financial stability map. *Engineering Applications of Artificial Intelligence, 26*(5–6), 1532–1539.
67. Scherbina, A., & Schlusche, B. (2015). *Economic linkages inferred from news stories and the predictability of stock returns.* AEI Economics Working Papers 873600, American Enterprise Institute. https://ideas.repec.org/p/aei/rpaper/873600.html
68. Schmitt, X., Kubler, S., Robert, J., Papadakis, M., & Traon, Y.L. (2019). A replicable comparison study of NER software: StanfordNLP, NLTK, OpenNLP, SpaCy, Gate. In *International Conference on Social Networks Analysis, Management and Security (SNAMS)* (pp. 338–343). Piscataway: IEEE.
69. Sen, S., Swoap, A. B., Li, Q., Boatman, B., Dippenaar, I., Gold, R., et al. (2017). Cartograph: Unlocking spatial visualization through semantic enhancement. In *Proceedings of the International Conference on Intelligent User Interfaces (IUI)* (pp. 179–190). Geneva: ACM Press.
70. Shen, W., Wang, J., & Han, J. (2015). Entity linking with a knowledge base: Issues, techniques, and solutions. *IEEE Transactions on Knowledge and Data Engineering, 27*(2), 443–460.
71. Smirnova, A., & Cudré-Mauroux, P. (2019). Relation extraction using distant supervision: A survey. *ACM Computing Surveys, 51*(5), 106:1–106:35.
72. Smith, A. E., & Humphreys, M. S. (2006). Evaluation of unsupervised semantic mapping of natural language with leximancer concept mapping. *Behavior Research Methods, 38*(2), 262–279.
73. Soares, L. B., FitzGerald, N., Ling, J., & Kwiatkowski, T. (2019). Matching the blanks: Distributional similarity for relation learning. In *Proceedings of the Annual Meeting of the Association for Computational Linguistics (ACL)* (pp. 2895–2905). Stroudsburg, PA, USA: Association for Computational Linguistics.
74. Socher, R., Chen, D., Manning, C. D., & Ng, A. Y. (2013). Reasoning with neural tensor networks for knowledge base completion. In *Proceedings of the Conference on Neural Information Processing Systems (NIPS)* (pp. 926–934).
75. Spitz, A., Almasian, S., & Gertz, M. (2019). TopExNet: Entity-centric network topic exploration in news streams. In *Proceedings of the International Conference on Web Search and Data Mining (WSDM)* (pp. 798–801). New York, NY, USA: ACM Press.
76. Suchanek, F. M., Kasneci, G., & Weikum, G. (2007). YAGO: A core of semantic knowledge. In *Proceedings of the International World Wide Web Conference (WWW)* (pp. 697–706).
77. Vrandečić, D., & Krötzsch, M. (2014). Wikidata: A free collaborative knowledgebase. *Communications of the ACM, 57*(10), 78–85.
78. Wang, L., Cao, Z., de Melo, G., & Liu, Z. (2016). Relation classification via multi-level attention CNNs. In *Proceedings of the Annual Meeting of the Association for Computational Linguistics (ACL)* (pp. 1298–1307). New York, NY, USA: ACM Press.
79. Wang, Q., Mao, Z., Wang, B., & Guo, L. (2017). Knowledge graph embedding: A survey of approaches and applications. *IEEE Transactions on Knowledge and Data Engineering, 29*(12), 2724–2743.
80. Yadav, V., & Bethard, S. (2018). A survey on recent advances in named entity recognition from deep learning models. In *Proceedings of the International Conference on Computational Linguistics (COLING)* (pp. 2145–2158). Stroudsburg, PA, USA: Association for Computational Linguistics.

81. Yamada, I., Shindo, H., Takeda, H., & Takefuji, Y. (2016). Joint learning of the embedding of words and entities for named entity disambiguation. In *Proceedings of the Conference on Computational Natural Language Learning (CoNLL)* (pp. 250–259). Stroudsburg, PA, USA: Association for Computational Linguistics.

82. Zheng, H., & Schwenkler, G. (2020). *The network of firms implied by the news.* ESRB Working Paper Series 108, European Systemic Risk Board.

83. Zuo, Z., Kasneci, G., Grütze, T., & Naumann, F. (2014). BEL: Bagging for entity linking. In J. Hajic & J. Tsujii (Eds.), *Proceedings of the International Conference on Computational Linguistics (COLING)* (pp. 2075–2086). Stroudsburg, PA, USA: Association for Computational Linguistics.

84. Zuo, Z., Loster, M., Krestel, R., & Naumann, F. (2017). Uncovering business relationships: Context-sensitive relationship extraction for difficult relationship types. In *Lernen, Wissen, Daten, Analysen (LWDA) Conference Proceedings, CEUR Workshop Proceedings* (vol. 1917, p. 271). CEUR-WS.org.

85. Zwicklbauer, S., Seifert, C., & Granitzer, M. (2016). Robust and collective entity disambiguation through semantic embeddings. In *Proceedings of the ACM Conference on Information Retrieval (SIGIR)* (pp. 425–434). New York, NY, USA: ACM Press.

Quantifying News Narratives to Predict Movements in Market Risk

Thomas Dierckx, Jesse Davis, and Wim Schoutens

Abstract The theory of *Narrative Economics* suggests that narratives present in media influence market participants and drive economic events. In this chapter, we investigate how financial news narratives relate to movements in the CBOE Volatility Index. To this end, we first introduce an uncharted dataset where news articles are described by a set of financial keywords. We then perform topic modeling to extract news themes, comparing the canonical latent Dirichlet analysis to a technique combining *doc2vec* and Gaussian mixture models. Finally, using the state-of-the-art XGBoost (Extreme Gradient Boosted Trees) machine learning algorithm, we show that the obtained news features outperform a simple baseline when predicting CBOE Volatility Index movements on different time horizons.

1 Introduction

Nowadays market participants must cope with new sources of information that yield large amounts of unstructured data on a daily basis. These include sources such as online new articles and social media. Typically, this kind of information comes in the form of text catered to human consumption. However, humans struggle to identify relevant complex patterns that are hidden in enormous collections of data. Therefore, investors, regulators, and institutions would benefit from more sophisticated automated approaches that are able to extract meaningful insights from such information. This need has become increasingly relevant since the inception of

T. Dierckx
Department of Statistics, KU Leuven, Leuven, Belgium
e-mail: thomas.dierckx@kuleuven.be

J. Davis
Department of Computer Science, KU Leuven, Leuven, Belgium
e-mail: jesse.davis@kuleuven.be

W. Schoutens (✉)
Department of Mathematics, KU Leuven, Leuven, Belgium
e-mail: wim.schoutens@kuleuven.be

© The Author(s) 2021
S. Consoli et al. (eds.), *Data Science for Economics and Finance*,
https://doi.org/10.1007/978-3-030-66891-4_12

Narrative Economics [23]. This theory proposes that the presence of narratives in media influence the belief systems of market participants and even directly affect future economic performance. Consequently, it would be useful to apply advanced data science techniques to discern possible narratives in these information sources and assess how they influence the market.

Currently, two distinct paradigms exist that show potential for this task. First, *topic modeling* algorithms analyze the text corpora in order to automatically discover hidden themes, or topics, present in the data. At a high level, topic models identify a set of topics in a document collection by exploiting the statistical properties of language to group together similar words. They then describe a document by assessing the mixture of topics present in the document. That is, they determine the proportion of each topic present in the given document. Second, *Text Embedding* techniques infer vector representations for the semantic meaning of text. While extremely popular in artificial intelligence, their use is less prevalent in economics. One potential reason is that topic models tend to produce human-interpretable models as they associate probabilities with (groups of) words. In contrast, humans have more difficulties capturing the meaning of the vectors of real values produced by embedding methods.

In the context of narratives, preceding work in the domain of *topic modeling* has already shown that certain latent themes extracted from press releases and news articles can be predictive for future abnormal stock returns [10, 9] and volatility [3]. Similarly, researchers have explored this using *Text Embedding* on news articles to predict bankruptcy [16] and abnormal returns [25, 1].

The contribution of this chapter is multifaceted. First, we noticed that most research involving *topic modeling* is constrained by the intricate nature of natural language. Aspects such as rich vocabularies, ambiguous phrasing, and complex morphological and syntactical structures make it difficult to capture information present in a text article. Consequently, various imperfect preprocessing steps such as stopword removal, stemming, and phrase detection have to be utilized. This study therefore refrains from applying quantification techniques on raw news articles. Instead, we introduce an unprecedented corpus of historical news metadata using the *Financial Times* news API, where each news article is represented by the set of financial sub-topics it covers. Second, at the time of writing, this study offers the first attempt to investigate the interplay between narratives and implied volatility. We hypothesize that the presence of financial news narratives can instill fear in market participants, altering their perception of market risk and consequently causing movements in the CBOE Volatility Index, also known as the *fear index*. In order to test this hypothesis, we first extract latent themes from the news corpus using two different topic modeling approaches. We employ the canonical latent Dirichlet analysis but also an alternative methodology using the modern *doc2vec* and Gaussian mixture models. Finally, using the state-of-the-art XGBoost (Extreme Gradient Boosted Trees) machine learning algorithm, we model the interplay between the obtained news features and the CBOE Volatility Index. We show that we can predict movements for different time horizons, providing empirical evidence for the validity of our hypothesis.

The remainder of this chapter is structured as follows: Section 2 outlines the preliminary material necessary to understand the applied methodology in our study, which in turn is detailed in Sect. 3. Section 4 then presents the experimental results together with a discussion, and finally Sect. 5 offers a conclusion for our conducted research.

2 Preliminaries

Our approach for extracting news narratives from our news dataset builds on several techniques, and this section provides the necessary background to understand our methodology. Section 2.1 describes existing topic modeling methodologies. Section 2.2 presents the Gradient Boosted Trees machine learning model. Lastly, Sect. 2.3 defines the notion of market risk and its relation to the CBOE Volatility Index.

2.1 Topic Modeling

Topic models are machine learning algorithms that are able to discover and extract latent themes, or *topics*, from large and otherwise unstructured collections of documents. The algorithms exploit statistical relationships among words in documents in order to group them into topics. In turn, the obtained topic models can be used to automatically categorize or summarize documents up to scale that would be unfeasible to do manually.

This study considers two different approaches of topic modeling. Section 2.1.1 details the popular latent Dirichlet analysis (LDA). Sections 2.1.2 and 2.1.3 describe the paragraph vector technique and Gaussian mixture models, respectively. Note that only the former is an actual topic modeling algorithm. However, the Methodology section (Sect. 3) will introduce a topic modeling procedure by combining paragraph vector and Gaussian mixture models.

2.1.1 Latent Dirichlet Analysis

Latent Dirichlet analysis (LDA) [4] belongs to the family of generative probabilistic processes. It defines topics to be random distributions over the finite vocabulary present in a corpus. The method hinges on the assumption that every document exhibits a random mixture of such topics and that the entire corpus was generated by the following imaginary two-step process:

1. For every document d in corpus D, there's a random distribution θ_d over K topics where each entry $\theta_{d,k}$ represents the proportion of topic k in document d.

2. For each word w in document d, draw a topic z from θ_d and sample a term from its distribution over a fixed vocabulary given by β_z.

The goal of any topic modeling is to automatically discover hidden topic structures in the corpus. To this end, LDA inverts the previously outlined imaginary generative process and attempts to find the hidden topic structure that *likely* produced the given collection of documents. Mathematically, the following posterior distribution is to be inferred:

$$P(\beta_{1:K}, \theta_{1:D}, z_{1:D} \mid w_{1:D}) = \frac{P(\beta_{1:K}, \theta_{1:D}, z_{1:D}, w_{1:D})}{p(w_{1:D})}. \tag{1}$$

Unfortunately, Eq. 1 is generally deemed computationally intractable. Indeed, the denominator denotes the probability of seeing the observed corpus under any possible topic model. Since the number of possible topic models is exponentially large, it is computational intractable to compute this probability [4]. Consequently, practical implementations resort to approximate inference techniques such as online variational Bayes algorithms [13].

The inference process is mainly governed by the hyper-parameters K and Dirichlet priors α and η. The parameter K indicates the number of latent topics to be extracted from the corpus. The priors control the document-topic distribution θ and topic-word distribution β, respectively. Choosing the right values for these hyper-parameters poses intricate challenges due to the unsupervised nature of the training process. Indeed, there is no prior knowledge as to how many and what kind of hidden topic structures reside within a corpus. Most research assesses model quality based on manual and subjective inspection (e.g., [3, 9, 10]). They examine the most probable terms per inferred topic and subsequently gauge them for human interpretability. Because this is a very time-intensive procedure and requires domain expertise, an alternative approach is to use quantitative evaluation metrics. For instance, the popular perplexity metric [26] gauges the predictive likelihood of held-out data given the learned topic model. However, the metric has been shown to be negatively correlated with human interpretable topics [6]. Newer and better measures have been proposed in the domain of topic coherence. Here, topic quality is based on the idea that a topic is coherent if all or most of its words are related [2]. While multiple measures have been proposed to quantify this concept, the coherence method named C_v has been shown to achieve the highest correlation with human interpretability of the topics [20].

2.1.2 Paragraph Vector

Paragraph vector [15], commonly known as *doc2vec*, is an unsupervised framework that learns vector representations for semantics contained in chunks of text such as sentences, paragraphs, and documents. It is a simple extension to the popular

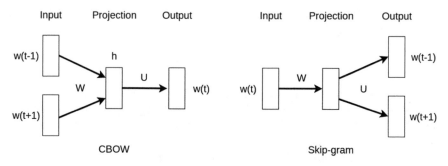

Fig. 1 The two *word2vec* approaches CBOW (left) and skip-gram (right) and their neural network architectures [17] for word predictions. The variables W and U represent matrices that respectively contain the input and output layer weights of the neural network. Function h is an aggregation function for the CBOW method to combine the multiple of input words w

word2vec model [17], which is a canonical approach for learning vector representations for individual words.

Word2vec builds on the distributional hypothesis in linguistics, which states that words occurring in the same context carry similar meaning [12]. There are two canonical approaches for learning a vector representation of a word: *continuous bag of words* (CBOW) and *skip-gram*. Both methods employ a shallow neural network but differ in input and output. CBOW attempts to predict which word is missing given its context, i.e., the surrounding words. In contrast, the *skip-gram* model inverts the prediction task and given a single word attempts to predict which words surround it. In the process of training a model for this prediction task, the network learns vector representations for words, mapping words with similar meaning to nearby points in a vector space. The architectures of both approaches are illustrated in Fig. 1. The remainder of this section continues to formally describe the CBOW method. The mathematical intuition of skip-gram is similar and can be inferred from the ensuing equations.

Formally, given a sequence of words $w_1, w_2, \ldots, w_N$, the objective of the *continuous bag of words* framework is to minimize the average log probability given by:

$$-\frac{1}{N} \sum_{n=k}^{N-k} \log p(w_n \mid w_{n-k}, \ldots, w_{n+k}) \tag{2}$$

where k denotes the number of context words to be considered on either side. Note that the value $2k + 1$ is often referred to as the window size. The prediction of the probability is typically computed using a softmax function, i.e.:

$$\log p(w_n \mid w_{n-k}, \ldots, w_{n+k}) = \frac{e^{y_{w_t}}}{\sum_i e^{y_i}} \tag{3}$$

with y_i being the unnormalized log probability for each output word i, which in turn is specified by:

$$y = b + Uh(w_{n-k}, \ldots, w_{n+k}; W) \tag{4}$$

where matrix W contains the weights between the input and hidden layers, matrix U contains the weights between the hidden and output layers, b is an optional bias vector, and lastly h is a function that aggregates the multiple of input vectors into one, typically by concatenation or summation.

The word vectors are learned by performing predictions, as outlined by Eqs. 3 and 4, for each word in the corpus. Errors made while predicting words will then cause the weights W and U of the network to be updated by the backpropagation algorithm [21]. After this training process converges, the weights W between the input and hidden layer represent the learned word vectors, which span a vector space where words with similar meaning tend to cluster. The two key hyper-parameters that govern this learning process are the word sequence length n and the word vector dimension d. Currently no measures exist to quantify the quality of a learned embedding, so practitioners are limited to performing a manual, subjective inspection of the learned representation.

Paragraph vector, or *doc2vec*, is a simple extension to *word2vec* which only differs in input. In addition to word vectors, this technique associates a vector with a chunk of text, or paragraph, to aid in predicting the target words. Note that *word2vec* builds word vectors by sampling word contexts from the entire corpus. In contrast, *doc2vec* only samples locally and restricts the contexts to be within the paragraph. Evidently, *doc2vec* not only learns corpus-wide word vectors but also vector representations for paragraphs. Note that the original frameworks depicted in Fig. 1 remain the same aside from some subtle modifications. The *continuous bag of words* extension now has an additional paragraph vector to predict the target word, whereas *skip-gram* now exclusively uses a paragraph vector instead of a word vector for predictions. These extensions are respectively called *distributed memory* (PV-DM) and *distributed bag of words* (PV-DBOW).

2.1.3 Gaussian Mixture Models

Cluster analysis attempts to identify groups of similar objects within the data. Often, clustering techniques make hard assignments where an object is assigned to exactly one cluster. However, this can be undesirable at times. For example, consider the scenario where the true clusters overlap, or the data points are spread out in such a way that they could belong to multiple clusters. Gaussian mixture models (GMM) that fit a mixture of Gaussian distributions on data overcome this problem by performing *soft* clustering where points are assigned a probability of belonging to each cluster.

A Gaussian mixture model [19] is a parametric probability density function that assumes data points are generated from a mixture of different multivariate

Gaussian distributions. Each distribution is completely determined by its mean μ and covariance matrix Σ, and therefore, a group of data points x with dimension D is modeled by the following Gaussian density function:

$$\mathcal{N}(x \mid \mu, \Sigma) = \frac{1}{(2\pi)^{D/2}|\Sigma|^{1/2}} \exp\left(-\frac{1}{2}(x - \mu)^T \Sigma^{-1}(x - \mu) \right). \tag{5}$$

The Gaussian mixture model, which is a weighted sum of Gaussian component densities, is consequently given by:

$$p(x) = \sum_{k=1}^{K} \pi_k \mathcal{N}(x \mid \mu_k, \Sigma_k) \tag{6}$$

$$\sum_{k=1}^{K} \pi_k = 1. \tag{7}$$

The training process is comprised of finding the optimal values for the weights π_k, means μ_k, and covariances Σ_k of each Gaussian component. Inferring these parameters is usually done using the expectation-maximization algorithm [14]. Note that Eqs. 6 and 7 require knowing k, which is the number of Gaussian components present in the data. However, in practice this is a hyper-parameter that must be tuned. A popular method to assess how well a Gaussian mixture model fits the data is by using the Bayesian Information Criterion [22], where the model with the lowest score is deemed best. This criterion is formally defined as:

$$BIC = \ln(n)k - 2\ln(\hat{L}). \tag{8}$$

where $\hat{L}$ is the maximized value of the likelihood function of the model, n the sample size, and k is the number of parameters estimated by the model. Increasing the number of components in the model will typically yield a higher likelihood of the used training data. However, this can also lead to overfitting. The Bayesian Information Criterion accounts for this phenomenon by introducing the term $\ln(n)k$ that penalizes a model based on the number of parameters it contains.

2.2 Gradient Boosted Trees

In the domain of machine learning, algorithms infer models on a given data in order to predict a supposed dependent variable. One of the most simple algorithms is CART [5], which builds a decision tree model. However, a single tree's prediction performance usually does not suffice in practice. Instead, ensembles of trees are built where the prediction is made by multiple trees together. To this end, the Gradient Boosted Trees algorithm [11] builds a sequence of small decision trees where each tree attempts to correct the mistake of the previous one. Mathematically, a Gradient Boosted Trees model can be specified as:

$$\hat{y}_i = \sum_{k=1}^{K} f_k(x_i), \quad f_k \in F \tag{9}$$

where K is the number of trees and f is a function in the set F of all possible CARTs. As with any machine learning model, the training process involves finding the set of parameters θ that best fit the training data x_i and labels y_i. An objective function is therefore maximized containing both a measure for training loss and a regularization term. This can be formalized as:

$$\text{obj}(\theta) = \sum_{i=1}^{n} l(y_i, \hat{y}_i^{(t)}) + \sum_{i=1}^{t} \Omega(f_i) \tag{10}$$

where l is a loss function, such as the mean squared error, t the amount of learned trees at a given step in the building process, and Ω the regularization term that controls the complexity of the model to avoid overfitting. One way to define the complexity of a tree model is by:

$$\Omega(f) = \gamma T + \frac{1}{2}\lambda \sum_{j=1}^{T} w_j^2 \tag{11}$$

with w the vector of scores on leaves, T the number of leaves, and hyper-parameters γ and λ.

2.3 Market Risk and the CBOE Volatility Index (VIX)

In the world of derivatives, options are one of the most prominent types of financial instruments available. A prime example is the European call option, giving the holder the right to buy stock for a pre-determined price K at time T. Options are exposed to risk for the duration of the contract. To quantify this risk, the expected price fluctuations of the underlying asset are considered over the course of the option contract. A measure that gauges this phenomenon is implied volatility and varies with the strike price and duration of the option contract. A famous example of such a measure in practice is the CBOE Volatility Index. This index, better known as VIX, is a measure of expected price fluctuations in the S&P 500 Index options over the next 30 days. It is therefore often referred to as the *fear index* and is considered to be a reflection of investor sentiment on the market.

3   Methodology

The main goal of this study is to explore the following question:

Are narratives present in financial news articles predictive of future movements in the CBOE Volatility Index?

In order to investigate the interplay between narratives and implied volatility, we have collected a novel news dataset which has not yet been explored by existing research. Instead of containing the raw text of news articles, our dataset simply describes each article using a set of keywords denoting financial sub-topics. Our analysis of the collected news data involves multiple steps. First, because there are both many keywords and semantic overlaps among different ones, we use topic modeling to group together similar keywords. We do this using both the canonical Latent Dirichlet analysis and an alternative approach based on embedding methods, which have received less attention in the economics literature. Second, we train a machine-learned model using these narrative features to predict whether the CBOE Volatility Index will increase or decrease for different time steps into the future.

The next sections explain our applied methodology in more detail. Sect. 3.1 describes how we constructed an innovative news dataset for our study. Section 3.2 then rationalizes our choice for topic modeling algorithms and details both proposed approaches. Section 3.3 then elaborates on how we applied machine learning on the obtained narrative features to predict movements in the CBOE Volatility Index. Lastly, Sect. 3.4 describes the time series cross-validation method we used to evaluate our predictions.

3.1   News Data Acquisition and Preparation

We used the *Financial Times* news API to collect keyword metadata of news articles published on global economy spanning the years 2010 and 2019. Every article is accompanied by a set of keywords where each keyword denotes a financial sub-topic the article covers. Keywords include terms such as *Central Banks, Oil,* and *UK Politics*. In total, more than 39,000 articles were obtained covering a variety of news genres such as opinions, market reports, newsletters, and actual news. We discarded every article that was not of the news genre, which yielded a corpus of roughly 26,000 articles. An example of the constructed dataset can be seen in Fig. 2.

```
2013-08-14:  ['Emerging markets', 'Global Economy']
2013-08-14:  ['Central banks', 'Markets', 'UK business & economy', '...']
2013-08-14:  ['Support services', 'French economy', 'Global Economy', '...']
2013-08-14:  ['Central banks', 'US quantitative easing', 'US economy', '...']
2013-08-14:  ['Central banks', 'Austerity Europe', 'Global Economy', '...']
```

Fig. 2 An example slice of the constructed temporally ordered dataset where a news article is represented by its set of keywords

We investigated the characteristics of the dataset and found 677 unique financial keywords. Not all keywords are as equivalently frequent as the average and median keyword frequency is respectively 114 and 12 articles. Infrequent keywords are probably less important and too specific. We therefore decided to remove the keywords that had occurred less than five times, which corresponds to the 32nd percentile. In addition, we found that keywords *Global Economy* and *World* are respectively present in 100 and 70% of all keywords sets. As their commonality implies weak differentiation power, we omitted both keywords from the entire dataset. Ultimately, 425 unique keywords remain in the dataset. The average keyword set is 6 terms long and more than 16,000 unique sets exist.

Note that in the following sections, terms like *article*, *keyword set*, and *document* will be used interchangeably and are therefore equivalent in meaning.

3.2 Narrative Extraction and Topic Modeling

There are several obvious approaches for extracting narratives and transforming the news corpus into a numerical feature matrix. The most straightforward way is to simply consider the provide keywords about financial sub-topics and represent each article as a binary vector of dimension 1×425, with 1 binary feature denoting the presence/absence of each of the 425 unique keywords. However, this approach yields a sparse feature space and more importantly neglects the semantics associated with each keyword. For example, consider the scenario where three sets are principally equal except for respectively those containing the terms *Federal Reserve*, *Inflation*, and *Climate*. Using the aforementioned approach, this scenario would yield three vectors that are equal in dissimilarity. In contrast, a human reader would use semantic information and consider the first two sets to be closely related. Naturally, incorporating semantic information is advantageous in the context of extracting narratives. We therefore employ topic modeling techniques that group keywords into abstract themes or *latent topics* based on co-occurrence statistics. This way, a keyword set can be represented as a vector of dimension $1 \times K$, denoting the present proportion of each latent topic k_i. In doing so, keyword sets become more comparable on a semantic level, solving the previously outlined problem. Figure 3 demonstrates the result of this approach, where an over-simplified scenario is depicted using the three keyword sets from the previous example. The keyword sets containing the keywords *Federal Reserve* and *Inflation* are now clearly mathematically more similar, suggesting the persistence of some narrative during that time.

To conclude formally, given a series of N news articles each represented by a keyword set, we first transform every article into a vector representing a mixture of K latent topics. This yields a temporally ordered feature matrix X of dimension $N \times K$ where each entry $x_{n,k}$ represents the proportion of topic k in article n. We then aggregate the feature vectors of articles published on the same day by summation,

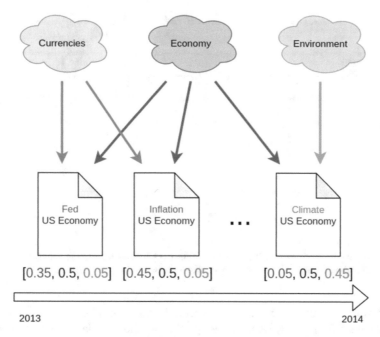

Fig. 3 An illustration of keyword sets being expressed as combinations of their latent themes. In this scenario, the three existing latent themes (clouds) make the documents directly comparable. As a consequence, more *similar* documents are closer to each other in a vector space

producing a new feature matrix X' of dimension $T \times K$, where each entry $x_{t,k}$ now represents the proportion of topic k on day t.

The following sections present how we employed two different approaches to achieve this transformation.

3.2.1 Approach 1: Narrative Extraction Using Latent Dirichlet Analysis

In our study, we utilized the Python library *Gensim* [18] to build LDA topic models. As explained in Sect. 2.1.1, the learning process is primarily controlled by three hyper-parameters K, α, and β. In the interest of finding the optimal hyper-parameter setting, we trained 50 different LDA models on all news articles published between the years 2010 and 2017 by varying the hyper-parameter K from 20 to 70. Prior distributions α and β were automatically inferred by the algorithm employed in *Gensim*. Subsequently, we evaluated the obtained models based on the proposed topic coherence measure C_v [20]. Figure 4 shows the coherence values for different values of K.

Note that the model achieving the highest score is not necessarily the best. Indeed, as the number of parameters in a model increases, so does the risk of overfitting. To alleviate this, we employ the elbow method [24] and identify the

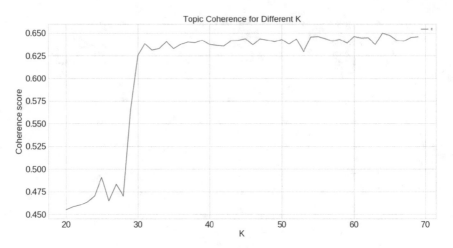

Fig. 4 Topic coherence score achieved by different LDA models for varying values of k. Results were obtained by training on news articles published between the years 2010 and 2017

smallest number of k topics where the score begins to level off. We observed this phenomenon for $k = 31$, where the graph (Fig. 4) shows a clear angle or so-called elbow. Although a somewhat subjective method, this likely yields an appropriate value for K that captures enough information without overfitting on the given data.

Finally, we can transform N given news articles into a temporally ordered feature matrix X of dimension $N \times 31$ using the best performing topic model *LDA(31)*. In turn, we aggregate the feature vectors of articles published on the same day by summation, transforming matrix X into matrix X' of dimension $T \times 31$.

3.2.2 Approach 2: Narrative Extraction Using Vector Embedding and Gaussian Mixture Models

As LDA analyzes documents as bag of words, it does not incorporate word order information. This subtly implies that each keyword co-occurrence within a keyword set is of equal importance. In contrast, vector embedding approaches such as *word2vec* and *doc2vec* consider co-occurrence more locally by using the word's context (i.e., its neighborhood of surrounding words). In an attempt to leverage this mechanism, we introduced order in the originally unordered keyword sets. Keywords belonging to the same financial article are often related to a certain degree. Indeed, take, for example, an article about Brexit that contains the keywords *Economy*, *UK Politics*, and *Brexit*. Not only do the keywords seem related, they tend to represent financial concepts with varying degrees of granularity. In practice, because keyword sets are unordered, more specialized concepts can end up in the vicinity of more general concepts. Evidently, these concepts will be less related, which might introduce noise for vector embedding approaches looking at a word's

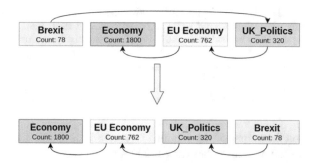

Fig. 5 An illustration of ordering a keyword set based on total corpus frequency. The arrow is an indicator of subsumption by a supposed parent keyword

context. We therefore argue that by ordering the keywords based on total frequency across the corpus, more specific terms will be placed closer to their subsuming keyword. This way, relevant terms are likely to be brought closer together. An example of this phenomenon is demonstrated in Fig. 5.

Note that the scenario depicted in Fig. 5 is ideal, and in practice the proposed ordering will also introduce noise by placing incoherent topics in each other's vicinity. The counts used for ordering were based on news articles published between 2010 and 2017.

For the purpose of topic modeling, we combined *doc2vec* with Gaussian mixture models. First, *doc2vec* is trained on a collection of ordered keyword sets, generating a vector space where similar sets are typically projected in each other's vicinity. Next, a Gaussian mixture model is fitted on this vector space to find k clusters or *latent topics*. In doing so, each document can then be expressed as a mixture of different clusters. *doc2vec* allows retrieving the original document associated with a certain vector. This way, we can compute word frequencies for each cluster, which in turn allows us to interpret them.

In practice, we built *doc2vec* models using the Python library *Gensim*. Recall that sliding window size w and vector dimension d are both important hyper-parameters to the training process. Unlike LDA, there is no quantifiable way to assess the effectiveness of an obtained vector space. We therefore built six *doc2vec* models using both *PV-DBOW* and *PV-DM*, choosing different sliding window sizes $w \in \{2, 5, 8\}$ for a constant $d = 25$. Most research utilizing these techniques tends to use arbitrary vector dimensions without experimental validation (i.e., [17, 15, 8]), suggesting that performance isn't very sensitive to this hyper-parameter. Our decision for the dimension hyper-parameter was ultimately also arbitrary, but chosen to be on the low end given that we are analyzing a relatively small corpus with a limited vocabulary. Each of the obtained vector spaces is then fitted with a Gaussian mixture model to cluster the vector space into k different topics. For each vector space, we found the optimal value for k by fitting 50 different Gaussian mixture models with $k \in \{20, 70\}$. We then applied the elbow technique, introduced

Table 1 The optimal number of Gaussian mixture components for each vector space obtained by using *doc2vec* with vector dimension $d = 25$ and window size $w \in \{2, 5, 8\}$. The results were found by applying the elbow method on the BIC of the Gaussian mixture models

	2	5	8
PV-DBOW	32	38	40
PV-DM	34	36	30

in Sect. 3.2.1, on the graphs of the obtained Bayesian Information Criterion scores. Table 1 presents the optimal values for k found for each vector space.

For each configuration, we can now transform the N given news articles into a temporally ordered feature matrix X of dimension $N \times K$ by first obtaining the vector representation for each article using *doc2vec* and subsequently classifying it with the associated Gaussian mixture model. Again, feature vectors of articles published on the same day are aggregated by summation, transforming matrix X into matrix X' of dimension $T \times K$.

3.3   Predicting Movements in Market Risk with Machine Learning

In our study, we took the CBOE Volatility Index as a proxy for market risk. Instead of solely studying 1-day-ahead predictions, we chose to predict longer-term trends in market risk as well. Consequently, we opted to predict whether the CBOE Volatility Index closes up or down in exactly 1, 2, 4, 6, and 8 trading days.

We downloaded historical price data of VIX through Yahoo Finance. Data points represent end-of-day close prices and have a daily granularity. To construct the actual target feature, we define the n-day-ahead difference in market implied volatility on day i as $y_i^* = (ivolatility_{i+n} - ivolatility_i)$ where $ivolatility_i$ denotes the end-of-day market-implied volatility on day i. We consider the movements to be upward whenever $y_i^* > 0$ and downward whenever $y_i^* \leq 0$. The final target feature is therefore a binary feature obtained by applying case equation 12.

$$y_i = \begin{cases} 1, & \text{if } y_i^* > 0. \\ 0, & \text{otherwise.} \end{cases} \tag{12}$$

In order to predict our target variable, we chose to employ XGBoost's implementation of Gradient Boosted Trees [7]. The implementation is fast and has been dominating Kaggle data science competitions since its inception. Moreover, because forest classifiers are robust to large feature spaces and scaling issues, we do not have to perform standardization or feature selection prior to utilization. Ultimately, we used eight distinctive XGBoost configurations in each experiment, with $max\_depth \in \{4, 5, 6, 7\}$, and $n\_estimators \in \{200, 400\}$. These models were trained on a temporally ordered feature matrix X^* of dimension $T \times (K + 1)$, obtained by concatenating the feature matrix X comprised of narrative features of dimension $T \times K$ together with the CBOE Volatility Index' close prices. Note that special care was taken to not introduce data leakage when using topic models to obtain the narrative feature matrix X. To this end, each prediction for given day t was made using feature vectors obtained by a topic model that was trained on news articles published strictly before day t.

3.4 Evaluation on Time Series

The Gradient Boosted Trees are evaluated using cross-validation, where data is repeatedly split into non-overlapping train and test sets. This way models are trained on one set and afterward evaluated on a test set comprised of unseen data to give a more robust estimate of the achieved generalization. However, special care needs to be taken when dealing with time series data. Classical cross-validation methods assume observations to be independent. This assumption does not hold for time series data, which inherently contains temporal dependencies among observations. We therefore split the data into training and test sets which take the temporal order into account to avoid data leakage. To be more concrete, we employ Walk Forward Validation (or Rolling Window Analysis) where a sliding window of t previous trading days is used to train the models and where trading day t_{t+1+m} is used for the out-of-sample test prediction. Note that special care needs to be taken when choosing a value for m. For example, if we want to perform an out-of-sample prediction for our target variable 2 days into the future given information on day t_i, we need to leave out day t_{i-1} from the train set in order to avoid data leakage. Indeed, the training data point t_{i-1} not only contains the information of narratives present on the said day but also whether the target variable has moved up or down by day t_{i+1}. It is evident that in reality we do not possess information on our target variable on day t_{i+1} at the time of our prediction on day t_i. Consequently, m has to be chosen so that $m \geq d - 1$ where d denotes how many time steps into the future the target variable is predicted.

Table 2 illustrates an example of this method where t_i denotes the feature vector corresponding to trading day i and predictions are made 2 days into the future. Note that in this scenario, when given a total of n observations and a sliding window of length t, you can construct a maximum of $n - (t + m)$ different train-test splits. Moreover, models need to be retrained during each iteration of the evaluation process, as is the case with any cross-validation method.

Table 2 Example of Walk Forward Validation where t_i represents the feature vector of trading day i. In this example, a sliding window of size three is taken to learn a model that predicts a target variable 2 days into the future. During the first iteration, we use the feature vectors of the first 3 consecutive trading days to train a model (underlined) and subsequently test the said model on the 5th day (bold), leaving out the 4th day to avoid data leakage as described in Sect. 3.4. This process is repeated j times where, after each iteration, the sliding window is shifted in time by 1 trading day

Iteration	Variable roles
1	$\underline{t_1\ t_2\ t_3}\ \cancel{t_4}\ \mathbf{t_5}\ t_6\ \cdots\ t_n$
2	$t_1\ \underline{t_2\ t_3\ t_4}\ \cancel{t_5}\ \mathbf{t_6}\ \cdots\ t_n$
$\vdots$	$\vdots$
j	$t_1\ \cdots\ \underline{t_{n-4}\ t_{n-3}\ t_{n-2}}\ \cancel{t_{n-1}}\ \mathbf{t_n}$

4 Experimental Results and Discussion

In this section, we present our experimental methodology and findings from our study. The study consists of two parts. First, we examined the soundness of our two proposed strategies for performing topic modeling on keyword sets. To this end, we contrasted the predictive performance of each strategy to a simple baseline for different prediction horizons. Second, we investigated the interplay between the prediction horizon and each feature setup on predictive performance.

4.1 Feature Setups and Predictive Performance

We examined whether feature matrices containing narrative features (obtained by the methodologies proposed in Sects. 3.2.1 and 3.2.2) achieve a better predictive accuracy compared to a simple baseline configuration that solely uses the daily CBOE Volatility Index' closing values as the predictive feature. To this end, we investigated the predictive performance for predicting CBOE Volatility Index movements for 1, 2, 4, 6, and 8 days ahead.

The Gradient Boosted Trees were trained on a sliding window of 504 trading days (2 years), where the out-of-sample test case was picked in function of the prediction horizon and according to the method outlined in Sect. 3.4. Because the optimal hyper-parameters for both our topic modeling approaches were found by utilizing news articles published between 01/01/2010 and 31/12/2017 (Sect. 3.2), we constrained our out-of-sample test set to the years 2018 and 2019 to avoid data leakage. Consequently, the trained Gradient Boosted Trees models were evaluated on 498 different out-of-sample movement predictions for the CBOE Volatility Index. Each proposed feature setup had a unique temporally ordered feature matrix of dimension $1002 \times C_i$, where C_i denotes the number of features for a particular setup i. We chose to quantify the performance of our predictions by measuring the predictive accuracy. Note that the target variable is fairly balanced with about 52% down movements and 48% up movements.

First, to examine the baseline configuration, predictions and evaluations were done using a temporally ordered feature matrix X_{vix} of dimension 1002×1 where each entry x_t represents the CBOE Volatility Index closing value for trading day t. Second, to study the performance of the feature matrix obtained by the latent Dirichlet analysis method outlined in Sect. 3.2.1, predictions and evaluations were done using a temporally ordered feature matrix X_{lda} of dimension $1002 \times (31 + 1)$. This feature matrix contains 31 topic features and an additional feature representing daily CBOE Volatility Index closing values. Lastly, to investigate the performance of the feature matrices obtained by using *doc2vec* and Gaussian mixture models outlined in Sect. 3.2.2, predictions and evaluations were done using six different temporally ordered features matrices X_{d2v}^i of dimension $1002 \times (K_i + 1)$ where K_i denotes the amount of topic features associated with one of the six proposed configurations. Note again that an additional feature representing daily CBOE Volatility Index closing values was added to the feature matrices.

Table 3 presents the best accuracy scores obtained by the Gradient Boosted Trees for different prediction horizons, following the methodology outlined in Sects. 3.3 and 3.4. First, Table 3 shows that for each prediction horizon except for the last one, there exists a feature setup that improves the predictive performance compared to the baseline. Second, for the scenario where movements are predicted 4 days into the future, all feature setups manage to outperform the baseline. In addition, all *doc2vec* feature setups manage to outperform the baseline and latent Dirichlet analysis feature setups for 6-day-ahead predictions. Third, the number of feature setups that outperform the baseline (bold numerals) increases as we predict further into the future. However, this trend does not hold when predicting 8 days into the future. Lastly, the *doc2vec* scenario, where PV-DM is used with a window size of two, seems to perform best overall except for the scenario where movements are predicted 2 days ahead.

Table 3 This table shows different feature setups and their best accuracy score obtained by Gradient Boosted Trees while predicting t-days ahead CBOE Volatility Index movements during 2018–2019 for $t \in \{1, 2, 4, 6, 8\}$. It demonstrates the contrast between simply using VIX closing values as a predictive feature (baseline) and feature matrices augmented with narrative features using respectively latent Dirichlet analysis (Sect. 3.2.1) and a combination of *doc2vec* and Gaussian mixture models (Sect. 3.2.2). Bold numerals indicate whether a particular setting outperforms the baseline, where underlined numerals indicate the best performing setting for the given prediction horizon

	$t = 1$	$t = 2$	$t = 4$	$t = 6$	$t = 8$
Baseline	54.0	51.5	53.4	54.4	<u>56.1</u>
LDA(31)	**55.7**	**52.4**	**54.7**	52.2	55.4
D2V(PV-DM, 2)	<u>**57.3**</u>	51.6	<u>**59.1**</u>	<u>**57.7**</u>	53.8
D2V(PV-DM, 5)	53.5	**53.7**	**57.8**	**57.3**	55.2
D2V(PV-DM, 8)	53.4	**53.8**	**57.5**	**57.0**	55.6
D2V(PV-DB, 2)	53.1	<u>**54.0**</u>	**55.0**	**55.5**	55.2
D2V(PV-DB, 5)	**55.0**	**52.3**	**57.3**	**56.2**	55.3
D2V(PV-DB, 8)	**54.2**	**52.5**	**57.0**	**55.6**	55.7

In conclusion, the narrative features contribute to an increased predictive performance compared to baseline. The *doc2vec* approach seems to yield the best performing models overall, consistently outperforming both the baseline and latent Dirichlet analysis feature setups for 4- and 6-day-ahead predictions. Lastly, the results suggest that the prediction horizon has an effect on predictive performance. The next section will investigate this further.

4.2 The Effect of Different Prediction Horizons

The results shown in Sect. 4.1 suggest that the prediction horizon influences the predictive performance for all different feature setups. In this part of the study, we investigated this phenomenon more in depth by examining to what degree feature setups outperform the baseline in function of different prediction horizons. The results are displayed in Fig. 6, where a bar chart is used to illustrate this interplay. Note that for both *doc2vec* scenarios using respectively PV-DM and PV-DBOW, the

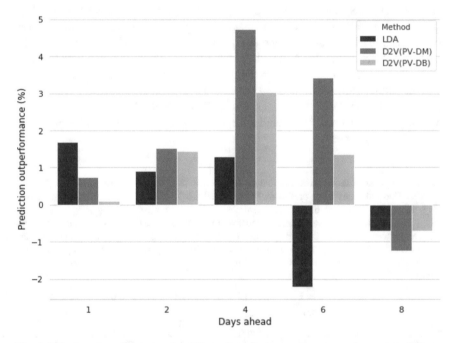

Fig. 6 This bar chart illustrates the effect of predictive performance when using different prediction horizons for different feature setups. The height of a bar denotes the outperformance of the given method compared to the baseline method of just using VIX closing values as the predictive feature. Note that for both D2V (PV-DM) and D2V (PV-DB), the accuracy scores were averaged across the different window size configurations prior to computing the prediction outperformance

accuracy scores were averaged across the different window size configurations prior to comparing the prediction performance compared to the baseline method.

First, Fig. 6 shows that for 1-day-ahead predictions, the narrative features obtained by using latent Dirichlet analysis perform better than *doc2vec* when performances are averaged across the different window sizes. However, note that the results from Sect. 4.1 show that the best performance for 1-day-ahead prediction is still achieved by an individual *doc2vec* feature setup. Nonetheless, this indicates that the performance of *doc2vec* feature setups is sensitive to the window size hyper-parameter. Second, a clear trend is noticeable looking at the outperformance achieved by both *doc2vec* PV-DM and PV-DBOW scenarios for different prediction horizons. Indeed, the performance for both scenarios increases by extending the prediction horizon. Moreover, the PV-DM method seems to consistently beat the PV-DBOW method. Third, the optimal prediction horizon for the *doc2vec* feature setups seems to be around 4 days, after which the performance starts to decline. Lastly, no feature setup is able to outperform the baseline model on a prediction horizon of 8 days.

In conclusion, we can state that the predictive performance of both latent Dirichlet analysis and *doc2vec* behaves differently. The best performance is achieved by *doc2vec* for a prediction horizon of 4 days, after which the performance starts to decline. This may suggest that the narrative features present in news only influence market participants for a short period of time, with market reaction peaking about 4 days into the future. Note that our study provides no evidence for causality.

5   Conclusion

Our study provides empirical evidence in favor of the theory of Narrative Economics by showing that quantified narratives extracted from news articles, described by sets of financial keywords, are predictive of future movements in the CBOE Volatility Index for different time horizons. We successfully demonstrate how both latent Dirichlet analysis and *doc2vec* combined with Gaussian mixture models can be used as effective topic modeling methods. However, overall we find that the *doc2vec* approach works better for this application. In addition, we show that the predictive power of extracted narrative features fluctuates in function of prediction horizon. Configurations using narrative features are able to outperform the baseline on 1-day, 2-day, 4-day, and 6-day-ahead predictions, but not on 8-day-ahead predictions. We believe this suggests that the narrative features present in news only influence market participants for a short period of time. Moreover, we show that the best predictive performance is achieved when predicting 4-day-ahead movements. This may suggest that market participants not always react instantaneously to narratives present in financial news, or that it takes time for this reaction to be reflected in the market.

References

1. Akita, R., Yoshihara, A., Matsubara, T., & Uehara, K. (2016, June). Deep learning for stock prediction using numerical and textual information. In *2016 IEEE/ACIS 15th International Conference on Computer and Information Science (ICIS)* (pp. 1–6).
2. Aletras, N., & Stevenson, M. (2013). Evaluating topic coherence using distributional semantics. In *Evaluating Topic Coherence Using Distributional Semantics Proceedings of the 10th International Conference on Computational Semantics (IWCS 2013)* (pp. 13–22). Association for Computational Linguistics.
3. Atkins, A., Niranjan, M., & Gerding, E. (2018). Financial news predicts stock market volatility better than close price. *The Journal of Finance and Data Science, 4*(2), 120–137.
4. Blei, D. M., Ng, A. Y., & Jordan, M. I. (2003). Latent dirichlet allocation. *J. Mach. Learn. Res., 3*, 993–1022.
5. Breiman, L., Friedman, J. H., Olshen, R. A., & Stone, C. J. (1984). *Classification and regression trees.* Monterey, CA: Wadsworth and Brooks. ISBN 9780412048418
6. Chang, J., Boyd-Graber, J., Gerrish, S., Wang, C., & Blei, D. M. (2009). Reading tea leaves: How humans interpret topic models. In *Proceedings of the 22nd International Conference on Neural Information Processing Systems, NIPS'09, Red Hook, NY* (pp. 288–296). Red Hook: Curran Associates Inc.
7. Chen, T., & Guestrin, C. (2016). XGBoost: A scalable tree boosting system. In *Proceedings of the 22nd ACM SIGKDD International Conference on Knowledge Discovery and Data Mining - KDD '16, 13–17-August-2016* (pp. 785–794).
8. Devlin, J., Chang, M.-W., Lee, K., & Toutanova, K. (2018). BERT: Pre-training of deep bidirectional transformers for language understanding. In *NAACL HLT 2019 - 2019 Conference of the North American Chapter of the Association for Computational Linguistics: Human Language Technologies - Proceedings of the Conference* (Vol. 1, pp. 4171–4186).
9. Feuerriegel, S., & Gordon, J. (2018). Long-term stock index forecasting based on text mining of regulatory disclosures. *Decision Support Systems, 112*, 88–97.
10. Feuerriegel, S., & Pröllochs, N. (2018). Investor reaction to financial disclosures across topics: An application of latent dirichlet allocation. *Decision Sciences.* Article in Press. https://doi.org/10.1111/deci.12346
11. Friedman, J. H. (2001). Greedy function approximation: A gradient boosting machine. *The Annals of Statistics, 29*(5), 1189–1232.
12. Harris, Z. S. (1954). Distributional structure. *WORD, 10*(2–3), 146–162.
13. Hoffman, M., Bach, F. R., & Blei, D. M. (2010). Online learning for latent dirichlet allocation. In J. D. Lafferty, C. K. I. Williams, J. Shawe-Taylor, R. S. Zemel, & A. Culotta (Eds.), *Advances in Neural Information Processing Systems 23* (pp. 856–864). Red Hook: Curran Associates, Inc.
14. Jin, X., & Han, J. (2011). Expectation-Maximization Algorithm. In C. Sammut & G. I. Webb (Eds.), *Encyclopedia of machine learning* (pp. 387–387). Boston: Springer. https://doi.org/10.1007/978-0-387-30164-8_291
15. Le, Q., & Mikolov, T. (2014). Distributed representations of sentences and documents. In *31st International Conference on Machine Learning, ICML 2014* (Vol. 4, pp. 2931–2939).
16. Mai, F., Tian, S., Lee, C., & Ma, L. (2019). Deep learning models for bankruptcy prediction using textual disclosures. *European Journal of Operational Research, 274*(2), 743–758.
17. Mikolov, T., Chen, K., Corrado, G., & Dean, J. (2013). Efficient estimation of word representations in vector space. In *1st International Conference on Learning Representations, ICLR 2013 - Workshop Track Proceedings.* Available at: https://arxiv.org/abs/1301.3781
18. Řehůřek, R. (2021). *Gensim-topic modelling for humans.* Last accessed on 12 March, 2021. Available at: https://radimrehurek.com/gensim/
19. Reynolds, D. (2015). Gaussian mixture models. In S. Z. Li & A. K. Jain (Eds.), *Encyclopedia of Biometrics.* Boston: Springer. https://doi.org/10.1007/978-1-4899-7488-4_196

20. Röder, M., Both, A., & Hinneburg, A. (2015). Exploring the space of topic coherence measures. In *Proceedings of the Eighth ACM International Conference on Web Search and Data Mining, WSDM '15, New York, NY* (pp. 399–408). New York: Association for Computing Machinery.
21. Rumelhart, D. E., Hinton, G. E., & Williams, R. J. (1986). *Learning internal representations by error propagation* (pp. 318–362). Cambridge: MIT Press.
22. Schwarz, G. (1978). Estimating the dimension of a model. *Annals of Statistics, 6*(2), 461–464.
23. Shiller, R. J. (2019). *Narrative economics: How stories go viral and drive major economic events*. Princeton: Princeton University Press.
24. Thorndike, R. (1953). Who belongs in the family? *Psychometrika, 18*(4), 267–276.
25. Vargas, M., Lima, B., & Evsukoff, A. (2017). Deep learning for stock market prediction from financial news articles. In *2017 IEEE International Conference on Computational Intelligence and Virtual Environments for Measurement Systems and Applications (CIVEMSA 2017)* (pp. 60–65).
26. Wallach, H. M., Murray, I., Salakhutdinov, R., & Mimno, D. (2009). Evaluation methods for topic models. In *Proceedings of the 26th Annual International Conference on Machine Learning, ICML '09, New York, NY* (pp. 1105–1112). New York: Association for Computing Machinery.

Do the Hype of the Benefits from Using New Data Science Tools Extend to Forecasting Extremely Volatile Assets?

Steven F. Lehrer, Tian Xie, and Guanxi Yi

Abstract This chapter first provides an illustration of the benefits of using machine learning for forecasting relative to traditional econometric strategies. We consider the short-term volatility of the Bitcoin market by realized volatility observations. Our analysis highlights the importance of accounting for nonlinearities to explain the gains of machine learning algorithms and examines the robustness of our findings to the selection of hyperparameters. This provides an illustration of how different machine learning estimators improve the development of forecast models by relaxing the functional form assumptions that are made explicit when writing up an econometric model. Our second contribution is to illustrate how deep learning can be used to measure market-level sentiment from a 10% random sample of Twitter users. This sentiment variable significantly improves forecast accuracy for every econometric estimator and machine algorithm considered in our forecasting application. This provides an illustration of the benefits of new tools from the natural language processing literature at creating variables that can improve the accuracy of forecasting models.

1 Introduction

Over the past few years, the hype surrounding words ranging from big data to data science to machine learning has increased from already high levels. This hype arises

S. F. Lehrer (✉)
Queen's University, Kingston, ON, Canada
NBER, Cambridge, MA, USA
e-mail: lehrers@queensu.ca

T. Xie
Shanghai University of Finance and Economics, Shanghai, China
e-mail: xietian@shufe.edu.cn

G. Yi
Digital Asset Strategies, LLP, Santa Monica, CA, USA
e-mail: Guanxi@das.fund

© The Author(s) 2021
S. Consoli et al. (eds.), *Data Science for Economics and Finance*,
https://doi.org/10.1007/978-3-030-66891-4_13

in part from three sets of discoveries. Machine learning tools have repeatedly been shown in the academic literature to outperform statistical and econometric techniques for forecasting.[1] Further, tools developed in the natural language processing literature that are used to extract population sentiment measures have also been found to help forecast the value of financial indices. This set of finding is consistent with arguments in the behavioral finance literature (see [23], among others) that the sentiment of investors can influence stock market activity. Last, issues surrounding data security and privacy have grown among the population as a whole, leading governments to consider blockchain technology for uses beyond what it was initially developed for.

Blockchain technology was originally developed for the cryptocurrency Bitcoin, an asset that can be continuously traded and whose value has been quite volatile. This volatility may present further challenges for forecasts by either machine learning algorithms or econometric strategies. Adding to these challenges is that unlike almost every other financial asset, Bitcoin is traded on both the weekend and holidays. As such, modeling the estimated daily realized variance of Bitcoin in US dollars presents an additional challenge. Many measures of conventional economic and financial data commonly used as predictors are not collected at the same points in time. However, since the behavioral finance literature has linked population sentiment measures to the price of different financial assets, we propose measuring and incorporating social media sentiment as an explanatory variable in the forecasting model. As an explanatory predictor, social media sentiment can be measured continuously providing a chance to capture and forecast the variation in the prices at which trades for Bitcoin are made.

In this chapter, we consider forecasts of Bitcoin realized volatility to first provide an illustration of the benefits in terms of forecast accuracy of using machine learning relative to traditional econometric strategies. While prior work contrasting approaches to conduct a forecast found that machine learning does provide gains primarily from relaxing the functional form assumptions that are made explicit when writing up an econometric model, those studies did not consider predicting an outcome that exhibits a degree of volatility of the magnitude of Bitcoin.

Determining strategies that can improve volatility forecasts is of significant value since they have come to play a large role in decisions ranging from asset allocation to derivative pricing and risk management. That is, volatility forecasts are used by traders as a component of their valuation procedure of any risky asset's value (e.g., stock and bond prices), since the procedure requires assessing the level and riskiness of future payoffs. Further, their value to many investors arises when using a strategy that adjust their holdings to equate the risk stemming from the different investments included in a portfolio. As such, more accurate volatility forecasts can provide

---

[1] See [25, 26], for example, with data from the film industry that conducts horse races between various strategies. Medeiros et al. [31] use the random forest estimator to examine the benefits of machine learning for forecasting inflation. Last, Coulombe et al. [13] conclude that the benefits from machine learning over econometric approaches for macroeconomic forecasting arise since they capture important nonlinearities that arise in the context of uncertainty and financial frictions.

valuable actionable insights for market participants. Finally, additional motivation for determining how to obtain more accurate forecasts comes from the financial media who frequently report on market volatility since it is hypothesized to have an impact on public confidence and thereby can have a significant effect on the broader global economy.

There are many approaches that could be potentially used to undertake volatility forecasts, but each requires an estimate of volatility. At present, the most popular method used in practice to estimate volatility was introduced by Andersen and Bollerslev [1] who proposed using the realized variance, which is calculated as the cumulative sum of squared intraday returns over short time intervals during the trading day.[2] Realized volatility possesses a slowly decaying autocorrelation function, sometimes known as long memory.[3] Various econometric models have been proposed to capture the stylized facts of these high-frequency time series models including the autoregressive fractionally integrated moving average (ARFIMA) models of Andersen et al. [3] and the heterogeneous autoregressive (HAR) model proposed by Corsi [11]. Compared with the ARFIMA model, the HAR model rapidly gained popularity, in part due to its computational simplicity and excellent out-of-sample forecasting performance. [4]

In our empirical exercise, we first use well-established machine learning techniques within the HAR framework to explore the benefits of allowing for general nonlinearities with recursive partitioning methods as well as sparsity using the least absolute shrinkage and selection operator (LASSO) of Tibshirani [39]. We consider alternative ensemble recursive partitioning methods including bagging and random forest that each place equal weight on all observations when making a forecast, as well as boosting that places alternative weight based on the degree of fit. In total, we evaluate nine conventional econometric methods and five easy-to-implement machine learning methods to model and forecast the realized variance of Bitcoin measured in US dollars.

Studies in the financial econometric literature have reported that a number of different variables are potentially relevant for the forecasting of future volatility. A

---

[2]Traditional econometric approaches to model and forecast such as the parametric GARCH or stochastic volatility models include measures built on daily, weekly, and monthly frequency data. While popular, empirical studies indicate that they fail to capture all information in high-frequency data; see [1, 7, 20], among others.

[3]This phenomenon has been documented by Dacorogna et al. [15] and Andersen et al. [3] for the foreign exchange market and by Andersen et al. [2] for stock market returns.

[4]Corsi et al. [12] provide a comprehensive review of the development of HAR-type models and their various extensions. The HAR model provides an intuitive economic interpretation that agents with three frequencies of trading (daily, weekly, and monthly) perceive and respond to, which changes the corresponding components of volatility. Müller et al. [33] refer to this idea as the Heterogeneous Market Hypothesis. Nevertheless, the suitability of such a specification is not subject to enough verification. Craioveanu and Hillebrand [14] employ a parallel computing method to investigate all of the possible combinations of lags (chosen within a maximum lag of 250) for the last two terms in the additive model, and they compared their in-sample and out-of-sample fitting performance.

secondary goal of our empirical exercise is to determine if there are gains in forecast accuracy of realized volatility by incorporating a measure of social media sentiment. We contrast forecasts using models that both include and exclude social media sentiment. This additional exercise allows us to determine if this measure provides information that is not captured by either the asset-specific realized volatility histories or other explanatory variables that are often included in the information set.

Specifically, in our application social media sentiment is measured by adopting a deep learning algorithm introduced in [17]. We use a random sample of 10% of all tweets posted from users based in the United States from the Twitterverse collected at the minute level. This allows us to calculate a sentiment score that is an equal tweet weight average of the sentiment values of the words within each Tweet in our sample at the minute level.[5] It is well known that there are substantial intraday fluctuations in social media sentiment but its weekly and monthly aggregates are much less volatile. This intraday volatility may capture important information and presents an additional challenge when using this measure for forecasting since the Bitcoin realized variance is measured at the daily level, a much lower time frequency than the minute-level sentiment index that we refer to as the US Sentiment Index (USSI). Rather than make ad hoc assumptions on how to aggregate the USSI to the daily level, we follow Lehrer et al. [28] and adopt the heterogeneous mixed data sampling (H-MIDAS) method that constructs empirical weights to aggregate the high-frequency social media data to a lower frequency.

Our analysis illustrates that sentiment measures extracted from Twitter can significantly improve forecasting efficiency. The gains in forecast accuracy as pseudo R-squared increased by over 50% when social media sentiment was included in the information set for all of the machine learning and econometric strategies considered. Moreover, using four different criteria for forecast accuracy, we find that the machine learning techniques considered tend to outperform the econometric strategies and that these gains arise by incorporating nonlinearities. Among the 16 methods considered in our empirical exercise, both bagging and random forest yield the highest forecast accuracy. Results from the [18] test indicate that the improvements that each of these two algorithms offers are statistically significant at the 5% level, yet the difference between these two algorithms is indistinguishable.

For practitioners, our empirical exercise also contains exercises including examining the sensitivity of our findings to the choices of hyperparameters made when implementing any machine learning algorithm. This provides value since the settings of the hyperparameters with any machine learning algorithm can be thought of in an analogous manner to model selection in econometrics. For example,

---

[5]We note that the assumption of equal weight is strong. Mai et al. [29] find that social media sentiment is an important predictor in determining Bitcoin's valuation, but not all social media messages are of equal impact. Yet, our measure of social media is collected from all Twitter users, a more diverse group than users of cryptocurrency forums in [29]. Thus, if we find any effect, it is likely a lower bound since our measure of social media sentiment likely has classical measurement error.

with the random forest algorithm, numerous hyperparameters can be adjusted by the researcher including the number of observations drawn randomly for each tree and whether they are drawn with or without replacement, the number of variables drawn randomly for each split, the splitting rule, the minimum number of samples that a node must contain, and the number of trees. Further, Probst and Boulesteix provide evidence that the benefits from changing hyperparameters differ across machine learning algorithms and are higher with the support vector regression than the random forest algorithm we employ. In our analysis, the default values of the hyperparameters specified in software packages work reasonably well, but we stress a caveat that our investigation was not exhaustive so there remains a possibility that there are particular specific combinations of hyperparameters with each algorithm that may lead to changes in the ordering of forecast accuracy in the empirical horse race presented. Thus, there may be a set of hyperparameters where the winning algorithms have a distinguishable different effect from the others that it is being compared to.

This chapter is organized as follows. In the next section, we briefly describe Bitcoin. Sections 3 and 4 provide a more detailed overview of existing HAR strategies as well as conventional machine learning algorithms. Section 5 describes the data we utilize and explains how we measure and incorporate social media data into our empirical exercise. Section 6 presents our main empirical results that compare the forecasting performance of each method introduced in Sects. 3 and 4 in a rolling window exercise. To focus on whether social media sentiment data adds value, we contrast the results of incorporating the USSI variable in each strategy to excluding this variable from the model. For every estimator considered, we find that incorporating the USSI variable as a covariate leads to significant improvements in forecast accuracy. We examine the robustness of our results by considering (1) different experimental settings, (2) different hyperparameters, and (3) incorporating covariates on the value of mainstream assets, in Sect. 7. We find that our main conclusions are robust to both changes in the hyperparameters and various settings, as well as little benefits from incorporating mainstream asset markets when forecasting the realized volatility in the value of Bitcoin. Section 8 concludes by providing additional guidance to practitioners to ensure that they can gain the full value of the hype for machine learning and social media data in their applications.

2 What Is Bitcoin?

Bitcoin, the first and still one of the most popular applications of the blockchain technology by far, was introduced in 2008 by a person or group of people known by the pseudonym, Satoshi Nakamoto. Blockchain technology allows digital information to be distributed but not copied. Basically, a time-stamped series of immutable records of data are managed by a cluster of computers that are not owned by any single entity. Each of these blocks of data (i.e., block) is secured and bound to each other using cryptographic principles (i.e., chain). The blockchain network has no central authority and all information on the immutable ledger is shared. The information on the blockchain is transparent and each individual involved is accountable for their actions.

The group of participants who uphold the blockchain network ensure that it can neither be hacked or tampered with. Additional units of currency are created by the nodes of a peer-to-peer network using a generation algorithm that ensures decreasing supply that was designed to mimic the rate at which gold was mined. Specifically, when a user/miner discovers a new block, they are currently awarded 12.5 Bitcoins. However, the number of new Bitcoins generated per block is set to decrease geometrically, with a 50% reduction every 210,000 blocks. The amount of time it takes to find a new block can vary based on mining power and the network difficulty.[6] This process is why it can be treated by investors as an asset and ensures that causes of inflation such as printing more currency or imposing capital controls by a central authority cannot take place. The latter monetary policy actions motivated the use of Bitcoin, the first cryptocurrency as a replacement for fiat currencies.

Bitcoin is distinguished from other major asset classes by its basis of value, governance, and applications. Bitcoin can be converted to a fiat currency using a cryptocurrency exchange, such as Coinbase or Kraken, among other online options. These online marketplaces are similar to the platforms that traders use to buy stock. In September 2015, the Commodity Futures Trading Commission (CFTC) in the United States officially designated Bitcoin as a commodity. Furthermore, the Chicago Mercantile Exchange in December 2017 launched a Bitcoin future (XBT) option, using Bitcoin as the underlying asset. Although there are emerging crypto-focused funds and other institutional investors,[7] this market remains retail investor dominated.[8]

---

[6]Mining is challenging since new blocks and miners are paid any transaction fees as well as a "subsidy" of newly created coins. For the new block to be considered valid, it must contain a proof of work that is verified by other Bitcoin nodes each time they receive a block. By downloading and verifying the blockchain, Bitcoin nodes are able to reach consensus about the ordering of events in Bitcoin. Any currency that is generated by a malicious user that does not follow the rules will be rejected by the network and thus is worthless. To make each new block more challenging to mine, the rate at which a new block can be found is recalculated every 2016 blocks increasing the difficulty.

[7]For example, the legendary former Legg Mason' Chief Investment Officer Bill Miller's fund has been reported to have 50% exposure to crypto-assets. There is also a growing set of decentralized exchanges, including IDEX, 0x, etc., but their market shares remain low today. Furthermore, given the SEC's recent charge against EtherDelta, a well-known Ethereum-based decentralized exchange, the future of decentralized exchanges faces significant uncertainties.

[8]Apart from Bitcoin, there are more than 1600 other alter coin or cryptocurrencies listed over 200 different exchanges. However, Bitcoin still maintains roughly 50% market dominance. At the end of December 2018, the market capitalization of Bitcoin is roughly 65 billion USD with 3800 USD per token. On December 17, 2017, it reached 330 billion USD cap peak with almost 19,000 USD per Bitcoin according to *Coinmarketcap.com*.

There is substantial volatility in BTC/USD, and the sharp price fluctuations in this digital currency greatly exceed that of most other fiat currencies. Much research has explored why Bitcoin is so volatile; our interest is strictly to examine different empirical strategies to forecast this volatility, which greatly exceeds that of other assets including most stocks and bonds.

3 Bitcoin Data and HAR-Type Strategies to Forecast Volatility

The price of Bitcoin is often reported to experience wild fluctuations. We follow Xie [42] who evaluates model averaging estimators with data on the Bitcoin price in US dollars (henceforth BTC/USD) at a 5-min. frequency between May 20, 2015, and Aug 20, 2017. This data was obtained from Poloniex, one of the largest US-based digital asset exchanges. Following Andersen and Bollerslev [1], we estimate the daily realized volatility at day t (RV_t) by summing the corresponding M equally spaced intra-daily squared returns $r_{t,j}$. Here, the subscript t indexes the day, and j indexes the time interval within day t:

$$RV_t \equiv \sum_{j=1}^{M} r_{t,j}^2 \qquad (1)$$

where $t = 1, 2, \ldots, n$, $j = 1, 2, \ldots, M$, and $r_{t,j}$ is the difference between log-prices $p_{t,j}$ ($r_{t,j} = p_{t,j} - p_{t,j-1}$). Poloniex is an active exchange that is always in operation, every minute of each day in the year. We define a trading day using Eastern Standard Time and with data calculate realized volatility of BTC/USD for 775 days. The evolution of the RV data over this full sample period is presented in Fig. 1.

In this section, we introduce some HAR-type strategies that are popular in modeling volatility. The standard HAR model of Corsi [11] postulates that the h-step-ahead daily RV_{t+h} can be modeled by[9]

$$logRV_{t+h} = \beta_0 + \beta_d logRV_t^{(1)} + \beta_w logRV_t^{(5)} + \beta_m logRV_t^{(22)} + e_{t+h}, \qquad (2)$$

---

[9]Using the log to transform the realized variance is standard in the literature, motivated by avoiding imposing positive constraints and considering the residuals of the below regression to have heteroskedasticity related to the level of the process, as mentioned by Patton and Sheppard [34]. An alternative is to implement weighted least squares (WLS) on RV, which does not suit well our purpose of using the least squares model averaging method.

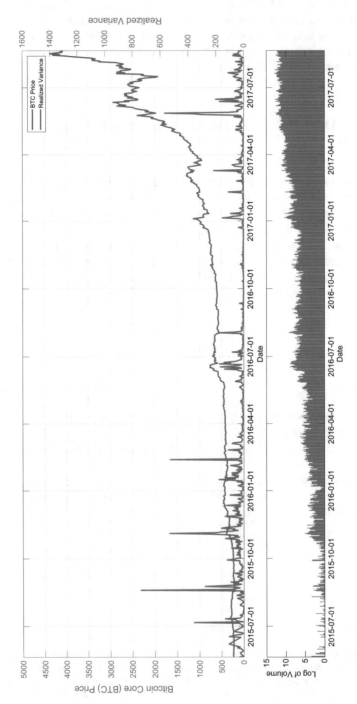

Fig. 1 BTC/USD price, realized variance, and volume on Poloniex

where the βs are the coefficients and $\{e_t\}_t$ is a zero mean innovation process. The explanatory variables take the general form of $log\mathrm{RV}_t^{(l)}$ that is defined as the l period averages of daily log RV:

$$log\mathrm{RV}_t^{(l)} \equiv l^{-1} \sum_{s=1}^{l} log\mathrm{RV}_{t-s}.$$

Another popular formulation of the HAR model in Eq. (2) ignores the logarithmic form and considers

$$\mathrm{RV}_{t+h} = \beta_0 + \beta_d \mathrm{RV}_t^{(1)} + \beta_w \mathrm{RV}_t^{(5)} + \beta_m \mathrm{RV}_t^{(22)} + e_{t+h}, \tag{3}$$

where $\mathrm{RV}_t^{(l)} \equiv l^{-1} \sum_{s=1}^{l} \mathrm{RV}_{t-s}$.

In an important paper, Andersen et al. [4] extend the standard HAR model from two perspectives. First, they added a daily jump component (J_t) to Eq. (3). The extended model is denoted as the HAR-J model:

$$\mathrm{RV}_{t+h} = \beta_0 + \beta_d \mathrm{RV}_t^{(1)} + \beta_w \mathrm{RV}_t^{(5)} + \beta_m \mathrm{RV}_t^{(22)} + \beta^j \mathrm{J}_t + e_{t+h}, \tag{4}$$

where the empirical measurement of the squared jumps is $\mathrm{J}_t = \max(\mathrm{RV}_t - \mathrm{BPV}_t, 0)$ and the standardized realized bipower variation (BPV) is defined as

$$\mathrm{BPV}_t \equiv (2/\pi)^{-1} \sum_{j=2}^{M} |r_{t,j-1}||r_{t,j}|.$$

Second, through a decomposition of RV into the continuous sample path and the jump components based on the Z_t statistic [22], Andersen et al. [4] extend the HAR-J model by explicitly incorporating the two types of volatility components mentioned above. The Z_t statistic respectively identifies the "significant" jumps CJ_t and continuous sample path components CSP_t by

$$\mathrm{CSP}_t \equiv \mathbb{I}(Z_t \leq \Phi_\alpha) \cdot \mathrm{RV}_t + \mathbb{I}(Z_t > \Phi_\alpha) \cdot \mathrm{BPV}_t,$$

$$\mathrm{CJ}_t = \mathbb{I}(Z_t > \Phi_\alpha) \cdot (\mathrm{RV}_t - \mathrm{BPV}_t).$$

where Z_t is the ratio-statistic defined in [22] and Φ_α is the cumulative distribution function(CDF) of a standard Gaussian distribution with α level of significance. The daily, weekly, and monthly average components of CSP_t and CJ_t are then constructed in the same manner as $\mathrm{RV}^{(l)}$. The model specification for the continuous HAR-J, namely, HAR-CJ, is given by

$$\mathrm{RV}_{t+h} = \beta_0 + \beta_d^c \mathrm{CSP}_t^{(1)} + \beta_w^c \mathrm{CSP}_t^{(5)} + \beta_m^c \mathrm{CSP}_t^{(22)} + \beta_d^j \mathrm{CJ}_t^{(1)} + \beta_w^j \mathrm{CJ}_t^{(5)} + \beta_m^j \mathrm{CJ}_t^{(22)} + e_{t+h}. \tag{5}$$

Note that compared with the HAR-J model, the HAR-CJ model explicitly controls for the weekly and monthly components of continuous jumps. Thus, the HAR-J model can be treated as a special and restrictive case of the HAR-CJ model for

$$\beta_d = \beta_d^c + \beta_d^j, \beta^j = \beta_d^j, \beta_w = \beta_w^c + \beta_w^j, \text{ and } \beta_m = \beta_m^c + \beta_m^j.$$

To capture the role of the "leverage effect" in predicting volatility dynamics, Patton and Sheppard [34] develop a series of models using signed realized measures. The first model, denoted as HAR-RS-I, decomposes the daily RV in the standard HAR model (3) into two asymmetric semi-variances RS_t^+ and RS_t^-:

$$RV_{t+h} = \beta_0 + \beta_d^+ RS_t^+ + \beta_d^- RS_t^- + \beta_w RV_t^{(5)} + \beta_m RV_t^{(22)} + e_{t+h}, \tag{6}$$

where $RS_t^- = \sum_{j=1}^{M} r_{t,j}^2 \cdot \mathbb{I}(r_{t,j} < 0)$ and $RS_t^+ = \sum_{j=1}^{M} r_{t,j}^2 \cdot \mathbb{I}(r_{t,j} > 0)$. To verify whether the realized semi-variances add something beyond the classical leverage effect, Patton and Sheppard [34] augment the HAR-RS-I model with a term interacting the lagged RV with an indicator for negative lagged daily returns $RV_t^{(1)} \cdot \mathbb{I}(r_t < 0)$. The second model in Eq. (7) is denoted as HAR-RS-II:

$$RV_{t+h} = \beta_0 + \beta_1 RV_t^{(1)} \cdot \mathbb{I}(r_t < 0) + \beta_d^+ RS_t^+ + \beta_d^- RS_t^- + \beta_w RV_t^{(5)} + \beta_m RV_t^{(22)} + e_{t+h}, \tag{7}$$

where $RV_t^{(1)} \cdot \mathbb{I}(r_t < 0)$ is designed to capture the effect of negative daily returns. As in the HAR-CJ model, the third and fourth models in [34], denoted as HAR-SJ-I and HAR-SJ-II, respectively, disentangle the signed jump variations and the BPV from the volatility process:

$$RV_{t+h} = \beta_0 + \beta_d^j SJ_t + \beta_d^{bpv} BPV_t + \beta_w RV_t^{(5)} + \beta_m RV_t^{(22)} + e_{t+h}, \tag{8}$$

$$RV_{t+h} = \beta_0 + \beta_d^{j-} SJ_t^- + \beta_d^{j+} SJ_t^+ + \beta_d^{bpv} BPV_t + \beta_w RV_t^{(5)} + \beta_m RV_t^{(22)} + e_{t+h}, \tag{9}$$

where $SJ_t = RS_t^+ - RS_t^-$, $SJ_t^+ = SJ_t \cdot \mathbb{I}(SJ_t > 0)$, and $SJ_t^- = SJ_t \cdot \mathbb{I}(SJ_t < 0)$. The HAR-SJ-II model extends the HAR-SJ-I model by being more flexible to allow the effect of a positive jump variation to differ in unsystematic ways from the effect of a negative jump variation.

The models discussed above can be generalized using the following formulation in practice:

$$y_{t+h} = x_t \beta + e_{t+h}$$

for $t = 1, \ldots, n$, where y_{t+h} stands for RV_{t+h} and variable x_t collects all the explanatory variables such that

$$x_t \equiv \begin{cases} \left[1, RV_t^{(1)}, RV_t^{(5)}, RV_t^{(22)}\right] & \text{for model HAR in (3),} \\ \left[1, RV_t^{(1)}, RV_t^{(5)}, RV_t^{(22)}, J_t\right] & \text{for model HAR-J in (4),} \\ \left[1, CSP_t^{(1)}, CSP_t^{(5)}, CSP_t^{(22)}, CJ_t^{(1)}, CJ_t^{(5)}, CJ_t^{(22)}\right] & \text{for model HAR-CJ in (5),} \\ \left[1, RS_t^-, RS_t^+, RV_t^{(5)}, RV_t^{(22)}\right] & \text{for model HAR-RS-I in (6),} \\ \left[1, RV_t^{(1)}\mathbb{I}_{r_t<0}, RS_t^-, RS_t^+, RV_t^{(5)}, RV_t^{(22)}\right] & \text{for model HAR-RS-II in (7),} \\ \left[1, SJ_t, BPV_t, RV_t^{(5)}, RV_t^{(22)}\right] & \text{for model HAR-SJ-I in (8),} \\ \left[1, SJ_t^-, SJ_t^+, BPV_t, RV_t^{(5)}, RV_t^{(22)}\right] & \text{for model HAR-SJ-II in (9).} \end{cases}$$

Since y_{t+h} is infeasible in period t, in practice, we usually obtain the estimated coefficient $\hat{\beta}$ from the following model:

$$y_t = x_{t-h}\beta + e_t, \tag{10}$$

in which both the independent and dependent variables are feasible in period $t = 1, \ldots, n$. Once the estimated coefficients $\hat{\beta}$ are obtained, the h-step-ahead forecast can be estimated by

$$\hat{y}_{t+h} = x_t\hat{\beta} \text{ for } t = 1, \ldots, n.$$

4 Machine Learning Strategy to Forecast Volatility

Machine learning tools are increasingly being used in the forecasting literature.[10] In this section, we briefly describe five of the most popular machine learning algorithms that have been shown to outperform econometric strategies when conducting forecast. That said, as Lehrer and Xie [26] stress the "No Free Lunch" theorem of Wolpert and Macready [41] indicates that in practice, multiple algorithms should be considered in any application.[11]

The first strategy we consider was developed to assist in the selection of predictors in the main model. Consider the regression model in Eq. (10), which contains many explanatory variables. To reduce the dimensionality of the set of the explanatory variables, Tibshirani [39] proposed the LASSO estimator of $\hat{\beta}$ that

---

[10]For example, Gu et al. [19] perform a comparative analysis of machine learning methods for measuring asset risk premia. Ban et al. [6] adopt machine learning methods for portfolio optimization. Beyond academic research, the popularity of algorithm-based quantitative exchange-traded funds (ETF) has increased among investors, in part since as LaFon [24] points out they both offer lower management fees and volatility than traditional stock-picking funds.

[11]This is an impossibility theorem that rules out the possibility that a general-purpose universal optimization strategy exists. As such, researchers should examine the sensitivity of their findings to alternative strategies.

solves

$$\hat{\beta}^{LASSO} = \arg\min_{\beta} \frac{1}{2n} \sum_{t=1}^{n} (y_t - x_{t-h}\beta)^2 + \lambda \sum_{j=1}^{L} |\beta_j|, \tag{11}$$

where λ is a tuning parameter that controls the penalty term. Using the estimates of Eq. (11), the h-step-ahead forecast is constructed in an identical manner as OLS:

$$\hat{y}_{t+h}^{LASSO} = x_t \hat{\beta}^{LASSO}.$$

The LASSO has been used in many applications and a general finding is that it is more likely to offer benefits relative to the OLS estimator when either (1) the number of regressors exceeds the number of observations, since it involves shrinkage, or (2) the number of parameters is large relative to the sample size, necessitating some form of regularization.

Recursive partitioning methods do not model the relationship between the explanatory variables and the outcome being forecasted with a regression model such as Eq. (10). Breiman et al. [10] propose a strategy known as classification and regression trees (CART), in which classification is used to forecast qualitative outcomes including categorical responses of non-numeric symbols and texts, and regression trees focus on quantitative response variables. Given the extreme volatility in Bitcoin gives rise to a continuous variable, we use regression trees (RT).

Consider a sample of $\{y_t, x_{t-h}\}_{t=1}^{n}$. Intuitively, RT operates in a similar manner to forward stepwise regression. A fast divide and conquer greedy algorithm considers all possible splits in each explanatory variable to recursively partition the data. Formally, at node τ containing n_τ observations with mean outcome $\overline{y}(\tau)$ of the tree can only be split by one selected explanatory variable into two leaves, denoted as τ_L and τ_R. The split is made at the explanatory variable which will lead to the largest reduction of a predetermined loss function between the two regions.[12] This splitting process continues at each new node until the gain to any forecast adds little value relative to a predetermined boundary. Forecasts at each final leaf are the fitted value from a local constant regression model.

Among machine learning strategies, the popularity of RT is high since the results of the analysis are easy to interpret. The algorithm that determines the split allows partitions among the entire covariate set to be described by a single tree. This contrasts with econometric approaches that begin by assuming a linear parametric form to explain the same process and as with the LASSO build a statistical model to make forecasts by selecting which explanatory variables to include. The tree

---

[12] A best split is determined by a given loss function, for example, the reduction of the sum of squared residuals (SSR). A simple regression will yield a sum of squared residuals, SSR_0. Suppose we can split the original sample into two subsamples such that $n = n_1 + n_2$. The RT method finds the best split of a sample to minimize the SSR from the two subsamples. That is, the SSR values computed from each subsample should follow: $SSR_1 + SSR_2 \leq SSR_0$.

structure considers the full set of explanatory variables and further allows for nonlinear predictor interactions that could be missed by conventional econometric approaches. The tree is simply a top-down, flowchart-like model which represents how the dataset was partitioned into numerous final leaf nodes. The predictions of a RT can be represented by a series of discontinuous flat surfaces forming an overall rough shape, whereas as we describe below visualizations of forecasts from other machine learning methods are not intuitive.

If the data are stationary and ergodic, the RT method often demonstrates gains in forecasting accuracy relative to OLS. Intuitively, we expect the RT method to perform well since it looks to partition the sample into subgroups with heterogeneous features. With time series data, it is likely that these splits will coincide with jumps and structural breaks. However, with primarily cross-sectional data, the statistical learning literature has discovered that individual regression trees are not powerful predictors relative to ensemble methods since they exhibit large variance [21].

Ensemble methods combine estimates from multiple outputs. Bootstrap aggregating decision trees (aka bagging) proposed in [8] and random forest (RF) developed in [9] are randomization-based ensemble methods. In bagging trees (BAG), trees are built on random bootstrap copies of the original data. The BAG algorithm is summarized as below:

(i) Take a random sample with replacement from the data.
(ii) Construct a regression tree.
(iii) Use the regression tree to make forecast, $\hat{f}$.
(iv) Repeat steps (i) to (iii), $b = 1, \ldots, B$ times and obtain $\hat{f}^b$ for each b.
(v) Take a simple average of the B forecasts $\hat{f}_{\text{BAG}} = \frac{1}{B} \sum_{b=1}^{B} \hat{f}^b$ and consider the averaged value $\hat{f}_{\text{BAG}}$ as the final forecast.

Forecast accuracy generally increases with the number of bootstrap samples in the training process. However, more bootstrap samples increase computational time. RF can be regarded as a less computationally intensive modification of BAG. Similar to BAG, RF also constructs B new trees with (conventional or moving block) bootstrap samples from the original dataset. With RF, at each node of every tree only a random sample (without replacement) of q predictors out of the total K ($q < K$) predictors is considered to make a split. This process is repeated and the remaining steps (iii)–(v) of the BAG algorithm are followed. Only if $q = K$, RF is roughly equivalent to BAG. RF forecasts involve B trees like BAG, but these trees are less correlated with each other since fewer variables are considered for a split at each node. The final RF forecast is calculated as the simple average of forecasts from each of these B trees.

The RT method can respond to highly local features in the data and is quite flexible at capturing nonlinear relationships. The final machine learning strategy we consider refines how highly local features of the data are captured. This strategy is known as boosting trees and was introduced in [21, Chapter 10]. Observations responsible for the local variation are given more weight in the fitting process. If the algorithm continues to fit those observations poorly, we reapply the algorithm with increased weight placed on those observations.

We consider a simple least squares boosting that fits RT ensembles (BOOST). Regression trees partition the space of all joint predictor variable values into disjoint regions R_j, $j = 1, 2, \ldots, J$, as represented by the terminal nodes of the tree. A constant j is assigned to each such region and the predictive rule is $X \in R_j \Rightarrow f(X) = \gamma_j$, where X is the matrix with tth component x_{t-h}. Thus, a tree can be formally expressed as $T(X, \Theta) = \sum_{j=1}^{J} \gamma_j \mathbb{I}(X \in R_j)$, with parameters $\Theta = \{R_j, \gamma_j\}_{j=1}^{J}$. The parameters are found by minimizing the risk

$$\hat{\Theta} = \arg\min_{\Theta} \sum_{j=1}^{J} \sum_{x_{t-h} \in R_j} \mathcal{L}(y_t, \gamma_j),$$

where $\mathcal{L}(\cdot)$ is the loss function, for example, the sum of squared residuals (SSR).

The BOOST method is a sum of all trees:

$$f_M(X) = \sum_{m=1}^{M} T(X; \Theta_m)$$

induced in a forward stagewise manner. At each step in the forward stagewise procedure, one must solve

$$\hat{\Theta}_m = \arg\min_{\Theta_m} \sum_{i=1}^{n} L\left(y_t, f_{m-1}(x_{t-h}) + T(x_{t-h}; \Theta_m)\right). \tag{12}$$

for the region set and constants $\Theta_m = \{R_{jm}, \gamma_{jm}\}_1^{J_m}$ of the next tree, given the current model $f_{m-1}(X)$. For squared-error loss, the solution is quite straightforward. It is simply the regression tree that best predicts the current residuals $y_t - f_{m-1}(x_{t-h})$, and $\hat{\gamma}_{jm}$ is the mean of these residuals in each corresponding region.

A popular alternative to a tree-based procedure to solve regression problems developed in the machine learning literature is the support vector regression (SVR). SVR has been found in numerous applications including Lehrer and Xie [26] to perform well in settings where there a small number of observations (< 500). Support vector regression is an extension of the support vector machine classification method of Vapnik [40]. The key feature of this algorithm is that it solves for a best fitting hyperplane using a learning algorithm that infers the functional relationships in the underlying dataset by following the structural risk minimization induction principle of Vapnik [40]. Since it looks for a functional relationship, it can find nonlinearities that many econometric procedures may miss using a prior chosen mapping that transforms the original data into a higher dimensional space.

Support vector regression was introduced in [16] and the true data that one wishes to forecast was known to be generated as $y_t = f(x_t) + e_t$, where f is unknown to the researcher and e_t is the error term. The SVR framework approximates $f(x_t)$ in terms of a set of basis functions: $\{h_s(\cdot)\}_{s=1}^{S}$:

$$y_t = f(x_t) + e_t = \sum_{s=1}^{S} \beta_s h_s(x_t) + e_t,$$

where $h_s(\cdot)$ is implicit and can be infinite-dimensional. The coefficients $\beta = [\beta_1, \cdots, \beta_S]^{\top}$ are estimated through the minimization of

$$H(\beta) = \sum_{t=1}^{T} V_\epsilon\left(y_t - f(x_t)\right) + \lambda \sum_{s=1}^{S} \beta_s^2, \tag{13}$$

where the loss function

$$V_\epsilon(r) = \begin{cases} 0 & \text{if } |r| < \epsilon \\ |r| - \epsilon & \text{otherwise} \end{cases}$$

is called an ϵ-insensitive error measure that ignores errors of size less than ϵ. The parameter ϵ is usually decided beforehand and λ can be estimated by cross-validation.

Suykens and Vandewalle [38] proposed a modification to the classic SVR that eliminates the hyperparameter ϵ and replaces the original ϵ-insensitive loss function with a least squares loss function. This is known as the least squares SVR (LSSVR). The LSSVR considers minimizing

$$H(\boldsymbol{\beta}) = \sum_{t=1}^{T} (y_t - f(x_t))^2 + \lambda \sum_{s=1}^{S} \beta_s^2, \tag{14}$$

where a squared loss function replaces $V_e(\cdot)$ for the LSSVR.

Estimating the nonlinear algorithms (13) and (14) requires a kernel-based procedure that can be interpreted as mapping the data from the original input space into a potentially higher-dimensional "feature space," where linear methods may then be used for estimation. The use of kernels enables us to avoid paying the computational penalty implicit in the number of dimensions, since it is possible to evaluate the training data in the feature space through indirect evaluation of the inner products. As such, the kernel function is essential to the performance of SVR and LSSVR since it contains all the information available in the model and training data to perform supervised learning, with the sole exception of having measures of the outcome variable. Formally, we define the kernel function $K(x, x_t) = h(x)h(x_t)^{\top}$ as the linear dot product of the nonlinear mapping for any input variable x. In our analysis, we consider the Gaussian kernel (sometimes referred to as "radial basis function" and "Gaussian radial basis function" in the support vector literature):

$$K(x, x_t) = \exp\left(-\frac{\|x - x_t\|^2}{2\sigma_x^2}\right),$$

where the hyperparameters σ_x^2 and γ.

In our main analysis, we use a tenfold cross-validation to pick the tuning parameters for LASSO, SVR, and LSSVR. For tree-type machine learning methods, we set the basic hyperparameters of a regression tree at their default values. These include but not limited to: (1) the split criterion is SSR; (2) the maximum number of split is 10 for BOOST and $n - 1$ for others; (3) the minimum leaf size is 1; (4) the number of predictors for split is $K/3$ for RF and K for others; and (5) the number of learning cycles is $B = 100$ for ensemble learning methods. We examine the robustness to different values for the hyperparameters in Sect. 7.3.

5   Social Media Data

Substantial progress has been made in the machine learning literature on quickly converting text to data, generating real-time information on social media content. To measure social media sentiment, we selected an algorithm introduced in [17] that pre-trained a five-hidden-layer neural model on 124.6 million tweets containing emojis in order to learn better representations of the emotional context embedded in the tweet. This algorithm was developed to provide a means to learn representations of emotional content in texts and is available with pre-processing code, examples of usage, and benchmark datasets, among other features at github.com/bfelbo/deepmoji. The pre-training data is split into a training, validation, and test set, where the validation and test set are randomly sampled in such a way that each emoji is equally represented. This data includes all English Twitter messages without URLs within the period considered that contained an emoji. The fifth layer of the algorithm focuses on attention and takes inputs from the prior levels which uses a multi-class learners to decode the text and emojis itself. See [17] for further details. Thus, an emoji is viewed as a labeling system for emotional content.

The construction of the algorithm began by acquiring a dataset of 55 billion tweets, of which all tweets with emojis were used to train a deep learning model. That is, the text in the tweet was used to predict which emoji was included with what tweet. The premise of this algorithm is that if it could understand which emoji was included with a given sentence in the tweet, then it has a good understanding of the emotional content of that sentence. The goal of the algorithm is to understand the emotions underlying from the words that an individual tweets. The key feature of this algorithm compared to one that simply scores words themselves is that it is better able to detect irony and sarcasm. As such, the algorithm does not score individual emotion words in a Twitter message, but rather calculates a score based on the probability of each of 64 different emojis capturing the sentiment in the full Twitter message taking the structure of the sentence into consideration. Thus, each emoji has a fixed score and the sentiment of a message is a weighted average of the type of mood being conveyed, since messages containing multiple words are translated to a set of emojis to capture the emotion of the words within.

In brief, for a random sample of 10% of all tweets every minute, the score is calculated as an equal tweet weight average of the sentiment values of the words within them.[13] That is, we apply the pre-trained classifier of Felbo et al. [17] to score each of these tweets and note that there are computational challenges related to data storage when using very large datasets to undertake sentiment analysis. In our application, the number of tweets per hour generally varies between 120,000 and 200,000 tweets per hour in our 10% random sample. We denote the minute-level sentiment index as the U.S. Sentiment Index (USSI).

In other words, if there are 10,000 tweets each hour, we first convert each tweet to a set of emojis. Then we convert the emojis to numerical values based on a fixed mapping related to their emotional content. For each of the 10,000 tweets posted in that hour, we next calculate the average of these scores as the emotion content or sentiment of that individual tweet. We then calculate the equal weighted average of these tweet-specific scores to gain an hourly measure. Thus, each tweet is treated equally irrespective of whether one tweet contains more emojis than the other. This is then repeated for each hour of each day in our sample providing us with a large time series.

Similar to many other text mining tasks, this sentiment analysis was initially designed to deal with English text. It would be simple to apply an off-the-shelf machine translation tool in the spirit of Google translate to generate pseudo-parallel corpora and then learn bilingual representations for downstream sentiment classification task of tweets that were initially posted in different languages. That said, due to the ubiquitous usage of emojis across languages and their functionality of expressing sentiment, alternative emoji powered algorithms have been developed with other languages. These have smaller training datasets since most tweets are in English and it is an open question as to whether they perform better than applying the [17] algorithm to pseudo-tweets.

Note that the way we construct USSI does not necessarily focus on sentiment related to cyptocurrency only as in [29]. Sentiment, in- and off-market, has been a major factor affecting the price of financial asset [23]. Empirical works have documented that large national sentiment swing can cause large fluctuation in asset prices, for example, [5, 37]. It is therefore natural to assume that national sentiment can affect financial market volatility.

---

[13]This is a 10% random sample of all tweets since the USSI was designed to measure the real-time mood of the nation and the algorithm does not restrict the calculations to Twitter accounts that either mention any specific stock or are classified as being a market participant.

Data timing presents a serious challenge in using minutely measures of the USSI to forecast the daily Bitcoin RV. Since USSI is constructed at minute level, we convert the minute-level USSI to match the daily sampling frequency of Bitcoin RV using the heterogeneous mixed data sampling (H-MIDAS) method of Lehrer et al. [28].[14] This allows us to transform 1,172,747 minute-level observations for USSI variable via a step function to allow for heterogeneous effects of different high-frequency observations into 775 daily observations for the USSI at different forecast horizons. This step function produces a different weight on the hourly levels in the time series and can capture the relative importance of user's emotional content across the day since the type of users varies in a manner that may be related to BTC volatility. The estimated weights used in the H-MIDAS transformation for our application are presented in Fig. 2.

Last, Table 1 presents the summary statistics for the RV data and p-values from both the Jarque–Bera test for normality and the Augmented Dickey–Fuller (ADF) tests for unit root. We consider the first half sample, the second half sample, and full sample. Each of the series exhibits tremendous variability and a large range across the sample period. Further, none of the series are normally distributed or nonstationary at 5% level.

6   Empirical Exercise

To examine the relative prediction efficiency of different HAR estimators, we conduct an h-step-ahead rolling window exercise of forecasting the BTC/USD RV for different forecasting horizons.[15] Table 2 lists each estimator analyzed in the exercise. For all the HAR-type estimators in Panel A (except the HAR-Full model which uses all the lagged covariates from 1 to 30), we set $l = [1, 7, 30]$. For the machine learning methods in Panel B, the input data includes all covariates as the one for HAR-Full model. Throughout the experiment, the window length is fixed at $WL = 400$ observations. Our conclusions are robust to other window lengths as discussed in Sect. 7.1.

To examine if the sentiment data extracted from social media improves forecasts, we contrasted the forecast from models that exclude the USSI to models that include the USSI as a predictor. We denote methods incorporating the USSI variable with

---

[14]We provide full details on this strategy in the appendix. In practice, we need to select the lag index $l = [l_1, \ldots, l_p]$ and determine the weight set $\mathcal{W}$ before the estimation. In this study, we set $\mathcal{W} \equiv \{\boldsymbol{w} \in \mathbb{R}^p : \sum_{j=1}^{p} w_j = 1\}$ and use OLS to estimate $\widehat{\beta \boldsymbol{w}}$. We consider $h = 1, 2, 4$, and 7 as in the main exercise. For the lag index, we consider $l = [1 : 5 : 1440]$, given there are 1440 minutes per day.

[15]Additional results using both the GARCH(1, 1) and the ARFIMA(p, d, q) models are available upon request. These estimators performed poorly relative to the HAR model and as such are not included for space considerations.

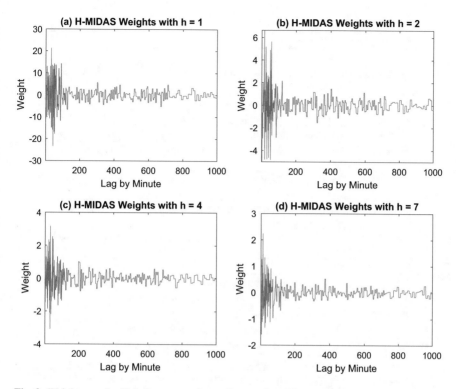

Fig. 2 Weights on the high-frequency observations under different lag indices. (**a**) H-MIDAS weights with h = 1. (**b**) H-MIDAS weights with h = 2. (**c**) H-MIDAS weights with h = 4. (**d**) H-MIDAS weights with h = 7

Table 1 Descriptive statistics

Statistics	Realized variance			USSI
	First half	Second half	Full sample	
Mean	43.4667	12.1959	27.8313	117.4024
Median	31.2213	7.0108	17.4019	125.8772
Maximum	197.6081	115.6538	197.6081	657.4327
Minimum	5.0327	0.5241	0.5241	−866.6793
Std. dev.	38.0177	15.6177	32.9815	179.1662
Skewness	2.1470	3.3633	2.6013	−0.8223
Kurtosis	7.8369	18.2259	11.2147	5.8747
Jarque–Bera	0.0000	0.0000	0.0000	0.0000
ADF test	0.0000	0.0000	0.0000	0.0000

∗ symbol in each table. The results of the prediction experiment are presented in Table 3. The estimation strategy is listed in the first column and the remaining columns present alternative criteria to evaluate the forecasting performance. The criteria include the mean squared forecast error (MSFE), quasi-likelihood (QLIKE),

Table 2 List of estimators

Panel A: conventional regression		
(1)	AR(1)	A simple autoregressive model
(2)	HAR-Full	The HAR model proposed in [11] with $l = [1, 2, \dots, 30]$, which is equivalent to AR(30)
(3)	HAR	The conventional HAR model proposed in [11] with $l = [1, 7, 30]$
(4)	HAR-J	The HAR model with jump component proposed in [4]
(5)	HAR-CJ	The HAR model with continuous jump component proposed in [4]
(6)	HAR-RS-I	The HAR model with semi-variance components (Type I) proposed in [34]
(7)	HAR-RS-II	The HAR model with semi-variance components (Type II) proposed in [34]
(8)	HAR-SJ-I	The HAR model with semi-variance and jump components (Type I) proposed in [34]
(9)	HAR-SJ-II	The HAR model with semi-variance and jump components (Type II) proposed in [34]
Panel B: machine learning strategy		
(10)	LASSO	The least absolute shrinkage and selection operator by Tibshirani [39]
(11)	RT	The regression tree method proposed by Breiman et al. [10]
(12)	BOOST	The boosting tree method described in [21]
(13)	BAG	The bagging tree method proposed by Breiman [8]
(14)	RF	The random forest method proposed by Breiman [9]
(15)	SVR	The support vector machine for regression by Drucker et al. [16]
(16)	LSSVR	The least squares support vector regression by Suykens and Vandewalle [38]

mean absolute forecast error (MAFE), and standard deviation of forecast error (SDFE) that are calculated as

$$\text{MSFE}(h) = \frac{1}{V} \sum_{j=1}^{V} e_{T_j,h}^2, \tag{15}$$

$$\text{QLIKE}(h) = \frac{1}{V} \sum_{j=1}^{V} \left(\log \hat{y}_{T_j,h} + \frac{y_{T_j,h}}{\hat{y}_{T_j,h}} \right), \tag{16}$$

$$\text{MAFE}(h) = \frac{1}{V} \sum_{j=1}^{V} |e_{T_j,h}|, \tag{17}$$

$$\text{SDFE}(h) = \sqrt{\frac{1}{V-1} \left(e_{T_j,h} - \frac{1}{V} \sum_{j=1}^{V} e_{T_j,h} \right)^2}, \tag{18}$$

Table 3 Forecasting performance of strategies in the main exercise

Method	MSFE	QLIKE	MAFE	SDFE	Pseudo R^2
Panel A: h = 1					
HAR	1666.8492	0.5356	17.0279	40.8271	0.3173
HAR-CJ	1690.4306	0.5299	17.1844	41.1148	0.3076
HAR-RS-II	2377.5159	0.5471	17.6936	48.7598	0.0262
LASSO	1726.2453	0.5649	17.4025	41.5481	0.2929
BOOST	3003.8597	2.3176	27.7473	54.8075	−0.2304
RF	1680.2756	0.4374	16.7922	40.9912	0.3118
BAG	1628.2674	0.4504	16.8285	40.3518	0.3331
SVR	2218.8594	1.3751	20.0765	47.1048	0.0912
LSSVR	1628.6800	0.4858	16.0397	40.3569	0.3329
HAR*	1459.7257	1.5488	19.2790	38.2064	0.4021
HAR-CJ*	1477.1162	1.7526	19.3398	38.4333	0.3950
HAR-RS-II*	2047.5427	1.5013	19.9458	45.2498	0.1613
LASSO*	1497.0621	1.8256	19.1215	38.6919	0.3868
BOOST*	1312.6693	2.4524	18.6123	36.2308	0.4623
RF*	1178.6862	0.3794	14.4059	34.3320	0.5172
BAG*	**1035.7081**	**0.3635**	**13.8235**	**32.1824**	**0.5758**
SVR*	2226.7603	1.4075	20.2407	47.1886	0.0879
LSSVR*	1494.0104	1.2801	16.4454	38.6524	0.3881
Panel B: h = 2					
HAR	2066.1864	0.6681	18.6000	45.4553	0.1558
HAR-CJ	2110.0401	0.6696	19.0773	45.9352	0.1379
HAR-RS-II	2028.5347	0.6838	18.8080	45.0393	0.1712
LASSO	2081.8131	0.6936	18.9990	45.6269	0.1494
BOOST	3615.6614	3.1268	28.7990	60.1304	−0.4772
RF	1880.7996	0.5376	17.1419	43.3682	0.2316
BAG	1994.2700	0.5733	17.8611	44.6572	0.1852
SVR	2224.9431	1.3804	20.1089	47.1693	0.0910
LSSVR	1872.4412	0.6192	16.5504	43.2717	0.2350
HAR*	1803.3278	1.5095	21.2684	42.4656	0.2632
HAR-CJ*	1832.2437	1.9863	21.4102	42.8047	0.2514
HAR-RS-II*	1783.0826	2.3170	21.4938	42.2266	0.2715
LASSO*	1817.9238	1.8877	20.8886	42.6371	0.2573
BOOST*	1832.3453	2.8026	21.2695	42.8059	0.2514
RF*	1511.0049	**0.4593**	15.5323	38.8716	0.3827
BAG*	**1428.6900**	0.4654	**15.1394**	**37.7980**	**0.4163**
SVR*	2232.1703	1.4105	20.2573	47.2458	0.0880
LSSVR*	1702.2016	1.0489	17.0578	41.2577	0.3045

(continued)

Table 3 (continued)

Method	MSFE	QLIKE	MAFE	SDFE	Pseudo R^2
Panel C: h = 4					
HAR	2064.3686	0.8043	19.5208	45.4353	0.1610
HAR-CJ	2100.3712	0.8181	20.0445	45.8298	0.1464
HAR-RS-II	2057.6179	0.8077	19.6796	45.3610	0.1638
LASSO	2068.0111	0.8231	19.8920	45.4754	0.1595
BOOST	2348.6453	4.6780	24.2304	48.4628	0.0455
RF	1936.6858	0.5980	17.5443	44.0078	0.2129
BAG	2035.9166	0.6470	17.9963	45.1211	0.1726
SVR	2235.8229	1.3882	20.1259	47.2845	0.0913
LSSVR	1963.1437	0.9329	17.3076	44.3074	0.2022
HAR*	1630.8296	2.5250	21.8847	40.3835	0.3372
HAR-CJ*	1641.7051	2.0302	22.0168	40.5180	0.3328
HAR-RS-II*	1638.4781	2.1343	21.9431	40.4781	0.3341
LASSO*	1636.6835	2.3301	21.5890	40.4559	0.3348
BOOST*	1447.7824	3.3492	20.7355	38.0497	0.4116
RF*	1205.4310	**0.4396**	**14.4692**	34.7193	0.5101
BAG*	**1075.4364**	0.4579	14.8433	**32.7938**	**0.5629**
SVR*	2241.9418	1.4129	20.2578	47.3491	0.0889
LSSVR*	1526.7558	1.3300	17.1047	39.0737	0.3795
Panel D: h = 7					
HAR	2108.7457	0.8738	19.9327	45.9211	0.1497
HAR-CJ	2119.8357	0.8872	20.2362	46.0417	0.1452
HAR-RS-II	2142.9983	0.9661	20.2572	46.2925	0.1359
LASSO	2100.7324	0.8939	20.2446	45.8337	0.1529
BOOST	2616.8282	2.9902	24.2636	51.1549	-0.0552
RF	1769.0548	0.5524	15.7001	42.0601	0.2867
BAG	1822.8425	0.5648	16.3405	42.6948	0.2650
SVR	2253.5470	1.4045	20.1991	47.4715	0.0913
LSSVR	2000.7088	0.8148	17.7411	44.7293	0.1933
HAR*	1703.6884	1.6255	22.3689	41.2758	0.3130
HAR-CJ*	1705.7788	1.7958	22.2928	41.3011	0.3122
HAR-RS-II*	1716.5970	1.5604	22.4318	41.4318	0.3078
LASSO*	1710.4945	4.1087	22.1347	41.3581	0.3103
BOOST*	1589.2483	2.8654	19.7297	39.8654	0.3592
RF*	1273.7997	**0.4656**	**14.4000**	35.6903	0.4864
BAG*	**1257.6470**	0.5070	15.1803	**35.4633**	**0.4929**
SVR*	2257.5369	1.4195	20.2793	47.5135	0.0897
LSSVR*	1561.7929	1.0831	18.0236	39.5195	0.3702

The best result under each criterion is highlighted in boldface

where $e_{T_j,h} = y_{T_j,h} - \hat{y}_{T_j,h}$ is the forecast error and $\hat{y}_{iT_j,h}$ is the h-day ahead forecast with information up to T_j that stands for the last observation in each of the V rolling windows. We also report the Pseudo R^2 of the Mincer–Zarnowitz regression [32] given by:

$$y_{T_j,h} = a + b\hat{y}_{T_j,h} + u_{T_j}, \text{ for } j = 1, 2, \ldots, V, \tag{19}$$

Each panel in Table 3 presents the result corresponding to a specific forecasting horizon. We consider various forecasting horizons $h = 1, 2, 4$, and 7.

To ease interpretation, we focus on the following representative methods: HAR, HAR-CJ, HAR-RS-II, LASSO, RF, BAG, and LSSVR with and without the USSI variable. Comparison results between all methods listed in Table 2 are available upon request. We find consistent ranking of modeling methods across all forecast horizons. The tree-based machine learning methods (BAG and RF) have superior performance than all others for each panel. Moreover, methods with USSI (indicated by $*$) always dominate those without USSI, which indicates the importance of incorporating social media sentiment data. We also discover that the conventional econometric methods have unstable performance, for example, the HAR-RS-II model without USSI has the worst performance when $h = 1$, but its performance improves when $h = 2$. The mixed performance of the linear models implies that this restrictive formulation may not be robust to model the highly volatile BTC/USD RV process.

To examine if the improvement from the BAG and RF methods is statistically significant, we perform the modified Giacomini–White test [18] of the null hypothesis that the *column method* performs equally well as the *row method* in terms of MAFE. The corresponding p values are presented in Table 4 for $h = 1, 2, 4, 7$. We see that the gains in forecast accuracy from BAG* and RF* relative to all other strategies are statistically significant, although results between BAG* and RF* are statistically indistinguishable.

7 Robustness Check

In this section, we perform four robustness checks of our main results. We first vary the window length for the rolling window exercise in Sect. 7.1. We next consider different sample periods in Sect. 7.2. We explore the use of different hyperparameters for the machine learning methods in Sect. 7.3. Our final robustness check examines if BTC/USD RV is correlated with other types of financial markets by including mainstream assets RV as additional covariates. Each of these robustness checks that are ported in the main text considers $h = 1$.[16]

---

[16]Although not reported due to space considerations, we investigated other forecasting horizons and our main findings are robust.

Table 4 Giacomini–White test results

	HAR	HAR-CJ	RS-II	LASSO	BOOST	RF	BAG	SVR	LSSVR	HAR*	HAR-CJ*	RS-II*	LASSO*	BOOST*	RF*	BAG*	SVR*
Panel A: h = 1																	
HAR	–	–	–	–	–	–	–	–	–	–	–	–	–	–	–	–	–
HAR-CJ	0.1609	–	–	–	–	–	–	–	–	–	–	–	–	–	–	–	–
HAR-RS-II	0.3155	0.4160	–	–	–	–	–	–	–	–	–	–	–	–	–	–	–
LASSO	0.0169	0.2295	0.6622	–	–	–	–	–	–	–	–	–	–	–	–	–	–
BOOST	0.0000	0.0000	0.0000	0.0000	–	–	–	–	–	–	–	–	–	–	–	–	–
RF	0.5722	0.4883	0.4241	0.3457	0.0000	–	–	–	–	–	–	–	–	–	–	–	–
BAG	0.9714	0.9095	0.6806	0.7371	0.0000	0.0811	–	–	–	–	–	–	–	–	–	–	–
SVR	0.0091	0.0178	0.1527	0.0211	0.0006	0.0002	0.0038	–	–	–	–	–	–	–	–	–	–
LSSVR	0.3449	0.2975	0.3122	0.1931	0.0000	0.3631	0.0946	0.0000	–	–	–	–	–	–	–	–	–
HAR*	0.0013	0.0033	0.1384	0.0091	0.0000	0.0134	0.0549	0.5354	0.0047	–	–	–	–	–	–	–	–
HAR-CJ*	0.0008	0.0020	0.1133	0.0065	0.0000	0.0135	0.0526	0.5773	0.0050	0.4450	–	–	–	–	–	–	–
HAR-RS-II*	0.0005	0.0006	0.0014	0.0025	0.0002	0.0265	0.0643	0.9384	0.0172	0.3066	0.3243	–	–	–	–	–	–
LASSO*	0.0017	0.0043	0.1688	0.0100	0.0000	0.0162	0.0662	0.4478	0.0059	0.2134	0.1287	0.2044	–	–	–	–	–
BOOST*	0.2661	0.3247	0.6207	0.4001	0.0000	0.1030	0.2551	0.3348	0.0500	0.6188	0.5917	0.4380	0.7039	–	–	–	–
RF*	0.0147	0.0129	0.0402	0.0061	0.0000	0.0009	0.0001	0.0000	0.0077	0.0000	0.0000	0.0003	0.0000	0.0001	–	–	–
BAG*	0.0104	0.0095	0.0322	0.0048	0.0000	0.0029	0.0004	0.0000	0.0085	0.0000	0.0000	0.0002	0.0000	0.0000	0.4810	–	–
SVR*	0.0061	0.0125	0.1269	0.0146	0.0008	0.0001	0.0024	0.0000	0.0000	0.4554	0.4960	0.8617	0.3744	0.2843	0.0000	0.0000	–
LSSVR*	0.5912	0.5127	0.4497	0.3794	0.0000	0.9770	0.3969	0.0001	0.3323	0.0031	0.0036	0.0202	0.0044	0.0772	0.0002	0.0005	0.0000
Panel B: h = 2																	
HAR	–	–	–	–	–	–	–	–	–	–	–	–	–	–	–	–	–
HAR-CJ	0.0012	–	–	–	–	–	–	–	–	–	–	–	–	–	–	–	–
HAR-RS-II	0.4668	0.4161	–	–	–	–	–	–	–	–	–	–	–	–	–	–	–
LASSO	0.0038	0.7427	0.5100	–	–	–	–	–	–	–	–	–	–	–	–	–	–
BOOST	0.0000	0.0000	0.0000	0.0000	–	–	–	–	–	–	–	–	–	–	–	–	–

	HAR	HAR-CJ	HAR-RS-II	LASSO	BOOST	RF	BAG	SVR	LSSVR	HAR*	HAR-CJ*	HAR-RS-II*	LASSO*	BOOST*	RF*	BAG*	SVR*	LSSVR*
RF	0.1446	0.0648	0.0850	0.0534	0.0000	–	–	–	–	–	–	–	–	–	–	–	–	–
BAG	0.5707	0.3177	0.4267	0.3205	0.0000	0.0287	–	–	–	–	–	–	–	–	–	–	–	–
SVR	0.1475	0.3618	0.1610	0.2555	0.0000	0.0054	0.0715	–	–	–	–	–	–	–	–	–	–	–
LSSVR	0.0596	0.0324	0.0291	0.0172	0.0000	0.3543	0.0981	0.0715	–	–	–	–	–	–	–	–	–	–
HAR*	0.0007	0.0091	0.0036	0.0041	0.0002	0.0002	0.0052	0.0981	0.0001	–	–	–	–	–	–	–	–	–
HAR-CJ*	0.0002	0.0034	0.0016	0.0017	0.0002	0.0002	0.0040	0.3019	0.0001	0.2366	–	–	–	–	–	–	–	–
HAR-RS-II*	0.0004	0.0054	0.0008	0.0020	0.0002	0.0000	0.0020	0.2221	0.0000	0.3331	0.7645	–	–	–	–	–	–	–
LASSO*	0.0029	0.0275	0.0103	0.0131	0.0001	0.0006	0.0110	0.5042	0.0001	0.0010	0.0015	0.0068	–	–	–	–	–	–
BOOST*	0.0872	0.1611	0.1083	0.1471	0.0002	0.0041	0.0297	0.4905	0.0022	0.9994	0.9240	0.8782	0.7968	–	–	–	–	–
RF*	0.0146	0.0065	0.0074	0.0043	0.0000	0.0316	0.0048	0.0002	0.3616	0.0002	0.0000	0.0000	0.0000	0.0000	–	–	–	–
BAG*	0.0127	0.0065	0.0078	0.0048	0.0000	0.0569	0.0157	0.0003	0.1931	0.0003	0.0000	0.0000	0.0000	0.0000	0.2217	–	–	–
SVR*	0.1131	0.2985	0.1201	0.1992	0.0000	0.0037	0.0549	0.0005	0.0000	0.4098	0.3609	0.2762	0.5892	0.5486	0.0001	0.0002	–	–
LSSVR*	0.1666	0.0934	0.1012	0.0688	0.0000	0.8533	0.3002	0.0015	0.2814	0.0000	0.0000	0.0000	0.0000	0.0031	0.0689	0.0344	0.0009	–

Panel C: h = 4

	HAR	HAR-CJ	HAR-RS-II	LASSO	BOOST	RF	BAG	SVR	LSSVR	HAR*	HAR-CJ*	HAR-RS-II*
HAR	–											
HAR-CJ	0.0224	–										
HAR-RS-II	0.6981	0.5251	–									
LASSO	0.0789	0.6490	0.6093	–								
BOOST	0.0033	0.0083	0.0048	0.0081	–							
RF	0.0106	0.0063	0.0006	0.0030	0.0000	–						
BAG	0.1467	0.0771	0.0527	0.0694	0.0000	0.0098	–					
SVR	0.4655	0.9347	0.5465	0.7681	0.0000	0.0322	0.0034	–				
LSSVR	0.0014	0.0012	0.0000	0.0001	0.0000	0.8528	0.3372	0.0014	–			
HAR*	0.0276	0.1108	0.0359	0.0621	0.1960	0.0004	0.0075	0.1806	0.0001	–		
HAR-CJ*	0.0157	0.0717	0.0240	0.0391	0.2175	0.0003	0.0057	0.1552	0.0001	0.3451	–	
HAR-RS-II*	0.0355	0.1292	0.0314	0.0696	0.2109	0.0003	0.0055	0.1622	0.0001	0.8194	0.8270	–

(continued)

Table 4 (continued)

	HAR	HAR-CJ	RS-II	LASSO	BOOST	RF	BAG	SVR	LSSVR	HAR*	HAR-CJ*	RS-II*	LASSO*	BOOST*	RF*	BAG*	SVR*
LASSO*	**0.0466**	0.1687	**0.0597**	0.0984	0.1420	**0.0008**	**0.0120**	0.2546	**0.0002**	**0.0013**	**0.0044**	0.1774	–	–	–	–	–
BOOST*	0.4973	0.7091	0.5281	0.6364	**0.0477**	**0.0252**	0.0991	0.7525	**0.0392**	0.4558	0.4091	0.4175	0.5780	–	–	–	–
RF*	**0.0000**	**0.0000**	**0.0000**	**0.0000**	**0.0000**	**0.0031**	**0.0004**	**0.0000**	**0.0041**	**0.0000**	**0.0000**	**0.0000**	**0.0000**	**0.0000**	–	–	–
BAG*	**0.0002**	**0.0001**	**0.0000**	**0.0000**	**0.0000**	**0.0272**	**0.0058**	**0.0001**	**0.0265**	**0.0000**	**0.0000**	**0.0000**	**0.0000**	**0.0000**	0.6820	–	–
SVR*	0.3783	0.8311	0.4400	0.6475	**0.0393**	**0.0025**	0.0577	**0.0117**	**0.0010**	0.2179	0.1882	0.1972	0.3025	0.8057	**0.0000**	**0.0001**	–
LSSVR*	**0.0113**	**0.0069**	**0.0010**	**0.0026**	**0.0000**	0.9283	0.3331	**0.0057**	0.7714	**0.0000**	**0.0000**	**0.0000**	**0.0000**	**0.0078**	**0.0004**	**0.0052**	**0.0042**

Panel D: h = 7

	HAR	HAR-CJ	RS-II	LASSO	BOOST	RF	BAG	SVR	LSSVR	HAR*	HAR-CJ*	RS-II*	LASSO*	BOOST*	RF*	BAG*	SVR*
HAR	–	–	–	–	–	–	–	–	–	–	–	–	–	–	–	–	–
HAR-CJ	0.1065	–	–	–	–	–	–	–	–	–	–	–	–	–	–	–	–
HAR-RS-II	0.1331	0.9319	–	–	–	–	–	–	–	–	–	–	–	–	–	–	–
LASSO	0.2138	0.9811	0.9725	–	–	–	–	–	–	–	–	–	–	–	–	–	–
BOOST	0.0533	0.0770	0.0790	0.0687	–	–	–	–	–	–	–	–	–	–	–	–	–
RF	**0.0000**	**0.0000**	**0.0000**	**0.0000**	**0.0000**	–	–	–	–	–	–	–	–	–	–	–	–
BAG	**0.0001**	**0.0001**	**0.0001**	**0.0000**	**0.0000**	**0.0032**	–	–	–	–	–	–	–	–	–	–	–
SVR	0.7526	0.9693	0.9494	0.9552	0.0607	**0.0000**	**0.0005**	–	–	–	–	–	–	–	–	–	–
LSSVR	**0.0022**	**0.0012**	**0.0035**	**0.0002**	**0.0009**	**0.0001**	**0.0087**	**0.0138**	–	–	–	–	–	–	–	–	–
HAR*	**0.0412**	0.0756	0.0913	0.0672	0.4768	**0.0000**	**0.0000**	0.1501	**0.0004**	–	–	–	–	–	–	–	–
HAR-CJ*	**0.0445**	0.0781	0.0954	0.0748	0.4590	**0.0000**	**0.0000**	0.1712	**0.0005**	0.6254	–	–	–	–	–	–	–
HAR-RS-II*	**0.0302**	0.0568	0.0667	0.0539	0.4943	**0.0000**	**0.0000**	0.1388	**0.0004**	0.7164	0.4060	–	–	–	–	–	–
LASSO*	0.0571	0.1043	0.1253	0.0918	0.4178	**0.0000**	**0.0000**	0.1812	**0.0004**	0.1073	0.4999	0.2253	–	–	–	–	–
BOOST*	0.9169	0.7966	0.7895	0.7876	**0.0118**	**0.0040**	**0.0199**	0.8192	0.2342	0.1673	0.1791	0.1613	0.2063	–	–	–	–
RF*	**0.0000**	**0.0000**	**0.0000**	**0.0000**	**0.0000**	0.1657	**0.0276**	**0.0002**	**0.0009**	**0.0000**	**0.0000**	**0.0000**	**0.0000**	**0.0000**	–	–	–
BAG*	**0.0008**	**0.0004**	**0.0005**	**0.0003**	**0.0000**	0.8017	0.3219	**0.0023**	**0.0316**	**0.0000**	**0.0000**	**0.0000**	**0.0000**	**0.0000**	0.0517	–	–
SVR*	0.6853	0.9647	0.9809	0.9663	0.0673	**0.0000**	**0.0004**	**0.0373**	**0.0121**	0.1670	0.1894	0.1547	0.2013	0.7898	**0.0002**	**0.0021**	–
LSSVR*	**0.0423**	**0.0192**	**0.0301**	**0.0168**	**0.0056**	**0.0027**	**0.0389**	0.0967	0.7300	**0.0000**	**0.0000**	**0.0000**	**0.0000**	0.2758	**0.0000**	**0.0013**	0.0871

p-values smaller than 5% are highlighted in boldface

7.1 Different Window Lengths

In the main exercise, we set the window length $WL = 400$. In this section, we also tried other window lengths $WL = 300$ and 500. Table 5 shows the forecasting performance of all the estimators for various window lengths. In all the cases BAG* and RF* yield smallest MSFE, MAFE, and SDFE and the largest Pseudo R^2. We examine the statistical significance of the improvement on forecasting accuracy in Table 6. The small p-values on testing BAG* and RF* against other strategies indicate that the forecasting accuracy improvement is statistically significant at the 5% level.

7.2 Different Sample Periods

In this section, we partition the entire sample period in half: the first subsample period runs from May 20, 2015, to July 29, 2016, and the second subsample period runs from July 30, 2016, to Aug 20, 2017. We carry out the similar out-of-sample analysis with $WL = 200$ for the two subsamples in Table 7 Panels A and B, respectively. We also examine the statistical significance in Table 8. The previous conclusions remain basically unchanged under the subsamples.

7.3 Different Tuning Parameters

In this section, we examine the effect of different tuning parameters for the machine learning methods. We consider a different set of tuning parameters: $B = 20$ for RF and BAG, and $\lambda = 0.5$ for LASSO, SVR, and LSSVR. The machine learning methods with the second set of tuning parameters are labeled as RF2, BAG2, and LASSO2. We replicate the main empirical exercise in Sect. 6 and compare the performance of machine learning methods with different tuning parameters.

The results are presented in Tables 9 and 10. Changes in the considered tuning parameters generally have marginal effects on the forecasting performance, although the results for the second tuning parameters are slightly worse than those under the default setting. Last, social media sentiment data plays a crucial role on improving the out-of-sample performance in each of these exercises.

7.4 Incorporating Mainstream Assets as Extra Covariates

In this section, we examine if the mainstream asset class has spillover effect on BTC/USD RV. We include the RVs of the S&P and NASDAQ indices ETFs (ticker

Table 5 Forecasting performance by different window lengths ($h = 1$)

Method	MSFE	QLIKE	MAFE	SDFE	Pseudo R^2
Panel A: $WL = 300$					
HAR	1626.1783	0.4658	17.3249	40.3259	0.3036
HAR-CJ	1691.6375	0.4806	17.3407	41.1295	0.2756
HAR-RS-II	2427.8630	0.4611	17.8985	49.2733	-0.0397
LASSO	1676.9910	0.4912	17.6299	40.9511	0.2819
BOOST	3902.5683	5.0682	30.7322	62.4705	-0.6712
RF	1725.6296	0.4611	18.2421	41.5407	0.2610
BAG	1633.2346	0.4540	17.5508	40.4133	0.3006
SVR	2017.5537	1.3343	19.3042	44.9172	0.1360
LSSVR	1632.6040	0.4961	17.3568	40.4055	0.3009
HAR*	1473.7240	1.8110	19.4883	38.3891	0.3689
HAR-CJ*	1526.2976	2.4053	19.6475	39.0679	0.3464
HAR-RS-II*	2159.5044	1.6874	20.1350	46.4705	0.0752
LASSO*	1510.2217	2.0658	19.4269	38.8616	0.3533
BOOST*	1531.6126	5.0383	20.4951	39.1358	0.3441
RF*	1277.5211	0.3751	15.7195	35.7424	0.4529
BAG*	**1182.1547**	**0.3602**	**14.7103**	**34.3825**	**0.4938**
SVR*	2022.3680	1.3688	19.4026	44.9707	0.1340
LSSVR*	1492.9071	1.8484	17.1765	38.6382	0.3607
Panel B: $WL = 500$					
HAR	2149.6161	0.5193	20.8155	46.3640	0.3510
HAR-CJ	2219.6210	0.5281	20.1791	47.1129	0.3298
HAR-RS-II	2851.7670	0.5199	21.5077	53.4019	0.1390
LASSO	2205.3996	0.5226	20.7104	46.9617	0.3341
BOOST	3106.4917	4.1749	29.2914	55.7359	0.0621
RF	2144.2577	0.4679	20.7959	46.3061	0.3526
BAG	2256.8494	0.4779	21.5526	47.5063	0.3186
SVR	2870.1779	1.2920	22.2445	53.5740	0.1334
LSSVR	2216.1386	0.4999	19.2678	47.0759	0.3309
HAR*	1686.7126	1.5249	21.6946	41.0696	0.4907
HAR-CJ*	1737.9884	1.5219	21.5992	41.6892	0.4753
HAR-RS-II*	2228.9633	2.0233	22.6721	47.2119	0.3270
LASSO*	1731.5366	1.6110	21.5009	41.6117	0.4772
BOOST*	1595.2616	4.8013	23.3670	39.9407	0.5184
RF*	1380.9952	0.3759	16.9718	37.1617	0.5830
BAG*	**1115.9729**	**0.3669**	**16.1018**	**33.4062**	**0.6631**
SVR*	2879.3386	1.3206	22.3949	53.6595	0.1307
LSSVR*	1890.4027	2.3489	19.2429	43.4788	0.4292

The best result under each criterion is highlighted in boldface

Table 6 Giacomini–White test results by different window lengths ($h = 1$)

	HAR	HAR-CJ	RS-II	LASSO	BOOST	RF	BAG	SVR	LSSVR	HAR*	HAR-CJ*	RS-II*	LASSO*	BOOST*	RF*	BAG*	SVR*
Panel A: W L = 300																	
HAR	–																
HAR-CJ	0.9338	–															
HAR-RS-II	0.4818	0.4462	–														
LASSO	0.0307	0.2119	0.7466	–													
BOOST	0.0000	0.0000	0.0000	0.0000	–												
RF	0.3721	0.4172	0.8416	0.5419	0.0000	–											
BAG	0.8110	0.8383	0.8332	0.9316	0.0000	0.0477	–										
SVR	0.0736	0.1037	0.4429	0.1229	0.0000	0.2329	0.0387	–									
LSSVR	0.9751	0.9885	0.7617	0.7860	0.0000	0.1028	0.7148	0.0085	–								
HAR*	0.0011	0.0025	0.1650	0.0059	0.0000	0.2636	0.0697	0.8791	0.0594	–							
HAR-CJ*	0.0003	0.0006	0.1028	0.0024	0.0000	0.2279	0.0615	0.7862	0.0548	0.2951	–						
HAR-RS-II*	0.0025	0.0015	0.0010	0.0079	0.0000	0.2672	0.1192	0.6477	0.1185	0.4168	0.4972	–					
LASSO*	0.0010	0.0023	0.1759	0.0046	0.0000	0.2727	0.0687	0.9174	0.0599	0.6547	0.2794	0.3800	–				
BOOST*	0.0090	0.0128	0.1289	0.0188	0.0000	0.0483	0.0073	0.3769	0.0100	0.4046	0.4957	0.8260	0.3761	–			
RF*	0.0932	0.1171	0.1826	0.0480	0.0000	0.0000	0.0002	0.0000	0.0012	0.0002	0.0003	0.0073	0.0002	0.0000	–		
BAG*	0.0230	0.0346	0.0877	0.0110	0.0000	0.0000	0.0002	0.0000	0.0009	0.0000	0.0000	0.0020	0.0000	0.0000	0.1109	–	
SVR*	0.0607	0.0877	0.4118	0.1027	0.0000	0.1932	0.0293	0.0000	0.0058	0.9436	0.8467	0.6872	0.9837	0.4177	0.0000	0.0000	–
LSSVR*	0.8887	0.8863	0.6880	0.6641	0.0000	0.0908	0.5488	0.0079	0.6627	0.0164	0.0173	0.0767	0.0160	0.0045	0.0016	0.0008	0.0055
Panel B: W L = 500																	
HAR	–																
HAR-CJ	0.0007	–															
HAR-RS-II	0.5914	0.3132	–														
LASSO	0.6862	0.0706	0.5393	–													

(continued)

Table 6 (continued)

	HAR	HAR-CJ	RS-II	LASSO	BOOST	RF	BAG	SVR	LSSVR	HAR*	HAR-CJ*	RS-II*	LASSO*	BOOST*	RF*	BAG*	SVR*
BOOST	**0.0001**	**0.0000**	**0.0058**	**0.0001**	–	–	–	–	–	–	–	–	–	–	–	–	–
RF	0.9827	0.5018	0.7248	0.9271	**0.0000**	–	–	–	–	–	–	–	–	–	–	–	–
BAG	0.3567	0.0966	0.9803	0.3041	**0.0001**	0.0950	–	–	–	–	–	–	–	–	–	–	–
SVR	0.1930	**0.0459**	0.7289	0.1745	**0.0037**	0.2397	0.5826	–	–	–	–	–	–	–	–	–	–
LSSVR	0.0867	0.3051	0.2939	0.1249	**0.0000**	**0.0194**	**0.0034**	**0.0018**	–	–	–	–	–	–	–	–	–
HAR*	0.3361	0.1173	0.9117	0.3096	**0.0006**	0.4461	0.8991	0.7049	**0.0397**	–	–	–	–	–	–	–	–
HAR-CJ*	0.3795	0.1278	0.9562	0.3464	**0.0006**	0.4930	0.9666	0.6494	**0.0453**	0.3877	–	–	–	–	–	–	–
HAR-RS-II*	0.1598	0.0691	0.1972	0.1440	**0.0133**	0.3375	0.5270	0.8385	0.0968	0.4019	0.3561	–	–	–	–	–	–
LASSO*	0.4266	0.1472	0.9967	0.3629	**0.0005**	0.5362	0.9615	0.5975	0.0501	0.3700	0.6529	0.3185	–	–	–	–	–
BOOST*	0.1798	0.1015	0.4720	0.1766	**0.0070**	0.1471	0.3213	0.6014	**0.0325**	0.3494	0.3291	0.7675	0.3065	–	–	–	–
RF*	**0.0011**	**0.0095**	**0.0333**	**0.0024**	**0.0000**	**0.0001**	**0.0000**	**0.0008**	**0.0325**	**0.0000**	**0.0000**	**0.0019**	**0.0000**	**0.0002**	–	–	–
BAG*	**0.0002**	**0.0018**	**0.0115**	**0.0005**	**0.0000**	**0.0001**	**0.0000**	**0.0002**	**0.0081**	**0.0000**	**0.0000**	**0.0003**	**0.0000**	**0.0000**	0.1547	–	–
SVR*	0.1520	**0.0330**	0.6768	0.1374	**0.0046**	0.1969	0.5057	**0.0059**	**0.0012**	0.6298	0.5754	0.8949	0.5260	0.6515	**0.0006**	**0.0001**	–
LSSVR*	0.1008	0.3379	0.2743	0.1405	**0.0000**	0.0646	**0.0103**	**0.0103**	0.9700	**0.0008**	**0.0012**	0.0516	**0.0015**	**0.0237**	**0.0060**	**0.0014**	**0.0073**

p-values smaller than 5% are highlighted in boldface

Table 7 Forecasting performance by different sample periods ($h = 1$)

Method	MSFE	QLIKE	MAFE	SDFE	Pseudo R^2
Panel A: first half					
HAR	2124.1237	0.4650	22.9310	46.0882	0.2335
HAR-CJ	2355.2555	0.4492	21.7508	48.5310	0.1500
HAR-RS-II	2603.1374	0.4914	24.0043	51.0210	0.0606
LASSO	2138.6650	0.4666	23.3848	46.2457	0.2282
BOOST	3867.2799	2.0069	32.0598	62.1875	-0.3956
RF	2099.9254	0.3727	19.6797	45.8249	0.2422
BAG	2106.6674	0.4048	19.5280	45.8984	0.2398
SVR	2153.3053	0.5631	22.7778	46.4037	0.2229
LSSVR	2040.2006	0.3860	21.1425	45.1686	0.2637
HAR*	1489.5345	2.9636	26.6309	38.5945	0.4625
HAR-CJ*	1541.1336	7.4995	26.9735	39.2573	0.4438
HAR-RS-II*	1711.2464	1.9009	27.7648	41.3672	0.3825
LASSO*	1448.5859	1.9891	26.4592	38.0603	0.4772
BOOST*	1273.8670	1.4514	22.1323	35.6913	0.5403
RF*	1201.8716	**0.2606**	16.8897	34.6680	0.5663
BAG*	**840.0199**	0.2629	**15.4812**	**28.9831**	**0.6969**
SVR*	2153.5420	0.5633	22.7812	46.4063	0.2228
LSSVR*	1331.7041	2.7236	19.8550	36.4925	0.5194
Panel B: second half					
HAR	3412.6612	0.4790	23.4856	58.4180	0.2370
HAR-CJ	3591.3391	0.4739	24.8167	59.9278	0.1970
HAR-RS-II	5357.5796	0.4995	25.1334	73.1955	−0.1979
LASSO	3575.5839	0.5118	24.0981	59.7962	0.2005
BOOST	6151.3787	4.0402	41.1825	78.4307	−0.3754
RF	3314.1729	0.5416	25.1547	57.5689	0.2590
BAG	3152.0846	0.5716	24.3284	56.1434	0.2952
SVR	3917.5789	1.9247	23.9854	62.5906	0.1241
LSSVR	3187.9434	0.5683	24.3457	56.4619	0.2872
HAR*	2747.1766	1.4813	24.0375	52.4135	0.3858
HAR-CJ*	2908.1546	1.4502	24.5958	53.9273	0.3498
HAR-RS-II*	4324.7752	2.3995	25.4931	65.7630	0.0330
LASSO*	2869.5404	0.7703	24.2617	53.5681	0.3584
BOOST*	2624.4054	5.9681	30.0566	51.2290	0.4132
RF*	2337.9213	**0.3759**	21.4734	48.3521	0.4773
BAG*	**2110.7631**	0.3847	**20.6086**	**45.9430**	**0.5281**
SVR*	3924.9867	1.9806	24.0556	62.6497	0.1224
LSSVR*	2952.6849	0.5104	24.0650	54.3386	0.3398

The best result under each criterion is highlighted in boldface

Table 8 Giacomini-White test results by different sample periods ($h = 1$)

	HAR	HAR-CJ	RS-II	LASSO	BOOST	RF	BAG	SVR	LSSVR	HAR*	HAR-CJ*	RS-II*	LASSO*	BOOST*	RF*	BAG*	SVR*
Panel A: first half																	
HAR	–	–	–	–	–	–	–	–	–	–	–	–	–	–	–	–	–
HAR-CJ	0.1354	–	–	–	–	–	–	–	–	–	–	–	–	–	–	–	–
HAR-RS-II	0.3850	**0.0068**	–	–	–	–	–	–	–	–	–	–	–	–	–	–	–
LASSO	0.0646	0.0526	0.6283	–	–	–	–	–	–	–	–	–	–	–	–	–	–
BOOST	**0.0005**	**0.0002**	**0.0055**	**0.0008**	–	–	–	–	–	–	–	–	–	–	–	–	–
RF	**0.0004**	0.0688	**0.0035**	**0.0001**	**0.0000**	–	–	–	–	–	–	–	–	–	–	–	–
BAG	**0.0004**	0.0605	**0.0035**	**0.0001**	**0.0000**	0.6653	–	–	–	–	–	–	–	–	–	–	–
SVR	0.8702	0.4822	0.5117	0.5202	**0.0011**	**0.0157**	**0.0157**	–	–	–	–	–	–	–	–	–	–
LSSVR	**0.0001**	0.5021	**0.0349**	**0.0000**	**0.0000**	**0.0478**	**0.0510**	0.1174	–	–	–	–	–	–	–	–	–
HAR*	0.0680	**0.0375**	0.3060	0.1094	0.0758	**0.0012**	**0.0011**	0.0599	**0.0073**	–	–	–	–	–	–	–	–
HAR-CJ*	**0.0359**	**0.0146**	0.1972	0.0632	0.0923	**0.0004**	**0.0004**	**0.0393**	**0.0032**	0.4765	–	–	–	–	–	–	–
HAR-RS-II*	**0.0171**	**0.0036**	0.0654	**0.0313**	0.1620	**0.0002**	**0.0002**	**0.0275**	**0.0014**	0.2676	0.2651	–	–	–	–	–	–
LASSO*	0.0778	**0.0410**	0.3285	0.1230	0.0656	**0.0014**	**0.0012**	0.0695	**0.0084**	0.5238	0.3100	0.1910	–	–	–	–	–
BOOST*	0.7816	0.9011	0.5709	0.6643	**0.0066**	0.3808	0.3515	0.8233	0.7283	**0.0441**	**0.0338**	**0.0231**	**0.0489**	–	–	–	–
RF*	**0.0000**	**0.0053**	**0.0005**	**0.0000**	**0.0000**	**0.0243**	**0.0322**	**0.0002**	**0.0015**	**0.0000**	**0.0000**	**0.0000**	**0.0000**	**0.0164**	–	–	–
BAG*	**0.0014**	**0.0156**	**0.0027**	**0.0007**	**0.0000**	0.0601	0.0693	**0.0019**	**0.0122**	**0.0000**	**0.0000**	**0.0000**	**0.0000**	**0.0001**	0.1855	–	–
SVR*	0.8731	0.4808	0.5130	0.5227	**0.0011**	**0.0156**	**0.0116**	0.4224	0.1169	0.0601	**0.0394**	**0.0276**	0.0698	0.8224	**0.0002**	**0.0019**	–
LSSVR*	**0.0169**	0.2273	**0.0259**	**0.0066**	**0.0000**	0.8966	0.8164	**0.0438**	0.2843	**0.0000**	**0.0000**	**0.0000**	**0.0000**	0.3193	**0.0034**	**0.0047**	**0.0436**
Panel B: second half																	
HAR	–	–	–	–	–	–	–	–	–	–	–	–	–	–	–	–	–
HAR-CJ	**0.0005**	–	–	–	–	–	–	–	–	–	–	–	–	–	–	–	–
HAR-RS-II	0.4168	0.8623	–	–	–	–	–	–	–	–	–	–	–	–	–	–	–
LASSO	0.1433	0.1745	0.6130	–	–	–	–	–	–	–	–	–	–	–	–	–	–
BOOST	**0.0000**	**0.0000**	**0.0000**	**0.0000**	–	–	–	–	–	–	–	–	–	–	–	–	–

RF	0.5305	0.9051	0.9962	0.6828	**0.0000**	–	–	–	–	–	–	–	–	–	–	–	–
BAG	0.7330	0.8538	0.8522	0.9237	**0.0000**	0.2049	–	–	–	–	–	–	–	–	–	–	–
SVR	0.8590	0.7862	0.8069	0.9680	**0.0000**	0.4623	0.8289	–	–	–	–	–	–	–	–	–	–
LSSVR	0.7489	0.8699	0.8634	0.9255	**0.0000**	0.3092	0.9838	0.8023	–	–	–	–	–	–	–	–	–
HAR*	0.6384	0.5469	0.6862	0.9624	**0.0000**	0.6605	0.9023	0.9845	0.9018	–	–	–	–	–	–	–	–
HAR-CJ*	0.3223	0.8517	0.8323	0.6860	**0.0000**	0.8360	0.9156	0.8308	0.9254	0.0696	–	–	–	–	–	–	–
HAR-RS-II*	0.2815	0.6915	0.7691	0.4684	**0.0000**	0.9354	0.7722	0.7294	0.7861	0.4615	0.6166	–	–	–	–	–	–
LASSO*	0.4982	0.6525	0.7429	0.8880	**0.0000**	0.7126	0.9764	0.9170	0.9723	0.5963	0.5101	0.5357	–	–	–	–	–
BOOST*	**0.0119**	0.0521	0.2097	**0.0251**	**0.0029**	0.0575	**0.0232**	**0.0323**	**0.0237**	**0.0097**	**0.0232**	0.1910	**0.0136**	–	–	–	–
RF*	0.5825	0.3362	0.4912	0.4401	**0.0000**	**0.0104**	0.0576	0.2770	**0.0350**	0.3694	0.2905	0.3893	0.3121	**0.0003**	–	–	–
BAG*	0.3290	0.1772	0.3417	0.2432	**0.0000**	**0.0168**	**0.0390**	0.0942	**0.0288**	0.1591	0.1250	0.2463	0.1385	**0.0000**	0.2344	–	–
SVR*	0.8395	0.8039	0.8185	0.9879	**0.0000**	0.4896	0.8635	**0.0000**	0.8404	0.9946	0.8500	0.7416	0.9381	**0.0343**	0.2613	0.0878	–
LSSVR*	0.8282	0.7923	0.8140	0.9900	**0.0000**	0.2548	0.7907	0.9576	0.6384	0.9904	0.8291	0.7258	0.9287	**0.0125**	**0.0198**	**0.0242**	0.9950

p-values smaller than 5% are highlighted in boldface

Table 9 Forecasting performance by different tuning parameters ($h = 1$)

Method	MSFE	QLIKE	MAFE	SDFE	Pseudo R^2
Panel A: without sentiment					
LASSO	1726.2453	0.5649	17.4025	41.5481	0.2929
BOOST	3003.8597	2.3176	27.7473	54.8075	-0.2304
RF	1680.2756	**0.4374**	16.7922	40.9912	0.3118
BAG	**1628.2674**	0.4504	16.8285	**40.3518**	**0.3331**
SVR	2218.8594	1.3751	20.0765	47.1048	0.0912
LSSVR	1628.6800	0.4858	**16.0397**	40.3569	0.3329
LASSO2	1736.6334	0.5546	17.4325	41.6729	0.2887
BOOST2	2965.5740	2.1399	27.2208	54.4571	-0.2147
RF2	1765.2329	0.4706	17.2435	42.0147	0.2770
BAG2	1659.4408	0.4611	16.7576	40.7362	0.3203
SVR2	2218.8594	1.3751	20.0765	47.1048	0.0912
LSSVR2	1635.2935	0.4900	16.0911	40.4388	0.3302
Panel B: with sentiment					
LASSO*	1497.0621	1.8256	19.1215	38.6919	0.3868
BOOST*	1312.6693	2.4524	18.6123	36.2308	0.4623
RF*	1178.6862	0.3794	14.4059	34.3320	0.5172
BAG*	T1035.7081	**0.3635**	**13.8235**	32.1824	0.5758
SVR*	2226.7603	1.4075	20.2407	47.1886	0.0879
LSSVR*	1494.0104	1.2801	16.4454	38.6524	0.3881
LASSO2*	1501.9018	2.1237	19.3177	38.7544	0.3848
BOOST2*	1324.7603	14.1393	18.2779	36.3973	0.4574
RF2*	1250.0685	0.3932	14.8282	35.3563	0.4880
BAG2*	**1007.2093**	0.3842	13.9225	**31.7366**	**0.5874**
SVR2*	2226.7603	1.4075	20.2407	47.1886	0.0879
LSSVR2*	1504.4609	1.7125	16.4577	38.7874	0.3838

The best result under each criterion is highlighted in boldface

names: SPY and QQQ, respectively) and the CBOE Volatility Index (VIX) as extra covariates. For SPY and QQQ, we proxy daily spot variances by daily realized variance estimates. For the VIX, we collect the daily data from CBOE. The extra covariates are described in Table 11

The data range is from May 20, 2015, to August 18, 2017, with 536 total observations. Fewer observations are available since mainstream asset exchanges are closed on the weekends and holidays. We truncate the BTC/USD data accordingly. We compare forecasts from models with two groups of covariate data: one with only the USSI variable and the other which includes both the USSI variable and the mainstream RV data (SPY, QQQ, and VIX). Estimates that include the larger covariate set are denoted by the symbol $**$.

The rolling window forecasting results with $WL = 300$ are presented in Table 12. Comparing results across any strategy between Panels A and B, we do not observe obvious improvements in forecasting accuracy. This implies that

Table 10 Giacomini–White test results by different tuning parameters ($h = 1$)

	LASSO	BOOST	RF	BAG	SVR	LSSVR	LASSO2	BOOST2	RF2	BAG2	SVR2	LSSVR2	LASSO*	BOOST*	RF*	BAG*	SVR*	LSSVR*	LASSO2*	BOOST2*	RF2*	SVR2*	LSSVR2*
LASSO	–																						
BOOST	**0.0000**	–																					
RF	0.5830	**0.0000**	–																				
BAG	0.6383	**0.0000**	0.7623	–																			
SVR	**0.0211**	**0.0006**	**0.0013**	**0.0020**	–																		
LSSVR	0.1931	**0.0000**	0.1252	0.1172	**0.0000**	–																	
LASSO2	0.7415	**0.0000**	0.5644	0.6197	**0.0263**	0.1895	–																
BOOST2	**0.0000**	**0.0114**	**0.0000**	**0.0000**	**0.0013**	**0.0000**	**0.0000**	–															
RF2	0.5977	**0.0000**	0.9806	0.8041	**0.0013**	0.1549	0.5797	**0.0000**	–														
BAG2	0.9837	**0.0000**	0.1410	0.1615	**0.0097**	**0.0324**	0.9606	**0.0000**	0.2193	–													
SVR2	**0.0211**	**0.0006**	**0.0013**	**0.0020**	1.0000	**0.0000**	**0.0263**	**0.0013**	**0.0013**	**0.0097**	–												
LSSVR2	0.2158	**0.0000**	0.1568	0.1455	**0.0000**	**0.0290**	0.2118	**0.0000**	0.1867	**0.0424**	**0.0000**	–											
LASSO*	**0.0100**	**0.0000**	**0.0435**	**0.0458**	0.4478	**0.0059**	**0.0125**	**0.0001**	**0.0491**	0.1237	0.4478	**0.0072**	–										
BOOST*	0.4001	**0.0000**	0.1872	0.2141	0.3348	0.0500	0.4163	**0.0000**	0.1809	0.3648	0.3348	0.0557	0.7039	–									
RF*	**0.0107**	**0.0000**	**0.0016**	**0.0009**	**0.0000**	**0.0466**	**0.0105**	**0.0000**	**0.0018**	**0.0003**	**0.0000**	**0.0407**	**0.0000**	**0.0001**	–								
BAG*	**0.0042**	**0.0000**	**0.0013**	**0.0005**	**0.0000**	**0.0076**	**0.0045**	**0.0000**	**0.0012**	**0.0002**	**0.0000**	**0.0065**	**0.0000**	**0.0000**	**0.0451**	–							
SVR*	**0.0146**	**0.0008**	**0.0007**	**0.0012**	**0.0000**	**0.0000**	**0.0186**	**0.0017**	**0.0008**	**0.0063**	**0.0000**	**0.0000**	0.3744	0.2843	**0.0000**	**0.0000**	–						
LSSVR*	0.3794	**0.0000**	0.5488	0.4744	**0.0001**	0.3323	0.3702	**0.0000**	0.5667	0.2064	**0.0001**	0.3976	**0.0044**	0.0772	**0.0039**	**0.0004**	**0.0000**	–					
LASSO2*	**0.0047**	**0.0000**	**0.0308**	**0.0321**	0.5551	**0.0041**	**0.0058**	**0.0001**	**0.0355**	0.0910	0.5551	**0.0051**	**0.0000**	0.6014	**0.0000**	**0.0000**	0.4734	**0.0028**	–				
BOOST2*	0.5432	**0.0000**	0.2858	0.3220	0.2362	0.0894	0.5610	**0.0000**	0.2776	0.5100	0.2362	0.0982	0.5292	**0.0441**	**0.0005**	**0.0000**	0.1969	0.1358	0.4418	–			
RF2*	**0.0084**	**0.0000**	**0.0011**	**0.0008**	**0.0000**	**0.0275**	**0.0084**	**0.0000**	**0.0015**	**0.0003**	**0.0000**	**0.0238**	**0.0000**	**0.0001**	0.3103	0.2140	**0.0000**	**0.0013**	**0.0000**	**0.0005**	–		
BAG2*	**0.0065**	**0.0000**	**0.0016**	**0.0007**	**0.0000**	**0.0093**	**0.0067**	**0.0000**	**0.0015**	**0.0003**	**0.0000**	**0.0079**	**0.0000**	**0.0000**	0.1363	0.6847	**0.0000**	**0.0008**	**0.0000**	**0.0000**	0.3800		
SVR2*	**0.0146**	**0.0008**	**0.0007**	**0.0012**	**0.0000**	**0.0000**	**0.0186**	**0.0017**	**0.0008**	**0.0063**	**0.0000**	**0.0000**	0.3744	0.2843	**0.0000**	**0.0000**	1.0000	**0.0000**	0.4734	0.1969	**0.0000**	–	
LSSVR2*	0.3899	**0.0000**	0.5604	0.4858	**0.0000**	0.3061	0.3806	**0.0000**	0.5773	0.2131	**0.0000**	0.3684	**0.0053**	0.0807	**0.0041**	**0.0005**	**0.0000**	0.6459	**0.0034**	0.1408	**0.0014**	**0.0008**	**0.0000**

p-values smaller than 5% are highlighted in boldface

Table 11 Descriptive statistics

Statistics	SPY	QQQ	VIX
Mean	0.3839	0.7043	15.0144
Median	0.2034	0.3515	13.7300
Maximum	12.1637	70.6806	40.7400
Minimum	0.0143	0.0468	9.3600
Std. Dev.	0.6946	3.1108	4.5005
Skewness	10.1587	21.3288	1.6188
Kurtosis	158.5806	479.5436	6.3394
Jarque–Bera	0.0010	0.0010	0.0010
ADF Test	0.0010	0.0010	0.0010

Table 12 Forecasting performance

Method	MSFE	QLIKE	MAFE	SDFE	Pseudo R^2
Panel A: with sentiment					
HAR*	1265.3736	1.7581	21.7060	35.5721	0.4299
HAR-CJ*	1258.1112	1.4488	21.4721	35.4699	0.4332
HAR-RS-II*	1312.9602	1.6025	22.4346	36.2348	0.4085
LASSO*	1251.4556	1.7235	21.3984	35.3759	0.4362
BOOST*	1135.0482	9.2958	19.0763	33.6905	0.4886
RF*	1015.7416	0.3845	15.1202	31.8707	0.5424
BAG*	**884.8778**	**0.3674**	**14.3677**	**29.7469**	**0.6013**
SVR*	1934.5500	1.4254	21.1660	43.9835	0.1284
LSSVR*	1311.5350	1.2829	18.2171	36.2151	0.4091
Panel B: with sentiment and extra covariates					
HAR**	1298.6001	8.7030	21.6841	36.0361	0.4149
HAR-CJ**	1299.4404	1.4853	21.7684	36.0478	0.4145
HAR-RS-II**	1349.2130	2.0542	22.4713	36.7316	0.3921
LASSO**	1251.6195	1.3544	21.1397	35.3782	0.4361
BOOST**	1489.1772	4.9792	22.1760	38.5899	0.3291
RF**	1024.0401	0.3846	15.3587	32.0006	0.5386
BAG**	**885.8634**	**0.3687**	**14.3526**	**29.7635**	**0.6009**
SVR**	1934.5502	1.4254	21.1660	43.9835	0.1284
LSSVR**	1336.3343	1.2665	17.7219	36.5559	0.3979

The best result under each criterion is highlighted in boldface

mainstream asset markets RV does not affect BTC/USD volatility, which reinforces the fact that crypto-assets are sometimes considered as a hedging device for many investment companies.[17]

Last, we use the GW test to formally explore if there are no differences in forecast accuracy between the panels in Table 13. For each estimator, we present the *p*-

---

[17]PwC-Elwood [36] suggests that the capitalization of cryptocurrency hedge funds increases at a steady pace since 2016.

Table 13 Giacomini–White test results

	HAR*	HAR-CJ*	RS-II*	LASSO*	BOOST*	RF*	BAG*	SVR*	LSSVR*	HAR**	HAR-CJ**	RS-II**	LASSO**	BOOST**	RF**	BAG**	SVR**
HAR*	–																
HAR-CJ*	0.2800	–															
HAR-RS-II*	**0.0308**	**0.0370**	–														
LASSO*	0.1862	0.7852	**0.0159**	–													
BOOST*	0.1080	0.1468	**0.0382**	0.1496	–												
RF*	**0.0000**	**0.0000**	**0.0000**	**0.0000**	**0.0011**	–											
BAG*	**0.0000**	**0.0000**	**0.0000**	**0.0000**	**0.0000**	0.0761	–										
SVR*	0.7212	0.8405	0.3878	0.8732	0.2890	**0.0000**	**0.0000**	–									
LSSVR*	**0.0000**	**0.0001**	**0.0000**	**0.0000**	0.5719	**0.0002**	**0.0004**	**0.0077**	–								
HAR**	0.9044	0.4386	**0.0467**	0.2991	0.1156	**0.0000**	**0.0000**	0.7276	**0.0000**	–							
HAR-CJ**	0.8329	0.1081	0.1848	0.2450	0.1071	**0.0000**	**0.0000**	0.6919	**0.0001**	0.7446	–						
HAR-RS-II**	0.0597	0.0511	0.8469	**0.0244**	**0.0390**	**0.0000**	**0.0000**	0.3653	**0.0000**	**0.0270**	0.1627	–					
LASSO**	**0.0421**	0.2672	**0.0050**	0.0524	0.2018	**0.0000**	**0.0000**	0.9853	**0.0001**	**0.0277**	**0.0421**	**0.0042**	–				
BOOST**	0.8039	0.7097	0.8906	0.6749	**0.0234**	**0.0000**	**0.0000**	0.6384	**0.0246**	0.7977	0.8313	0.8769	0.5775	–			
RF**	**0.0000**	**0.0000**	**0.0000**	**0.0000**	**0.0016**	0.7798	0.0725	**0.0000**	**0.0001**	**0.0000**	**0.0000**	**0.0000**	**0.0000**	**0.0000**	–		
BAG**	**0.0000**	**0.0000**	**0.0000**	**0.0000**	**0.0002**	0.6173	**0.0045**	**0.0004**	**0.0054**	**0.0000**	**0.0000**	**0.0000**	**0.0000**	**0.0000**	0.5515	–	
SVR**	0.7212	0.8405	0.3878	0.8732	0.2890	**0.0000**	**0.0000**	0.4629	**0.0077**	0.7276	0.6919	0.3653	0.9853	0.6384	**0.0000**	**0.0004**	–
LSSVR**	**0.0000**	**0.0000**	**0.0000**	**0.0000**	0.3831	**0.0025**	**0.0022**	**0.0011**	**0.0038**	**0.0000**	**0.0000**	**0.0000**	**0.0000**	**0.0127**	**0.0015**	**0.0194**	**0.0011**

p-values smaller than 5% are highlighted in boldface

values from different covariate groups in bold. Each of these p-values exceeds 5%, which support our finding that mainstream asset RV data does not improve forecasts sharply, unlike the inclusion of social media data.

8 Conclusion

In this chapter, we compare the performance of numerous econometric and machine learning forecasting strategies to explain the short-term realized volatility of the Bitcoin market. Our results first complement a rapidly growing body of research that finds benefits from using machine learning techniques in the context of financial forecasting. Our application involves forecasting an asset that exhibits significantly more variation than much of the earlier literature which could present challenges in settings such as ours with fewer than 800 observations. Yet, our result further highlights that what drives the benefits of machine learning is the accounting for nonlinearities and there are much smaller gains from using regularization or cross-validation. Second, we find substantial benefits from using social media data in our forecasting exercise that hold irrespective of the estimator. These benefits are larger when we consider new econometric tools to more flexibly handle the difference in the timing of the sampling of social media and financial data.

Taken together, there are benefits from using both new data sources from the social web and predictive techniques developed in the machine learning literature for forecasting financial data. We suggest that the benefits from these tools will likely increase as researchers begin to understand why they work and what they measure. While our analysis suggests nonlinearities are important to account for, more work is needed to incorporate heterogeneity from heteroskedastic data in machine learning algorithms.[18] We observe significant differences between SVR and LSSVR so the change in loss function can explain a portion of the gains within machine learning relative to econometric strategies, but not to the same extent as nonlinearities, which the tree-based strategies also account for and use a similar loss function based on SSR.

Our investigation focused on the performance of what are currently the most popular algorithms considered by social scientists. There have been many advances developing powerful algorithms in the machine learning literature including deep learning procedures which consider more hidden layers than the neural network procedures considered in the econometrics literature between 1995 and 2015. Similarly, among tree-based procedures, we did not consider eXtreme gradient boosting which applies more penalties in the boosting equation when updating

---

[18]Lehrer and Xie [26] pointed out that all of the machine learning algorithms considered in this paper assume homoskesdastic data. In their study, they discuss the consequences of heteroskedasticity for these algorithms and the resulting predictions, as well as propose alternatives for this data.

trees and residual compared to the classic boosting method we employed. Both eXtreme gradient boosting and deep learning methods present significant challenges regarding interpretability relative to the algorithms we examined in the empirical exercise.

Further, machine learning algorithms were not developed for time series data and more work is needed to develop methods that can account for serial dependence, long memory, as well as the consequences of having heterogeneous investors.[19] That is, while time series forecasting is an important area of machine learning (see [19, 30], for recent overviews that consider both one-step-ahead and multi-horizon time series forecasting), concepts such as autocorrelation and stationarity which pervade developments in financial econometrics have received less attention. We believe there is potential for hybrid approaches in the spirit of Lehrer and Xie [25] with group LASSO estimators. Further, developing machine learning approaches that consider interpretability appears crucial for many forecasting exercises whose results need to be conveyed to business leaders who want to make data-driven decisions. Last, given the random sample of Twitter users from which we measure sentiment, there is likely measurement error in our sentiment and our estimate should be interpreted as a lower bound.

Given the empirical importance of incorporating social media data in our forecasting models, there is substantial scope for further work that generates new insights with finer measures of this data. For example, future work could consider extracting Twitter messages that only capture the views of market participants rather than the entire universe of Twitter users. Work is also needed to clearly identify bots and consider how best to handle fake Twitter accounts. Similarly, research could strive to understand shifting sentiment for different groups on social media in response to news events. This can help improve our understanding of how responses to unexpected news leads lead investors to reallocate across asset classes.[20]

In summary, we remain at the early stages of extracting the full set of benefits from machine learning tools used to measure sentiment and conduct predictive analytics. For example, the Bitcoin market is international but the tweets used to estimate sentiment in our analysis were initially written in English. Whether the findings are robust to the inclusion of Tweets posted in other languages represents

---

[19]Lehrer et al. [27] considered the use of model averaging with HAR models to account for heterogeneous investors.

[20]As an example, following the removal of Ivanka Trump's fashion line from their stores, President Trump issued a statement via Twitter:

> My daughter Ivanka has been treated so unfairly by @Nordstrom. She is a great person – always pushing me to do the right thing! Terrible!

The general public response to this Tweet was to disagree with President Trump's stance on Nordstrom so aggregate Twitter sentiment measures rose and the immediate negative effects from the Tweet on Nordstrom stock of a decline of 1% in the minute following the tweet were fleeting since the stock closed the session posting a gain of 4.1%. See http://www.marketwatch.com/story/nordstrom-recovers-from-trumps-terrible-tweet-in-just-4-minutes-2017-02-08 for more details on this episode.

an open question for future research. As our understanding of how to account for real-world features of data increases with these data science tools, the full hype of machine learning and data science may be realized.

Acknowledgments We wish to thank Yue Qiu, Jun Yu, and Tao Zeng, seminar participants at Singapore Management University, for helpful comments and suggestions. Xie's research is supported by the Natural Science Foundation of China (71701175), the Chinese Ministry of Education Project of Humanities and Social Sciences (17YJC790174), and the Fundamental Research Funds for the Central Universities. Contact Tian Xie (e mail: xietian@shufe.edu.cn) for any questions concerning the data and/or codes. The usual caveat applies.

Appendix: Data Resampling Techniques

Substantial progress has been made in the machine learning literature on quickly converting text to data, generating real-time information on social media content. In this study, we also explore the benefits of incorporating an aggregate measure of social media sentiment, the Wall Street Journal-IHS Markit US Sentiment Index (USSI) in forecasting the Bitcoin RV. However, data timing presents a serious challenge in using minutely measures of the USSI to forecast the daily Bitcoin RV. To convert minutely USSI measure to match the sampling frequency of Bitcoin RV, we hereby introduce a few popular data resampling techniques.

Let y_{t+h} be target h-step-ahead future a low-frequency variable (e.g., the daily realized variance) that is sampled at periods denoted by a time index t for $t = 1, \ldots, n$. Consider a higher-frequency (e.g., the USSI) predictor X_t^{hi} that is sampled m times within the period of t:

$$X_t^h \equiv \left[X_t^{hi}, X_{t-\frac{1}{m}}^{hi}, \ldots, X_{t-\frac{m-1}{m}}^{hi} \right]^\top. \tag{20}$$

A specific element among the high-frequency observations in X_t^{hi} is denoted by $X_{t-\frac{i}{m}}^{hi}$ for $i = 0, \ldots, m - 1$. Denoting $L^{i/m}$ as the lag operator, then $X_{t-\frac{i}{m}}^{hi}$ can be reexpressed as $X_{t-\frac{i}{m}}^{hi} = L^{i/m} X_t^{hi}$ for $i = 0, \ldots, m - 1$.

Since X_t^h on y_{t+h} is measured at different frequencies, we need to convert the higher-frequency data to match the lower-frequency data. A simple average of the high-frequency observations X_t^h:

$$\bar{X}_t = \frac{1}{m} \sum_{i=0}^{m-1} L^{i/m} X_t^h,$$

where $\bar{X}_t$ is likely the easiest way to estimate a low-frequency X_t that can match the frequency of y_{t+h}. With the variables y_{t+h} and $\bar{X}_t$ being measured in the same time domain, a regression approach is simply

$$y_{t+h} = \alpha + \gamma \bar{X}_t + \epsilon_t = \alpha + \frac{\gamma}{m} \sum_{i=0}^{m-1} L^{i/m} X_t^h + \epsilon_t, \tag{21}$$

where α is the intercept and γ is the slope coefficient on the time-averaged $\bar{X}_t$. This approach assumes that each element in X_t^h has an identical effect on explaining y_{t+h}.

These homogeneity assumptions may be quite strong in practice. One could assume that each of the slope coefficients for each element in X_t^{hi} is unique. Following Lehrer et al. [28], extending Model (21) to allow for heterogeneous effects of the high-frequency observations generates

$$y_{t+h} = \alpha + \sum_{i=0}^{m-1} \gamma_i L^{i/m} X_t^{hi} + \epsilon_t, \tag{22}$$

where γ_i represents a set of slope coefficients for all high-frequency observations $X_{t-\frac{i}{m}}^{hi}$.

Since γ_i is unknown, estimating these parameters can be problematic when m is a relatively large number. The heterogeneous mixed data sampling (H-MIDAS) method by Lehrer et al. [28] uses a step function to allow for heterogeneous effects of different high-frequency observations on the low-frequency dependent variable. A low-frequency $\bar{X}_t^{(l)}$ can be constructed following

$$\bar{X}_t^{(l)} \equiv \frac{1}{l} \sum_{i=0}^{l-1} L^{i/m} X_t^{hi} = \frac{1}{l} \sum_{i=0}^{l-1} X_{t-\frac{i}{m}}^{hi}, \tag{23}$$

where l is a predetermined number and $l \leq m$. Equation (23) implies that we compute variable $\bar{X}_t^{(l)}$ by a simple average of the first l observations in X_t^{hi} and ignored the remaining observations. We consider different values of l and group all $\bar{X}_t^{(l)}$ into $\tilde{X}_t$ such that

$$\tilde{X}_t = \left[\bar{X}_t^{(l_1)}, \bar{X}_t^{(l_2)}, \ldots, \bar{X}_t^{(l_p)} \right],$$

where we set $l_1 < l_2 < \cdots < l_p$. Consider a weight vector $\boldsymbol{w} = \begin{bmatrix} w_1, w_2, \ldots, w_p \end{bmatrix}^{\mathsf{T}}$ with $\sum_{j=1}^{p} w_j = 1$; we can construct regressor X_t^{new} as $X_t^{new} = \tilde{\boldsymbol{X}}_t \boldsymbol{w}$. The regression based on the H-MIDAS estimator can be expressed as

$$
y_{t+h} = \beta X_t^{new} + \epsilon_t = \beta \sum_{s=1}^{p} \sum_{j=s}^{p} \frac{w_j}{l_j} \sum_{i=l_{s-1}}^{l_s - 1} L^{i/m} X_t^h + \epsilon_t = \beta \sum_{s=1}^{p} \sum_{i=l_{s-1}}^{l_s - 1} w_s^* L^{i/m} X_t^h + \epsilon_t,
$$
(24)

where $l_0 = 0$ and $w_s^* = \sum_{j=s}^{p} \frac{w_j}{l_j}$.

The weights $\boldsymbol{w}$ play a crucial role in this procedure. We first estimate $\widehat{\beta \boldsymbol{w}}$ following

$$
\widehat{\beta \boldsymbol{w}} = \arg \min_{\boldsymbol{w} \in \mathcal{W}} \left\| y_{t+h} - \tilde{\boldsymbol{X}}_t \cdot \beta \boldsymbol{w} \right\|^2
$$

by any appropriate econometric method necessary, where $\mathcal{W}$ is some predetermined weight set. Once $\widehat{\beta \boldsymbol{w}}$ is obtained, we estimate the weight vector $\hat{\boldsymbol{w}}$ by rescaling following

$$
\hat{\boldsymbol{w}} = \frac{\widehat{\beta \boldsymbol{w}}}{\text{Sum}(\widehat{\beta \boldsymbol{w}})},
$$

since the coefficient β is a scalar.

References

1. Andersen, T. G., & Bollerslev, T. (1998). Answering the skeptics: yes, standard volatility models do provide accurate forecasts. *International Economic Review, 39*(4), 885–905.
2. Andersen, T., Bollerslev, T., Diebold, F., & Ebens, H. (2001). The distribution of realized stock return volatility. *Journal of Financial Economics, 61*(1), 43–76.
3. Andersen, T. G., Bollerslev, T., Diebold, F. X., & Labys, P. (2001). The distribution of realized exchange rate volatility. *Journal of the American Statistical Associatio, 96*(453), 42–55.
4. Andersen, T. G., Bollerslev, T., & Diebold, F. X. (2007). Roughing it up: including jump components in the measurement, modelling, and forecasting of return volatility. *The Review of Economics and Statistics, 89*(4), 701–720.
5. Baker, M., & Wurgler, J. (2007). Investor sentiment in the stock market. *Journal of Economic Perspectives, 21*(2), 129–152.
6. Ban, G.-Y., Karoui, N. E., & Lim, A. E. B. (2018). Machine learning and portfolio optimization. *Management Science, 64*(3), 1136–1154.
7. Blair, B. J., Poon, S.-H., & Taylor, S. J. (2001). Forecasting S&P 100 volatility: the incremental information content of implied volatilities and high-frequency index returns. *Journal of Econometrics, 105*(1), 5–26.
8. Breiman, L. (1996). Bagging predictors. *Machine Learning, 24*, 123–140.
9. Breiman, L. (2001). Random forests. *Machine Learning, 45*, 5–32.

10. Breiman, L., Friedman, J., Stone, C. J., & Olshen, R. A. (1984). *Classification and regression trees*. New York: Chapman and Hall/CRC.
11. Corsi, F. (2009). A simple approximate long-memory model of realized volatility. *Journal of Financial Econometrics, 7*(2), 174–196.
12. Corsi, F., Audrino, F., & Renó, R. (2012). HAR modelling for realized volatility forecasting. In *Handbook of volatility models and their applications* (pp. 363–382). Hoboken: : John Wiley & Sons.
13. Coulombe, P. G., Leroux, M., Stevanovic, D., & Surprenant, S. (2019). How is machine learning useful for macroeconomic forecasting? In *Cirano Working Papers, CIRANO*. https://economics.sas.upenn.edu/system/files/2019-03/GCLSS_MC_MacroFcst.pdf
14. Craioveanu, M., & Hillebrand, E. (2012). *Why it is ok to use the har-rv (1, 5, 21) model*. Technical Report 1201, University of Central Missouri. https://ideas.repec.org/p/umn/wpaper/1201.html
15. Dacorogna, M. M., Müller, U. A., Nagler, R. J., Olsen, R. B., & Pictet, O. V. (1993). A geographical model for the daily and weekly seasonal volatility in the foreign exchange market. *Journal of International Money and Finance, 12*(4), 413–438.
16. Drucker, H., Burges, C. J. C., Kaufman, L., Smola, A. J., & Vapnik, V. (1996). Support vector regression machines. In M. C. Mozer, M. I. Jordan, & T. Petsche (Eds.), *Advances in neural information processing systems* (Vol. 9, pp. 155–161). Cambridge: MIT Press.
17. Felbo, B., Mislove, A., Søgaard, A., Rahwan, I., & Lehmann, S. (2017). Using millions of emoji occurrences to learn any-domain representations for detecting sentiment, emotion and sarcasm. In *Proceedings of the 2017 Conference on Empirical Methods in Natural Language Processing* (pp. 1615–1625). Stroudsburg: Association for Computational Linguistics.
18. Giacomini, R., & White, H. (2006). Tests of conditional predictive ability. *Econometrica, 74*(6), 1545–1578.
19. Gu, S., Kelly, B., & Xiu, D. (2020). Empirical asset pricing via machine learning. *Review of Financial Studies, 33*(5), 2223–2273. Society for Financial Studies.
20. Hansen, P. R., & Lunde, A. (2005). A forecast comparison of volatility models: does anything beat a garch(1,1)? *Journal of Applied Econometrics, 20*(7), 873–889.
21. Hastie, T., Tibshirani, R., & Friedman, J. (2009). *The elements of statistical learning. Springer series in statistics*. New York, NY: Springer.
22. Huang, X., & Tauchen, G. (2005). The relative contribution of jumps to total price variance. *Journal of Financial Econometrics, 3*(4), 456–499.
23. Ke, Z. T., Kelly, B. T., & Xiu, D. (2019). Predicting returns with text data. In *NBER Working Papers 26186*. Cambridge: National Bureau of Economic Research, Inc.
24. LaFon, H. (2017). Should you jump on the smart beta bandwagon? https://money.usnews.com/investing/funds/articles/2017-08-24/are-quant-etfs-worth-buying
25. Lehrer, S. F., & Xie, T. (2017). Box office buzz: does social media data steal the show from model uncertainty when forecasting for hollywood? *Review of Economics and Statistics, 99*(5), 749–755.
26. Lehrer, S. F., & Xie, T. (2018). The bigger picture: Combining econometrics with analytics improve forecasts of movie success. In *NBER Working Papers 24755*. Cambridge: National Bureau of Economic Research.
27. Lehrer, S. F., Xie, T., & Zhang, X. (2019). *Does adding social media sentiment upstage admitting ignorance when forecasting volatility?* Technical Report, Queen's University, NY. Available at: http://econ.queensu.ca/faculty/lehrer/mahar.pdf
28. Lehrer, S. F., Xie, T., & Zeng, T. (2019). Does high frequency social media data improve forecasts of low frequency consumer confidence measures? In *NBER Working Papers 26505*. Cambridge: National Bureau of Economic Research.
29. Mai, F., Shan, J., Bai, Q., Wang, S., & Chiang, R. (2018). How does social media impact bitcoin value? A test of the silent majority hypothesis. *Journal of Management Information Systems, 35*, 19–52.
30. Makridakis, S., Spiliotis, E., & Assimakopoulos, V. (2018). Statistical and machine learning forecasting methods: concerns and ways forward. *PloS One, 13*(3), Article No. e0194889. https://doi.org/10.1371/journal.pone.0194889

31. Medeiros, M. C., Vasconcelos, G. F. R., Veiga, Á., & Zilberman, E. (2019). Forecasting inflation in a data-rich environment: The benefits of machine learning methods. *Journal of Business & Economic Statistics, 39*(1), 98–119. https://doi.org/10.1080/07350015.2019. 1637745

32. Mincer, J., & Zarnowitz, V. (1969). The evaluation of economic forecasts. In *Economic forecasts and expectations: Analysis of forecasting behavior and performance* (pp. 3–46). Cambridge: National Bureau of Economic Research, Inc.

33. Müller, U. A., Dacorogna, M. M., Davé, R. D., Pictet, O. V., Olsen, R. B., & Ward, J. (1993). *Fractals and intrinsic time – a challenge to econometricians.* Technical report SSRN 5370. https://ssrn.com/abstract=5370

34. Patton, A. J., & Sheppard, K. (2015). Good volatility, bad volatility: signed jumps and the persistence of volatility. *The Review of Economics and Statistics, 97*(3), 683–697.

35. Probst, P., Boulesteix, A., & Bischl, B. (2019). Tunability: Importance of hyperparameters of machine learning algorithms. *Journal of Machine Learning Research, 20,* 1–32.

36. PwC-Elwood. (2019). 2019 crypto hedge fund report. https://www.pwc.com/gx/en/financial-services/fintech/assets/pwc-elwood-2019-annual-crypto-hedge-fund-report.pdf

37. Schumaker, R. P., Zhang, Y., Huang, C.-N., & Chen, H. (2012). Evaluating sentiment in financial news articles. *Decision Support Systems, 53*(3), 458–464.

38. Suykens, J., & Vandewalle, J. (1999). Least squares support vector machine classifiers. *Neural Processing Letters, 9,* 293–300.

39. Tibshirani, R. (1996). Regression shrinkage and selection via the lasso. *Journal of the Royal Statistical Society, Series B, 58,* 267–288.

40. Vapnik, V. N. (1996). *The nature of statistical learning theory.* New York, NY: Springer-Verlag.

41. Wolpert, D. H., & Macready, W. G. (1997). No free lunch theorems for optimization. *IEEE Transactions on Evolutionary Computation, 1*(1), 67–82.

42. Xie, T. (2019). Forecast bitcoin volatility with least squares model averaging. *Econometrics, 7*(3), 40:1–40:20.

Network Analysis for Economics and Finance: An Application to Firm Ownership

Janina Engel, Michela Nardo, and Michela Rancan

Abstract In this chapter, we introduce network analysis as an approach to model data in economics and finance. First, we review the most recent empirical applications using network analysis in economics and finance. Second, we introduce the main network metrics that are useful to describe the overall network structure and characterize the position of a specific node in the network. Third, we model information on firm ownership as a network: firms are the nodes while ownership relationships are the linkages. Data are retrieved from Orbis including information of millions of firms and their shareholders at worldwide level. We describe the necessary steps to construct the highly complex international ownership network. We then analyze its structure and compute the main metrics. We find that it forms a giant component with a significant number of nodes connected to each other. Network statistics show that a limited number of shareholders control many firms, revealing a significant concentration of power. Finally, we show how these measures computed at different levels of granularity (i.e., sector of activity) can provide useful policy insights.

1 Introduction

Historically, networks have been studied extensively in graph theory, an area of mathematics. After many applications to a number of different subjects including statistical physics, health science, and sociology, over the last two decades, an extensive body of theoretical and empirical literature was developed also in economics and finance. Broadly speaking, a network is a system with nodes connected by linkages. A node can be, e.g., an individual, a firm, an industry, or

J. Engel · M. Nardo
European Commission Joint Research Centre, Brussels, Belgium
e-mail: janina.engel@tum.de; michela.nardo@ec.europa.eu

M. Rancan (✉)
Marche Polytechnic University, Ancona, Italy
e-mail: m.rancan@univpm.it

© The Author(s) 2021
S. Consoli et al. (eds.), *Data Science for Economics and Finance*,
https://doi.org/10.1007/978-3-030-66891-4_14

331

even a geographical area. Correspondingly, different types of relationships have been represented as linkages. Indeed, network has become such a prominent cross-disciplinary topic [10] because it is extremely helpful to model a variety of data, even when they are big data [67]. At the same time, network analysis provides the capacity to estimate effectively the main patterns of several complex systems [66]. It is a prominent tool to better understand today's interlinked world, including economic and financial phenomena. To mention just a few applications, networks have been used to explain the trade of goods and services [39], financial flows across countries [64], innovation diffusion among firms, or the adoption of new products [3]. Another flourishing area of research related to network is the one of social connections, which with the new forms of interaction like online communities (e.g., Facebook or LinkedIn) will be even more relevant in the future [63]. Indeed network analysis is a useful tool to understand strategic interactions and external-ities [47]. Another strand of literature, following the 2007–2008 financial crisis, has shown how introducing a network approach in financial models can feature the interconnected nature of financial systems and key aspects of risk measurement and management, such as credit risk [27], counterparty risk [65], and systemic risk [13]. This network is also central to understanding the bulk of relationships that involve firms [12, 74, 70]. In this chapter, we present an application explaining step by step how to construct the network and perform some analysis, in which links are based on ownership information. Firms' ownership structure is an appropriate tool to identify the concentration of power [7], and a network perspective is particularly powerful to uncover intricate relationships involving indirect ownership. In this context, the connectivity of a firm depends on the entities with direct shares, if these entities are themselves controlled by other shareholders, and whether they also have shares in other firms. Hence, some firms are embedded in tightly connected groups of firms and shareholders, others are relatively disconnected. The overall structure of relationships will tell whether a firm is central in the whole web of ownership system, which may have implications, for example, for foreign direct investment (FDI).

Besides this specific application, the relevance of the network view has been particularly successful in economics and finance thanks to the unique insights that this approach can provide. A variety of network measures, at a global scale, allow to investigate in depth the structure, even of networks including a large number of nodes and/or links, explaining what patterns of linkages facilitate the transmission of valuable information. Node centrality measures may well complement information provided by other node attributes or characteristics. They may enrich other settings, such as standard micro-econometric models, or they may explain why an idiosyn-cratic shock has different spillover effects on the overall system depending on the node that is hit. Moreover, the identification of key nodes (i.e., nodes that can reach many other nodes) can be important for designing effective policy intervention. In a highly interconnected world, for example, network analysis can be useful to map the investment behavior of multinational enterprises and analyze the power concentration of nodes in strategic sectors. It can also be deployed to describe the extension and geographical location of value chains and the changes we are currently observing with the reshoring of certain economic activities as well as the degree of dependence on foreign inputs for the production of critical technologies. The variety of contexts to which network tools can be applied and the insights that this modeling technique may provide make network science extremely relevant for policymaking. Policy makers and regulators face dynamic and interconnected socioeconomic systems. The ability to map and understand this complex web of technological, economic, and social relationships is therefore critical for taking policy decision and action—even more, in the next decades when policy makers will face societal and economic challenges such as inequality, population ageing, innovation challenges, and climate risk. Moreover, network analysis is a promising tool also for investigating the fastest-changing areas of non-traditional financial intermediation, such as peer-to-peer lending, decentralized trading, and adoption of new payment instruments.

This chapter introduces network analysis providing suggestions to model data as a network for beginners and describes the main network tools. It is organized as follows. Section 2 provides an overview of recent applications of network science in the area of economics and finance. Section 3 introduces formally the fundamental mathematical concepts of a network and some tools to perform a network analysis. Section 4 illustrates in detail the application of network analysis to firm ownership and Sect. 5 concludes.

2   Network Analysis in the Literature

In economics, a large body of literature using micro-data investigates the effects of social interactions. Social networks are important determinants to explain job opportunity [18], student school performance [20],[1] criminal behavior [18], risk sharing [36], investment decisions [51], CEO compensations of major corporations [52], corporate governance of firms [42], and investment decision of mutual fund managers [26]. However, social interaction effects are subject to significant identification challenges [61]. An empirical issue is to disentangle the network effect of each other's behaviors from individual and group characteristics. An additional challenge is that the network itself cannot be always considered as exogenous but depends on unobservable characteristics of the individuals. To address these issues several strategies have been exploited: variations in the set of peers having different observable characteristics; instrumental variable approaches using as instrument, for example, the architecture of network itself; or modeling the network formation. See [14] and [15] for a deep discussion of the econometric framework and the identification conditions. Network models have been applied to the study of markets with results on trading outcomes and price formation corroborated by evidence obtained in laboratory [24]. Spreading of information is so important in some

---

[1]The literature on peer effects in education is extensive; see [71] for a review.

markets that networks are useful to better understand even market panics (see, e.g., [55]). Other applications are relevant to explain growth and economic outcome. For example, [3] find that past innovation network structures determine the process of future technological and scientific progress. Moreover, networks determine how technological advances generate positive externalities for related fields. Empirical evidences are relevant also for regional innovation policies [41]. In addition, network concepts have been adopted in the context of input–output tables, in which nodes represent the individual industries of different countries and links denote the monetary flows between industries [22], and the characterization of different sectors as suppliers to other sectors to explain aggregate fluctuations [1].[2]

In the area of finance,[3] since the seminal work by [5] network models have been revealed suitable to address potential domino effects resulting from interconnected financial institutions. Besides the investigation of the network structure and its properties [28], this framework has been used to answer the question whether the failure of an institution may propagate additional losses in the banking system [75, 72, 65, 34]. Importantly it has been found that network topology influences contagion [43, 2]. In this stream of literature, financial institutions are usually modeled as the nodes, while direct exposures are represented by the linkages (in the case of banking institutions, linkages are the interbank loans). Some papers use detailed data containing the actual exposures and the counterparties involved in the transactions. However, those data are usually limited to the banking sector of a single country (as they are disclosed to supervisory authorities) or a specific market (e.g., overnight interbank lending [53]). Unfortunately, most of the time such a level of detail is not available, and thus various methods have been developed to estimate networks, which are nonetheless informative for micro- and macro-prudential analysis (see [8] for an evaluation of different estimation methodologies). The mapping of balance sheet exposures and the associated risks through networks is not limited to direct exposures but has been extended to several financial instruments and common asset holdings such as corporate default swaps (CDS) exposures [23], bailinable securities [50], syndicated loans [48, 17], and inferred from market price data [62, 13]. Along this line, when different financial instruments are considered at the same time, financial institutions are then interconnected in different market segments by multiple layer networks [11, 60, 68]. Network techniques are not limited to model interlinkages across financial institutions at micro level. Some works consider as a node the overall banking sector of a country to investigate more aggregated effects [31] and the features of the global banking network [64].

---

[2]A complementary body of literature uses network modeling in economic theory reaching important achievements in the area of network formation, games of networks, and strategic interaction. For example, general theoretical models of networks provide insights on how network characteristics may affect individual behavior, payoffs, efficiency and consumer surplus (see, e.g., [54, 44, 40]), the importance of identifying key nodes through centrality measures [9], and the production of public goods [35]. This stream of literature is beyond the scope of this contribution.

[3]Empirical evidences about networks in economics and finance are often closely related. Here we aim to highlight some peculiarities regarding financial networks.

Other papers have applied networks to cross-border linkages and interdependencies of the international financial system, such as the international trade flows [38, 39] and cross-border exposures by asset class (foreign direct investment, portfolio equity, debt, and foreign exchange reserves) [56]. Besides the different level of aggregation to which a node can be defined, also a heterogeneous set of agents can be modeled in a network framework. This is the approach undertaken in [13], where nodes are hedge funds, banks, broker/dealers, and insurance companies, and [21] that consider the institutional sectors of the economy (non-financial corporations, monetary financial institutions, other financial institutions, insurance corporations, government, households, and the rest of the world).

Both in economics and finance, the literature has modeled firms as nodes considering different types of relationships, such as production, supply, or ownership, which may create an intricate web of linkages. The network approach has brought significant insights in the organization of production and international investment. [12] exploit detailed data on production network in Japan, showing that geographic proximity is important to understand supplier–customer relationships. The authors, furthermore, document that while suppliers to well-connected firms have, on average, relatively few customers, suppliers to less connected firms have, on average, many customers (negative degree assortativity). [6] exploring the structure of national and multinational business groups find a positive relationship between a groups' hierarchical complexity and productivity. [32] provide empirical evidence that parent companies and affiliates tend to be located in proximity over a supply chain. Starting with the influential contribution of [58], an extensive body of literature in corporate finance investigates the various types of firm control. Important driving forces are country legal origin and investor protection rights [58, 59]. In a recent contribution, [7] describe extensively corporate control for a large number of firms, documenting persistent differences across countries in corporate control and the importance of various institutional features. [74] investigate the network structure of transnational corporations and show that it can be represented as a bow-tie structure with a relatively small number of entities in the core. In a related paper, [73] study the community structure of the global corporate network identifying the importance of the geographic location of firms. A strong concentration of corporate power is documented also in [70]. Importantly, they show that parent companies choose indirect control when located in countries with better financial institutions and more transparent forms of corporate governance. Formal and informal networks may have even performance consequences [49], affect governance mechanisms [42], and lead to distortions in director selection [57]. For example, in [42] social connections between executives and directors undermine independent corporate governance, having a negative impact on firm value. In the application of this chapter, we focus on ownership linkages following [74, 70], and we provide an overview of the worldwide network structure and the main patterns of control.

This section does not provide a comprehensive literature review, but rather it aims to give an overview of the variety of applications using network analysis and the type of insights it may suggest. In this way we hope to help the reader to think about his own data as a network.

3   Network Analysis

This section formally introduces graphs[4] and provides an overview of standard network metrics, proceeding from local to more global measures.

A graph $G = (V, E)$ consists of a set of nodes V and a set of edges $E \subseteq V^2$ connecting the nodes. A graph G can conveniently be represented by a matrix $W \in \mathbb{R}^{n \times n}$, where $n \in \mathbb{N}$ denotes the number of nodes in G and the matrix element w_{ij} represents the edge from node i to node j. Usually $w_{ij} = 0$ is used to indicate a non-existing edge. Graphs are furthermore described by the following characteristics (see also Figs. 1 and 2):

- A graph G is said to be **undirected**, if there is no defined edge direction. This especially means $w_{ij} = w_{ji}$ for all $i, j = 1, \ldots, n$, and the matrix W is symmetric. Conversely, if we distinguish between the edge w_{ij} (from node i to node j) and the edge w_{ji} (from node j to node i), the graph is said to be **directed**.[5]
- If a graph describes only the existence of edges, i.e., $W \in \{0, 1\}^{n \times n}$, the graph is said to be **unweighted**. In this case, the matrix W is called the **adjacency matrix**, commonly denoted by A. Note that for every graph W there exists an adjacency matrix $A \in \{0, 1\}^{n \times n}$, defined by $a_{ij} := \mathbb{1}_{\{w_{ij} \neq 0\}}$. Conversely, if the edges w_{ij} carry weights, i.e., $w_{ij} \in \mathbb{R}$, the graph G is said to be **weighted**.[6]

While a visual inspection can be very helpful for small networks, this approach quickly becomes difficult as the number of nodes increases. Especially in the era of big data, more sophisticated techniques are required. Thus, various network metrics

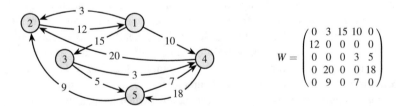

$$W = \begin{pmatrix} 0 & 3 & 15 & 10 & 0 \\ 12 & 0 & 0 & 0 & 0 \\ 0 & 0 & 0 & 3 & 5 \\ 0 & 20 & 0 & 0 & 18 \\ 0 & 9 & 0 & 7 & 0 \end{pmatrix}$$

Fig. 1 Example of a directed and weighted graph

---

[4]The terms "graph" and "network," as well as the terms "link" and "edge," and the terms "vertex" and "node" are used interchangeably throughout this chapter.

[5]For example, if countries are represented as nodes, the distance between them would be a set of undirected edges, while trade relationships would be directed edges, with w_{ij} representing the export from i to j and w_{ji} the import from i to j.

[6]Relationship in social media, such as Facebook or Twitter, can be represented as unweighted edges (i.e., whether two individuals are friends/followers) or weighted edges (i.e., the number of interactions in a given period).

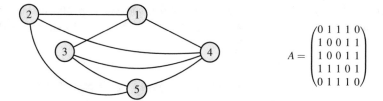

Fig. 2 Example of an undirected and unweighted graph

and measures have been developed to help describe and analyze complex networks. The most common ones are explained in the following.

Most networks do not exhibit self-loops, i.e., edges connecting a node with itself. For example, in social networks it makes no sense to model a person being friends with himself or in financial networks a bank lending money to itself. Therefore, in the following we consider networks without self-loops. It is however straightforward to adapt the presented network statistics to graphs containing self-loops. Moreover, we consider the usual case of networks comprising only positive weights, i.e., $W \in \mathbb{R}_{\geq 0}^{n \times n}$. Adaptations to graphs with negative weights are however also possible. Throughout this section let $W^{(\text{dir})}$ denote a directed graph and $W^{(\text{undir})}$ an undirected graph.

The network **density** $\rho \in [0, 1]$ is defined as the ratio of the number of existing edges and the number of possible edges, i.e., for $W^{(\text{dir})}$ and $W^{(\text{undir})}$, the density is given by:

$$\rho_{W^{(\text{dir})}} = \frac{\sum_{i=1}^{n} \sum_{j=1}^{n} \mathbb{1}_{\{w_{ij} > 0\}}}{n(n-1)}, \qquad \rho_{W^{(\text{undir})}} = \frac{\sum_{i=1}^{n} \sum_{j>i}^{n} \mathbb{1}_{\{w_{ij} > 0\}}}{n(n-1)/2}. \qquad (1)$$

The density of a network describes how tightly the nodes are connected. Regarding financial networks, the density can also serve as an indicator for diversification. The higher the density, the more edges, i.e., the more diversified the investments. For example, the graph pictured in Fig. 1 has a density of 0.5, indicating that half of all possible links, excluding self-loops, exist.

While the density summarizes the overall interconnectedness of the network, the **degree sequence** describes the connectivity of each node. The degree sequence $d = (d_1, \dots, d_n) \in \mathbb{N}_0^n$ of $W^{(\text{dir})}$ and $W^{(\text{undir})}$ is given for all $i = 1, \dots, n$ by:

$$d_{i, W^{(\text{dir})}} = \sum_{j=1}^{n} \mathbb{1}_{\{w_{ij} > 0\}} + \mathbb{1}_{\{w_{ji} > 0\}}, \qquad d_{i, W^{(\text{undir})}} = \sum_{j=1}^{n} \mathbb{1}_{\{w_{ij} > 0\}}. \qquad (2)$$

For a directed graph $W^{(\text{dir})}$ we can differentiate between incoming and outgoing edges and thus define the **in-degree sequence** $d^{(\text{in})}$ and the **out-degree sequence** $d^{(\text{out})}$ as:

$$d_{i,W^{(\text{dir})}}^{(\text{in})} = \sum_{j=1}^{n} \mathbb{1}_{\{w_{ji}>0\}}, \qquad d_{i,W^{(\text{dir})}}^{(\text{out})} = \sum_{j=1}^{n} \mathbb{1}_{\{w_{ij}>0\}}. \tag{3}$$

The degree sequence shows how homogeneously the edges are distributed among the nodes. Financial networks, for example, are well-known to include some well-connected big intermediaries and many small institutions and hence exhibit a heterogeneous degree sequence. For example, for the graph pictured in Fig. 1, we get the following in- and out-degree sequences, indicating that node 4 has the highest number of connections, 3 incoming edges, and 2 outgoing edges:

$$
\begin{aligned}
d_{W^{(\text{dir})}}^{(\text{in})} &= \left(d_{1,W^{(\text{dir})}}^{(\text{in})}, d_{2,W^{(\text{dir})}}^{(\text{in})}, \ldots, d_{5,W^{(\text{dir})}}^{(\text{in})} \right) = (1,3,1,3,2), \\
d_{W^{(\text{dir})}}^{(\text{out})} &= \left(d_{1,W^{(\text{dir})}}^{(\text{out})}, d_{2,W^{(\text{dir})}}^{(\text{out})}, \ldots, d_{5,W^{(\text{dir})}}^{(\text{out})} \right) = (3,1,2,2,2).
\end{aligned}
\tag{4}
$$

Similarly, for weighted graphs, the distribution of the weight among the nodes is described by the **strength sequence** $s = (s_1, \ldots, s_n) \in \mathbb{R}_{\geq 0}^{n}$ and is given for all $i = 1, \ldots, n$ by:

$$s_{i,W^{(\text{dir})}} = \sum_{j=1}^{n} w_{ij} + w_{ji}, \qquad s_{i,W^{(\text{undir})}} = \sum_{j=1}^{n} w_{ij}. \tag{5}$$

In addition, for the weighted and directed graph $W^{(\text{dir})}$, we can differentiate between the weight that flows into a node and the weight that flows out of it. Thus, the **in-strength sequence** $s^{(\text{in})}$ and the **out-strength sequence** $s^{(\text{out})}$ are defined for all $i = 1, \ldots, n$ as:

$$s_{i,W^{(\text{dir})}}^{(\text{in})} = \sum_{j=1}^{n} w_{ji}, \qquad s_{i,W^{(\text{dir})}}^{(\text{out})} = \sum_{j=1}^{n} w_{ij}. \tag{6}$$

For example, for the graph pictured in Fig. 1, we get the following in- and out-strength sequences:

$$
\begin{aligned}
s_{W^{(\text{dir})}}^{(\text{in})} &= \left(s_{1,W^{(\text{dir})}}^{(\text{in})}, s_{2,W^{(\text{dir})}}^{(\text{in})}, \ldots, s_{5,W^{(\text{dir})}}^{(\text{in})} \right) = (12,32,15,20,23), \\
s_{W^{(\text{dir})}}^{(\text{out})} &= \left(s_{1,W^{(\text{dir})}}^{(\text{out})}, s_{2,W^{(\text{dir})}}^{(\text{out})}, \ldots, s_{5,W^{(\text{dir})}}^{(\text{out})} \right) = (28,12,8,38,16).
\end{aligned}
\tag{7}
$$

Node 2 is absorbing more weight than all other nodes with an in-strength of 32, while node 4 is distributing more weight than all other nodes with an out-strength of 38.

The homogeneity of a graph in terms of its edges or weights is measured by the **assortativity**. Degree (resp. strength) assortativity is defined as Pearson's correlation coefficient of the degrees (resp. strengths) of connected nodes. Likewise, we can define the in- and out-degree assortativity and in- and out-strength assortativity. Negative assortativity, also called **disassortativity**, indicates that nodes with few edges (resp. low weight) tend to be connected with nodes with many edges (resp. high weight) and vice versa. This is, for example, the case for financial networks, where small banks and corporations maintain financial relationships (e.g., loans, derivatives) rather with big well-connected financial institutions than between themselves. Positive assortativity, on the other hand, indicates that nodes tend to be connected with nodes that have a similar degree (resp. similar weight). For example, the graph pictured in Fig. 1 has a degree disassortativity of -0.26 and a strength disassortativity of -0.24, indicating a slight heterogeneity of the connected nodes in terms of their degrees and strengths.

The importance of a node is assessed through centrality measures. The three most prominent centrality measures are betweenness, closeness, and eigenvector centrality and can likewise be defined for directed and undirected graphs. (Directed or undirected) **betweenness centrality** b_i of vertex i is defined as the sum of fractions of (resp. directed or undirected) shortest paths that pass through vertex i over all node pairs, i.e.:

$$b_i = \sum_{j,h=1}^{n} \frac{s_{jh}(i)}{s_{jh}}, \tag{8}$$

where $s_{jh}(i)$ is the number of shortest paths between vertices j and h that pass through vertex i, s_{jh} is the number of shortest paths between vertices j and h, and with the convention that $s_{jh}(i)/s_{jh} = 0$ if there is no path connecting vertices j and h. For example, the nodes of the graph pictured in Fig. 1 have betweenness centralities $b = (b_1, b_2, \ldots, b_5) = (5, 5, 1, 2, 1)$, i.e., nodes 1 and 2 are the most powerful nodes as they maintain the highest ratio of shortest paths passing through them.

(Directed or undirected) **closeness centrality** c_i of vertex i is defined as the inverse of the average shortest path (resp. directed or undirected) between vertex i and all other vertices, i.e.:

$$c_i = \frac{n-1}{\sum_{j \neq i} d_{ij}}, \tag{9}$$

where d_{ij} denotes the length of the shortest path from vertex i to vertex j. For example, the nodes of the graph pictured in Fig. 1 have closeness centralities $c = (c_1, c_2, \ldots, c_5) = (0.80, 0.50, 0.57, 0.57, 0.57)$. Note that in comparison to betweenness centrality, node 1 is closer to other nodes than node 2 as it has more outgoing edges.

Eigenvector centrality additionally accounts for the importance of a node's neighbors. Let λ denote the largest eigenvalue of the adjacency matrix a and e the corresponding eigenvector, i.e., $\lambda a = ae$ holds. The eigenvector centrality of vertex i is given by:

$$e_i = \frac{1}{\lambda} \sum_j a_{ij} e_j. \tag{10}$$

The closer a node is connected to other important nodes, the higher is its eigenvector centrality. For example, the nodes of the graph pictured in Fig. 2 (representing the undirected and unweighted version of the graph in Fig. 1) have eigenvector centralities $e = (e_1, e_2, \ldots, e_5) = (0.19, 0.19, 0.19, 0.24, 0.19)$, i.e., node 4 has the highest eigenvector centrality. Taking a look at the visualization in Fig. 2, this result is no surprise. In fact node 4 is the only node that is directly connected to all other nodes, naturally rendering it the most central node.

Another interesting network statistic is the **clustering coefficient**, which indicates the tendency to form triangles, i.e., the tendency of a node's neighbors to be also connected to each other. An intuitive example for a highly clustered network are friendship networks, as two people with a common friend are likely to be friends as well. Let a denote the adjacency matrix of an undirected graph. The clustering coefficient C_i of vertex i is defined as the ratio of realized to possible triangles formed by i:

$$C_i = \frac{\left(a^3\right)_{ii}}{d_i \left(d_i - 1\right)}, \tag{11}$$

where d_i denotes the degree of node i. For example, the nodes of the graph pictured in Fig. 2 have clustering coefficients $C = (C_1, C_2, \ldots, C_5) = (0.67, 0.67, 0.67, 0.67, 0.67)$. This can be easily verified via the visualization in Fig. 2. Nodes 1, 2, 3, and 5 form each part of 2 triangles and have 3 edges, which give rise to a maximum of 3 triangles ($C_1 = 2/3$). Node 4 forms part of 4 triangles and has 4 links, which would make 6 triangles possible ($C_4 = 4/6$). For an extension of the clustering coefficient to directed and weighted graphs, the reader is kindly referred to [37].

Furthermore, another important strand of literature works on community detection. Communities are broadly defined as groups of nodes that are densely connected within each group and sparsely between the groups. Identifying such groupings can provide valuable insight since nodes of the same community often have further features in common. For example, in social networks, communities are formed by families, sports clubs, and educationally or professionally linked colleagues; in biochemical networks, communities may constitute functional modules; and in citation networks, communities indicate a common research topic. Community detection is a difficult and often computationally intensive task. Many different approaches have been suggested, such as the minimum-cut method, modularity maximization, and the Girvan–Newman algorithm, which identifies communities by iteratively cutting the links with the highest betweenness centrality. Detailed information on community detection and comparison of different approaches are available in, e.g., [67] and to [30].

One may be interested in separating the nodes in communities that are tightly connected inside but with a few links between nodes that are part of different communities. Furthermore, we can identify network components that are of special interest. The most common components are the **largest weakly connected component (LWCC) and largest strongly connected component (LSCC)**. The LWCC is the largest subset of nodes, such that within the subset there exists an undirected path from each node to every other node. The LSCC is the largest subset of nodes, such that within the subset there exists a directed path from each node to every other node.

The concepts and measures we had presented are general and can be applied to any network.[7] However, their interpretation depends on the specific context of application. A knowledge of the underlined economic/financial phenomenon is also necessary before starting to model the raw data as a network. When building a network, a preliminary exploration of the data helps to control or mitigate errors that arise from working with data collected in real-world settings (i.e., missing data, measurement errors...). Data quality, or at least awareness of data limitations, is important to perform an accurate network analysis and to draw credible inferences and conclusions. When performing a network analysis, it is also important to remind that the line of investigation depends on the network under consideration: in some cases it could be more relevant to study deeply centrality measures while in others to detect communities.

For data processing and implementation of network measures, several software are available. R, Python, and MATLAB include tools and packages that allow the computation of the most popular measures and network analysis, while Gephi and Pajek are open-source options for exploring visually a network.

4 Network Analysis: An Application to Firm Ownership

In this section we present an application of network analysis to firm ownership, that is, the shareholders of firms. We first describe the data and how the network was built. Then we show the resulting network structure and comment the main results.

---

[7]A more extensive introduction to networks can be found, e.g., in [67].

4.1 Data

Data on firm ownership are retrieved from Orbis compiled by Bureau van Dijk (a Moody's Analytics Company). Orbis provides detailed firm's ownership information. Bureau van Dijk collects ownership information directly from multiple sources including the company (annual reports, web sites, private correspondence) and official regulatory bodies (when they are in charge of collecting this type of information) or from the associated information providers (who, in turn, have collected it either directly from the companies or via official bodies). It includes mergers and acquisitions when completed. Ownership data include for each firm the list of shareholders and their shares. They represent voting rights, rather than cash-flow rights, taking into account dual shares and other special types of share. In this application, we also consider the country of incorporation and the entity type.[8] In addition, we collect for each firm the primary sector of activity (NACE Revision 2 codes)[9] and, when available, financial data (in this application we restrict our interest to total assets, equity, and revenues). Indeed Orbis is widely used in the literature for the firms' balance sheets and income statements, which are available at an annual frequency. All data we used refer to year 2016.

4.2 Network Construction

In this application we aim to construct an *ownership network* that consists of a set of nodes representing different economic actors, as listed in footnote 8, and a set of directed weighted links denoting the shareholding positions between the nodes.[10] More precisely, a link from node A to node B with weight x means that A holds $x\%$ of the shares of B. This implies that the weights are restricted to the interval $[0, 100]$.[11]

---

[8] Orbis database provides information regarding the type of entity of most of the shareholders. The classification is as follows: insurance company (A); bank (B); industrial company (C); unnamed private shareholders (D); mutual and pension funds, nominee, trust, and trustee (E); financial company not elsewhere classified (F); foundation/research institute (J); individuals or families (I); self-ownership (H); other unnamed private shareholders (L); employees, managers, and directors (M); private equity firms (P); branch (Q); public authorities, states, and government (S); venture capital (V); hedge fund (Y); and public quoted companies (Z). The "type" is assigned according to the information collected from annual reports and other sources.

[9] NACE Rev. 2 is the revised classification of the official industry classification used in the European Union adopted at the end of 2006. The level of aggregation used in this contribution is the official sections from A to U. Extended names of sections are reported in Table 5 together with some summary statistics.

[10] For further applications of networks and graph techniques, the reader is kindly referred to [33, 69].

[11] Notice that the definition of nodes and edges and network construction are crucial steps, which depend on the specific purpose of the investigation. For example, in case one wanted to do some

Starting from the set of data available in Orbis, we extract listed firms.[12] This set of nodes can be viewed as the seed of the network. Any other seed of interest can of course be chosen likewise. Then, using the ownership information (the names of owners and their respective ownership shares) iteratively, the network is extended by integrating all nodes that are connected to the current network through outgoing or incoming links.[13] At this point, we consider all entities and both the direct and the total percentage figures provided in the Orbis database. This process stops when all outgoing and incoming links of all nodes lead to nodes which already form part of the network. To deal with missing and duplicated links, we subsequently perform the following adjustments: (1) in case Orbis lists multiple links with direct percentage figures from one shareholder to the same firm, these shares are aggregated into a single link; (2) in case direct percentage figures are missing, the total percentage figures are used; (3) in case both the direct and total percentage figures are missing, the link is removed; and (4) when shareholders of some nodes jointly own more than 100%, the concerned links are proportionally rescaled to 100%. From the resulting network, we extract the largest weakly connected component (LWCC) that comprises over 98% of the nodes w.r.t. the network derived so far.

The resulting sample includes more than 8.1 million observations, of which around 4.6 million observations are firms (57%).[14] The majority of firms are active in the sectors wholesale and retail trade; professional, scientific, and technical activities; and real estate activities (see Table 5). When looking at the size of sectors with respect to accounting the variables, the picture changes. In terms of total assets and equity, the main sectors are financial and insurance activities and manufacturing, while in terms of revenues, as expected, manufacturing and wholesale and retail trade have the largest share. We also report the average values, which again display a significant variation between sectors. Clearly the overall numbers hide a wide heterogeneity within sector, but some sectors are dominated by very large firms (e.g., mining and quarrying), while in others micro or small firms are prevalent (e.g., wholesale and retail trade). The remaining sample includes entities of various types, such as individuals, which do not have to report a balance sheet. Nodes are from

---

econometric analysis at firm level, it could have been more appropriate to exclude from the node definition all those entities that are not firms.

[12] We chose to study listed firms, as their ownership structure is often hidden behind a number of linkages forming a complex network. Unlisted firms, in contrast, are usually owned by a unique shareholder (identified by a GUO 50 in Orbis).

[13] All computations for constructing and analyzing the ownership network have been implemented in Python. Python is extremely useful for big data projects, such as analyzing complex networks comprising millions of nodes. Other common programming languages such as R and MATLAB are not able to manipulate huge amounts of data.

[14] Unfortunately balance sheet data are available only for a subsample corresponding to roughly 30% of the firms. Missing data are due to national differences in firm reporting obligations or Bureau van Dijk not having access to data in some countries. Still Orbis is considered one of the most comprehensive source for firms' data.

■	NO	(22.72%)
■	CN	(14.34%)
■	IT	(13.2%)
■	DE	(7.36%)
■	ES	(6.6%)
■	FR	(4.82%)
■	WW	(4.57%)
■	BE	(2.79%)
■	SI	(2.03%)
■	IN	(1.9%)
■	GB	(1.9%)

Fig. 3 Visualization of the IN component (see Sect. 4.4) and considering only the links with the weight of at least 1%. Countries that contain a substantial part of the nodes of this subgraph are highlighted in individual colors according to the legend on the right-hand side. This graph was produced with Gephi

Table 1 General characteristics of the ownership network

	Year 2016
Number of nodes (in million)	8.1
Number of links (in million)	10.4
Avg. weight per node	38.0%
Density	$1E - 6$

all over the world, but with a prevalence from developed countries and particularly from those having better reporting standards.

A visualization of the entire network with 8.1 million nodes is obviously not possible here. However, to still gain a better idea of the structure of the network, Fig. 3 visualizes part of the network, namely, the IN component (see Sect. 4.4).[15] It is interesting to note that the graph shows some clear clusters for certain countries.

4.3 Network Statistics

The resulting ownership network constitutes a complex network with millions of nodes and links. This section demonstrates how network statistics can help us gain insight into such opaque big data structure.

Table 1 summarizes the main network characteristics. The network includes more than 8.1 million of nodes and 10.4 million of links. This also implies that the ownership network is extremely sparse with a density of less 1E-6. The average share is around 38.0% but with a substantial heterogeneity across links.

---

[15]Gephi is one of the most commonly used open-source software for visualizing and exploring graphs.

Table 2 Centrality measures: summary statistics

	Mean	St. Dev.	Pt. 5	Pt. 50	Pt. 95
In-degree	1.27	5.17	0.00	1.00	4.00
Out-degree	1.27	85.81	0.00	1.00	4.00
In-strength	48.30	47.59	0.00	50.01	100.00
Out-strength	48.30	6432.56	0.00	0.10	183.33
Eigenvector (%)	0.01	0.03	0.00	0.00	0.04
Closeness	0.45	0.45	0.00	0.42	1.00

Table 2 shows the summary statistics of the network measures computed at node level. The ownership network is characterized by a high heterogeneity: there are firms wholly owned by a single shareholder (i.e., owning 100% of the shares) and firms with a dispersed ownership in which some shareholders own a tiny percentage. These features are reflected in the in-degree and in-strength. Correspondingly, there are shareholders with just a participation in a unique firm and others with shares in many different firms (see the out-degree and out-strength).

To gain further insights, we investigate the in-degree and out-degree distribution, that is, an analysis frequently used in complex networks. Common degree distributions identified in real-world networks are Poisson, exponential, or power-law distributions. Networks with power-law degree distribution, usually called scale-free networks, show many end nodes, other nodes with a low degree, and a handful of very well-connected nodes.[16] Since power laws show a linear relationship in logarithmic scales, it is common to visualize the degree distributions in the form of the complementary cumulative distribution function (CDF) in a logarithmic scale. Figure 4 displays the in- and out-degree distribution in panels (a) and (b), respectively. Both distributions show the typical behavior of scale-free networks, with the majority of nodes having a low degree and a few nodes having a large value. When considering the in-degree distribution, we can notice that there are 94% of the nodes with an in-degree equal or lower than 3. While this is partially explained by the presence of pure investors, when excluding these nodes from the distribution, the picture does not change much (90% of the nodes have an in-degree equal or lower than 3). This provides further evidence that the majority of firms are owned by very few shareholders, while a limited number of firms, mainly listed firms, are owned by many shareholders. A similar pattern is observed for the out-degree; indeed many shareholders invest in a limited number of firms, while few shareholders own shares in a large number of firms. This is the case of investment funds that aim to have a diversified portfolio.[17]

Concerning the centrality measures, the summary statistics in Table 2 suggest a high heterogeneity across nodes. It is also interesting to notice that centrality measures are positively correlated with financial data. Entities having high values of

---

[16]For more information on scale-free networks and power laws, see [4] and [25].

[17]A similar analysis can be performed also for the strength distribution; however, in this context, it is less informative.

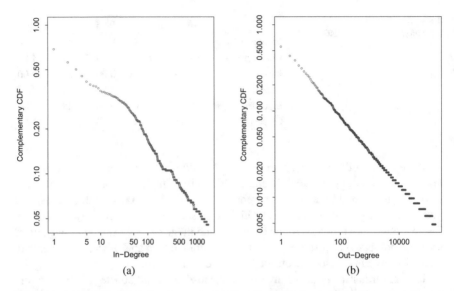

Fig. 4 Degree distribution in log–log scale. Panel **a** (Panel **b**) shows the in-degree (out-degree) distribution. The *y*-axis denotes the complementary cumulative distribution function

centrality are usually financial entities and institutional shareholders, such as mutual funds, banks, and private equity firms. In some cases, entities classified as states and governments have high values possibly due to state-owned enterprises, which in some countries are still quite diffused in certain sector of the economy.

4.4 Bow-Tie Structure

The ownership networks can be split into the components of a bow-tie structure (see, e.g., [74, 46]), as pictured in Fig. 5. Each component identifies a group of entities with a specific structure of interactions. In the center we have a set of closely interconnected firms forming the largest strongly connected component (LSCC). Next, we can identify all nodes that can be reached via a path of outgoing edges starting from the LSCC. These nodes constitute the OUT component and describe firms that are at least partially owned by the LSCC. Likewise, all nodes that can be reached via a path of incoming edges leading to the LSCC are grouped in the IN component. These nodes own at least partially the LSCC and thus indirectly also the OUT component. Nodes that lie on a path connecting the IN with the OUT component form the Tubes. All nodes that are connected through a path with nodes of the Tubes are also added to the Tubes component. The set of nodes that is reached via a path of outgoing edges starting from the IN component and not leading to the LSCC constitutes the IN-Tendrils. Analogously, nodes that are reached via a path on incoming edges leading to the OUT component and are not connecting the LSCC

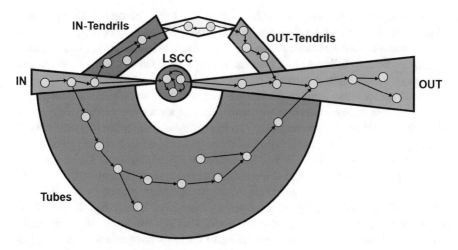

Fig. 5 Ownership networks: the bow-tie structure

Table 3 Size of the components of the bow-tie structure

Component	N (in %)
IN	0.20%
LSCC	0.03%
OUT	15.24%
Tube	59.49%
IN-Tendril	12.51%
OUT-Tendril	12.72%
IN- and OUT-Tendrils	0.20%

form the OUT-Tendrils. Again, nodes of the LWCC that are connected to the IN-Tendrils (resp. OUT-Tendrils) and are not part of any other component are added to the IN-Tendrils (resp. OUT-Tendrils). These nodes can construct a path from the OUT-Tendrils to the IN-Tendrils.[18]

Table 3 shows the distribution of the nodes of the ownership network among the components of the bow-tie structure. The biggest component is the Tube, which contains almost 59.49% of the nodes. Interestingly, the IN and the LSCC components include a very limited number of entities equal to only 0.20% and 0.03%, respectively, of the overall sample. The OUT component and the OUT-Tendrils, on the other side, show a fraction of, respectively, 15.24% and 12.72% on average. All other components hold less than 1% of the nodes. As expected, in the OUT component, most of the entities are firms (87%). Two components are key in terms of control of power in the network: the IN and the LSCC components. The IN component includes mainly individuals, for which even the country is not available

---

[18]Other networks characterized by a bow-tie architecture are the Web [16] and many biological systems [29].

in many instances, and large financial entities. The LSCC component has a similar distribution of entities from A to F with a slight prevalence of very large companies, banks, and mutual funds. These entities are more frequently located in United States and Great Britain, followed by China and Japan. Entities in this component are also the ones with the highest values of centrality.

Next, we focus on firms in the bow-tie structure and investigate the role played by each sector in the different components. Table 4 shows the number of firms and the total assets (both as percentage) by components. We can notice that the financial sector plays a key role in the IN and LSCC components, while it is less prominent in other components. Indeed, it is well-known that the financial sector is characterized by a limited number of financial institutions very large and internationalized. The network approach provides evidence of the key position played by the financial sector and, specifically, by some nodes in the global ownership. In the OUT-component other prominent sectors are manufacturing, wholesale and retail trade, and professional activities. The composition of the other components is more varied. As expected, sectors wholesale and retail trade and real activities are well-positioned in all the chain of control, while some other sectors (sections O to U) always play a limited role. Within each component, it would be possible to go deeper in the analysis separating sub-components or groups of nodes with specific characteristics.

Firm ownership has implications for a wide range of economic phenomena, which span from competition to foreign direct investments, where having a proper understanding of the ownership structure is of primary importance for policy makers. This is the case, for example, of the concentration of voting rights obtained by large investment funds holding within their grasp small stakes in many companies. According to the Financial Times, "BlackRock, Vanguard and State Street, the three biggest index-fund managers, control about 80 per cent of the US equity ETF market, about USD 1.6tn in total. Put together, the trio would be the largest shareholder of 88 per cent of all S&P 500 companies."[19] Our analysis of the network structure, jointly with the centrality measures, permits the identification of key nodes and concentration of power and therefore grants policy makers a proper assessment of the degree of influence exerted by these funds in the economy. Our findings at sectoral level also provide a rationale for having some sectors more regulated (i.e., the financial sector) than others. Moreover, the ownership network, in the context of policy support activities to the European Commission, has been used for supporting the new FDI screening regulation.[20] In the case of non-EU investments in Europe, the correct evaluation of the nationality of the investor is of particular importance. With the historical version covering the period 2007–2018, we tracked the change over time in ownership of EU companies owned by non-EU entities identifying the origin country of the controlling investor, as well as the sectors of activity targeted by non-EU investments. This approach constitutes an improvement

---

[19] *Financial Times*, "Common Ownership of shares faces regulatory scrutiny", January 22 2019.

[20] Regulation (EU) 2019/452 establishes a framework for the screening of foreign direct investments into the European Union.

Table 4 Bow-tie components and firm sectors

Sector	IN		LSCC		OUT		Tube		IN-Tendril		OUT-Tendril		IN- and OUT-Tendrils	
	N	TA	N	TA	N	TA	N	TA	N	TA	N	TA	N	TA
A	1.20%	0.29%	0.14%	0.01%	0.56%	0.19%	6.45%	0.76%	4.22%	0.96%	1.55%	0.54%	3.06%	1.36%
B	0.31%	0.02%	1.24%	2.04%	1.29%	4.62%	0.40%	2.54%	0.48%	2.08%	0.39%	1.03%	0.14%	12.29%
C	3.76%	15.43%	10.28%	8.18%	15.22%	17.91%	7.81%	13.49%	12.22%	13.04%	11.39%	14.86%	9.83%	29.47%
D	0.54%	11.30%	0.90%	2.72%	3.61%	4.35%	2.58%	5.98%	0.92%	7.43%	1.47%	2.52%	0.86%	5.56%
E	0.12%	0.00%	0.07%	0.10%	0.86%	0.27%	0.70%	1.15%	0.91%	0.47%	0.65%	0.47%	0.74%	0.50%
F	2.87%	0.60%	0.90%	1.14%	5.33%	2.53%	8.21%	4.23%	9.73%	4.07%	8.16%	4.43%	13.57%	2.69%
G	4.46%	2.89%	3.10%	1.77%	14.03%	5.70%	13.84%	5.93%	22.68%	6.38%	14.63%	8.50%	16.63%	12.26%
H	1.63%	4.07%	1.66%	0.77%	3.23%	2.95%	3.45%	5.64%	4.14%	6.06%	2.89%	4.34%	3.22%	0.76%
I	0.43%	0.02%	0.55%	0.04%	1.57%	0.36%	3.00%	0.44%	2.67%	0.84%	2.84%	0.75%	3.16%	1.20%
J	0.62%	0.51%	2.21%	3.02%	7.16%	4.48%	4.38%	1.69%	4.11%	2.29%	4.82%	1.78%	4.08%	0.47%
K	56.31%	58.12%	67.52%	78.64%	13.72%	38.38%	8.90%	38.93%	5.20%	38.92%	11.52%	40.47%	6.80%	8.76%
L	8.03%	1.42%	1.52%	0.21%	8.82%	2.51%	14.06%	6.54%	10.84%	6.20%	13.67%	7.40%	10.87%	3.90%
M	9.35%	3.49%	6.21%	0.76%	12.27%	9.66%	14.23%	7.27%	11.78%	3.48%	13.52%	7.35%	17.23%	10.60%
N	3.76%	1.22%	3.03%	0.55%	7.66%	4.75%	5.76%	2.95%	4.82%	2.34%	6.59%	3.34%	5.72%	9.68%
O	2.17%	0.03%	0.07%	0.00%	0.08%	0.06%	0.30%	0.98%	0.36%	2.97%	0.26%	0.49%	0.04%	0.00%
P	0.27%	0.16%	0.07%	0.00%	0.36%	0.03%	0.91%	0.11%	0.61%	0.26%	0.60%	0.10%	0.48%	0.05%
Q	0.54%	0.25%	0.07%	0.00%	1.42%	0.19%	1.79%	0.71%	1.38%	1.10%	1.42%	0.66%	0.90%	0.02%
R	0.27%	0.00%	0.14%	0.04%	0.72%	0.19%	1.36%	0.35%	1.02%	0.51%	1.39%	0.38%	0.50%	0.02%
S	3.26%	0.17%	0.34%	0.00%	1.76%	0.86%	1.78%	0.30%	1.78%	0.61%	1.69%	0.58%	1.50%	0.42%
T	0.00%	0.00%	0.00%	0.00%	0.33%	0.01%	0.09%	0.00%	0.13%	0.00%	0.53%	0.00%	0.62%	0.00%
U	0.08%	0.00%	0.00%	0.00%	0.01%	0.00%	0.01%	0.00%	0.00%	0.00%	0.02%	0.00%	0.00%	0.00%

with respect to the current practice, and it is crucial for depicting the network of international investments. Usually cross-border investments are measured using aggregated foreign direct investment statistics coming from national accounts that cover all cross-border transactions and positions between the reporting country and the first partner country. Official data, however, neglect the increasingly complex chain of controls of multinational enterprises and thus provide an incomplete and partial picture of international links, where the first partner country is often only one of the countries involved in the investment and in many cases not the origin. The centrality in the network of certain firms or sectors (using the more refined NACE classification at four-digit level) can be further used in support to the screening of foreign mergers and acquisitions in some key industries, such as IT, robotics, artificial intelligence, etc. Indeed, FDI screening is motivated by the protections of essential national or supra-national interests as requested by the new regulation on FDI screening that will enter into force in October 2020.

5 Conclusion

In light of today's massive and ubiquitous data collection, efficient techniques to manipulate these data and extract the relevant information become more and more important. One powerful approach is offered by network science, which finds increasing attention also in economics and finance. Our application of network analysis to ownership information demonstrates how network tools can help gain insights into large data. But it also shows how this approach provides a unique perspective on firm ownership, which can be particularly useful to inform policy makers. Extending the analysis to data covering several years, it would be possible to study the role of the evolving network structure over time on the macroeconomics dynamics and business cycle. Other avenues for future research concern the relationship of the economic performance of firms with their network positions and the shock transmission along the chain of ownership caused by firm bankruptcy. An important caveat of our analysis is that the results depend on the accuracy of the ownership data (i.e., incomplete information of the shareholders may result in a misleading network structure). But this is a common future to most of the network applications; indeed the goodness of the raw data from which the network is constructed is a key element. As better quality and detailed data will be available in the future, the more one can obtain novel findings and generalize results using a network approach. Overall, while each application of network analysis to real-world data has some challenges, we believe that the effort to implement it is worthy.

Appendix

See Table 5.

Table 5 Sample overview of firms

NACE section	Share (%)				Average value		
	N	TA	Equity	Revenues	TA	Equity	Revenues
A—Agriculture, forestry, and fishing	4.82%	0.31%	0.42%	0.54%	11.78	6.24	7.13
B—Mining and quarrying	0.56%	3.43%	4.46%	3.54%	524.90	268.60	212.80
C—Manufacturing	9.76%	15.50%	18.91%	34.37%	127.10	61.29	101.00
D—Electricity, gas, steam, and air conditioning supply	2.51%	4.87%	4.78%	6.43%	248.60	95.84	138.90
E—Water supply, sewerage, waste management	0.74%	0.42%	0.45%	0.40%	46.68	19.63	16.30
F—Construction	7.86%	2.69%	2.66%	4.17%	32.56	12.67	18.87
G—Wholesale and retail trade, repair of motor vehicles and motorcycles	14.71%	5.22%	5.30%	22.81%	31.67	12.65	50.42
H—Transportation and storage	3.42%	3.40%	3.88%	4.12%	92.86	41.73	42.11
I—Accommodation and food service activities	2.71%	0.34%	0.45%	0.47%	13.91	7.34	7.17
J—Information and communication	4.86%	3.36%	4.66%	4.26%	71.39	38.94	36.06
K—Financial and insurance activities	9.69%	44.65%	32.10%	10.41%	500.70	142.40	67.52
L—Real estate activities	12.83%	3.22%	3.53%	1.28%	24.18	10.40	3.76
M—Professional, scientific, and technical activities	13.62%	7.54%	11.55%	3.11%	64.92	39.08	11.97
N—Administrative and support service activities	6.06%	3.57%	4.93%	2.55%	65.79	35.67	22.89
O—Public administration and defense	0.26%	0.30%	0.39%	0.14%	192.90	99.34	38.21
P—Education	0.76%	0.05%	0.09%	0.07%	11.00	7.52	5.65
Q—Human health and social work activities	1.65%	0.31%	0.44%	0.60%	19.26	10.87	15.64
R—Arts, entertainment, and recreation	1.22%	0.21%	0.29%	0.33%	20.68	11.36	13.08
S—Other service activities	1.77%	0.59%	0.69%	0.40%	49.15	22.49	14.58
T—Activities of households as employers	0.17%	0.01%	0.00%	0.02%	2.41	0.56	17.64
U—Activities of extraterritorial organizations and bodies	0.01%	0.00%	0.00%	0.00%	20.54	4.64	4.33

References

1. Acemoglu, D., Carvalho. V. M., Ozdaglar, A., & Tahbaz-Salehi, A. (2012). The network origins of aggregate fluctuations. *Econometrica, 80*, 1977–2016.
2. Acemoglu, D., Ozdaglar, A., & Tahbaz-Salehi, A. (2015). Systemic risk and stability in financial networks. *American Economic Review, 105*, 564–608.
3. Acemoglu, D., Akcigit, U., & Kerr, W. R. (2016). Innovation network. *Proceedings of the National Academy of Sciences, 113*, 11483–11488.
4. Albert R., & Barabasi A. L. (2002). Statistical mechanics of complex networks. *Reviews of Modern Physics, 74*, 47–97.
5. Allen, F., & Gale, D. (2000). Financial contagion. *Journal of Political Economy, 108*, 1–33.
6. Altomonte, C., & Rungi, A. (2013). Business groups as hierarchies of firms: Determinants of vertical integration and performance. In *Working Paper Series 1554*. Frankfurt: European Central Bank.
7. Aminadav, G., & Papaioannou, E. (2020). Corporate control around the world. *The Journal of Finance, 75*(3), 1191–1246. https://doi.org/10.1111/jofi.12889
8. Anand, K., van Lelyveld, I., Banai, A., Friedrich, S., Garratt, R., Hałaj, G., et al. (2018). The missing links: A global study on uncovering financial network structures from partial data. *Journal of Financial Stability, 35*, 107–119.
9. Ballester, C., Calvó Armengol, A., & Zenou, Y. (2006). Who's who in networks. Wanted: The key player. *Econometrica, 74*, 1403–1417.
10. Barabási, A.-L., & Pósfai, M. (2016). *Network science*. Cambridge: Cambridge University Press.
11. Bargigli, L., Di Iasio, G., Infante, L., Lillo, F., & Pierobon, F. (2015). The multiplex structure of interbank networks. *Quantitative Finance, 15*, 673–691.
12. Bernard, A. B., Moxnes, A., & Saito, Y. U. (2019). Production networks, geography, and firm performance. *Journal of Political Economy, 127*, 639–688.
13. Billio, M., Getmansky, M., Lo, A., & Pelizzon, L. (2012). Econometric measures of connectedness and systemic risk in the finance and insurance sectors. *Journal of Financial Economics, 104*, 535–559.
14. Blume, L. E., Brock, W. A., Durlauf, S. N., & Ioannides, Y. M. (2011). Identification of social interactions. In *Handbook of social economics* (Vol. 1, pp. 853–964). North-Holland: Elsevier.
15. Blume, L. E., Brock, W. A., Durlauf, S. N., & Jayaraman, R. (2015). Linear social interactions models. *Journal of Political Economy, 123*, 444–496.
16. Broder, A., Kumar, R., Maghoul, F., Raghavan, P., Rajagopalan, S., Stata, R., et al. (2000). Graph structure in the Web. *Computer Networks, 33*, 309–320.
17. Cai, J., Eidam, F., Saunders, A., & Steffen, S. (2018). Syndication, interconnectedness, and systemic risk. *Journal of Financial Stability, 34*, 105–120.
18. Calvó-Armengol, A., & Zenou, Y. (2004). Social networks and crime decisions: The role of social structure in facilitating delinquent behavior. *International Economic Review, 45*, 939–958.
19. Calvó-Armengol, A., & Zenou, Y. (2005). Job matching, social network and word-of-mouth communication. *Journal of Urban Economics, 57*, 500–522.
20. Calvó-Armengol, A., Patacchini, E., & Zenou, Y. (2009). Peer effects and social networks in education. *The Review of Economic Studies, 76*, 1239–1267.
21. Castrén, O., & Rancan, M. (2014). Macro-Networks: An application to euro area financial accounts. *Journal of Banking & Finance, 46*, 43–58.
22. Cerina, F., Zhu, Z., Chessa, A., & Riccaboni, M. (2015). World input-output network. *PloS One, 10*, e0134025.
23. Cetina, J., Paddrik, M., & Rajan, S. (2018). Stressed to the core: Counterparty concentrations and systemic losses in CDS markets. *Journal of Financial Stability, 35*, 38–52.
24. Choi, S., Galeotti, A., & Goyal, S. (2017). Trading in networks: theory and experiments. *Journal of the European Economic Association, 15*, 784–817.

25. Clauset, A., Shalizi, C. R., & Newman, M. E. J. (2009). Power-law distributions in empirical data. *SIAM Review, 51*, 661–703.
26. Cohen, L., Frazzini, A., & Malloy, C. (2008). The small world of investing: Board connections and mutual fund returns. *Journal of Political Economy, 116*, 951–979.
27. Cossin, D., & Schellhorn, H. (2007). Credit risk in a network economy. *Management Science, 53*, 1604–1617.
28. Craig, B., & Von Peter, G. (2014). Interbank tiering and money center banks. *Journal of Financial Intermediation, 23*, 322–347.
29. Csete, M., & Doyle, J. (2004). Bow ties, metabolism and disease. *TRENDS in Biotechnology, 22*, 446–450.
30. Danon, L., Díaz-Guilera, A., Duch, J., & Arenas, A. (2005). Comparing community structure identification. *Journal of Statistical Mechanics: Theory and Experiment, 2005*(09), Article No. P09008. https://doi.org/10.1088/1742-5468/2005/09/P09008
31. Degryse, H., Elahi, M. A., & Penas, M. F. (2010). Cross border exposures and financial contagion. *International Review of Finance, 10*, 209–240.
32. Del Prete, D., & Rungi, A. (2017). Organizing the global value chain: A firm-level test. *Journal of International Economics, 109*, 16–30.
33. Ding, Y., Rousseau, R., & Wolfram, D. (2014). *Measuring scholarly impact: methods and practice*. Cham: Springer.
34. Eisenberg, L., & Noe, T. H. (2001). Systemic risk in financial systems. *Management Science, 47*, 236–249.
35. Elliott, M., & Golub, B. (2019). A network approach to public goods. *Journal of Political Economy, 127*, 730–776.
36. Fafchamps, M., & Lund, S. (2003). Risk-sharing networks in rural Philippines. *Journal of Development Economics, 71*, 261–287.
37. Fagiolo, G. (2007). Clustering in complex directed networks. *Physical Review E, 76*, 026107.
38. Fagiolo, G., Reyes, J., & Schiavo, S. (2009). World-trade web: Topological properties, dynamics, and evolution. *Physical Review E, 79*, 036115.
39. Fagiolo, G., Reyes, J., & Schiavo, S. (2010). The evolution of the world trade web: a weighted-network analysis. *Journal of Evolutionary Economics, 20*, 479–514.
40. Fainmesser, I. P., & Galeotti, A. (2015). Pricing network effects. *The Review of Economic Studies, 83*, 165–198.
41. Fleming, L., King III, C., & Juda, A. I. (2007). Small worlds and regional innovation. *Organization Science, 18*, 938–954.
42. Fracassi, C., & Tate, G. (2012). External networking and internal firm governance. *The Journal of Finance, 67*, 153–194.
43. Gai, P., & Kapadia, S. (2010). Contagion in financial networks. *Proceedings of the Royal Society A: Mathematical, Physical and Engineering Sciences, 466*, 2401–2423.
44. Galeotti, A., Goyal, S., Jackson, M. O., Vega-Redondo, F., & Yariv, L. (2010). Network games. *The Review of Economic Studies, 77*, 218–244.
45. Girvan, M., & Newman, M. E. (2002). Community structure in social and biological networks. *Proceedings of the National Academy of Sciences, 99*, 7821–7826.
46. Glattfelder, J. B., & Battiston, S. (2019). *The architecture of power: Patterns of disruption and stability in the global ownership network*. Working paper. Available at SSRN: 3314648
47. Goyal, S. (2012). *Connections: an introduction to the economics of networks*. Princeton: Princeton University Press.
48. Hale, G. (2012). Bank relationships, business cycles, and financial crises. *Journal of International Economics, 88*, 312–325.
49. Hochberg, Y. V., Ljungqvist, A., & Lu, Y. (2007). Whom you know matters: Venture capital networks and investment performance. *The Journal of Finance, 62*, 251–301.
50. Hüser, A. C., Halaj, G., Kok, C., Perales, C., & van der Kraaij, A. (2018). The systemic implications of bail-in: A multi-layered network approach. *Journal of Financial Stability, 38*, 81–97.

51. Hvide, H. K., & Östberg, P. (2015). Social interaction at work. *Journal of Financial Economics, 117*, 628–652.
52. Hwang, B. H., & Kim, S. (2009). It pays to have friends. *Journal of Financial Economics, 93*, 138–158.
53. Iori, G., De Masi, G., Precup, O., Gabbi, G., & Caldarelli, G. (2008). A network analysis of the Italian overnight money market. *Journal of Economic Dynamics and Control, 32*, 259–278.
54. Jackson, M. O., & Wolinsky, A. (1996). A strategic model of social and economic networks. *Journal of Economic Theory, 71*, 44–74.
55. Kelly, M., & Ó Gráda, C. (2000). Market contagion: Evidence from the panics of 1854 and 1857. *American Economic Review, 90*, 1110–1124.
56. Kubelec, C., & Sa, F. (2010). The geographical composition of national external balance sheets: 1980–2005. In *Bank of England Working Papers 384*. London: Bank of England.
57. Kuhnen, C. M. (2009). Business networks, corporate governance, and contracting in the mutual fund industry. *The Journal of Finance, 64*, 2185–2220.
58. La Porta, R., Lopez-de-Silanes, F., Shleifer, A., & Vishny, R. W. (1997). Legal determinants of external finance. *The Journal of Finance, 52*, 1131–1150.
59. La Porta, R., Lopez-de-Silanes, F., & Shleifer, A. (1997). Corporate ownership around the world. *The Journal of Finance, 54*, 471–517.
60. Langfield, S., Liu, Z., & Ota, T. (2014). Mapping the UK interbank system. *Journal of Banking & Finance, 45*, 288–303.
61. Manski, C. F. (1993). Identification of endogenous social effects: The reflection problem. *The Review of Economic Studies, 60*, 531–542.
62. Mantegna, R. N. (1999). Hierarchical structure in financial markets. *The European Physical Journal B-Condensed Matter and Complex Systems, 11*, 193–197.
63. Mayer, A. (2009). Online social networks in economics. *Decision Support Systems, 47*, 169–184.
64. Minoiu, C., & Reyes, J. A. (2013). A network analysis of global banking: 1978–2010. *Journal of Financial Stability, 9*, 168–184.
65. Mistrulli, P. E. (2011). Assessing financial contagion in the interbank market: Maximum entropy versus observed interbank lending patterns. *Journal of Banking & Finance, 35*, 1114–1127.
66. Newman, M. E. J. (2003). The structure and function of complex networks. *SIAM Review, 45*, 167–256.
67. Newman, M. E. J. (2010). *Networks: An introduction*. Oxford: Oxford University Press.
68. Poledna, S., Molina-Borboa, J. L., Martínez-Jaramillo, S., Van Der Leij, M., & Thurner, S. (2015). The multi-layer network nature of systemic risk and its implications for the costs of financial crises. *Journal of Financial Stability, 20*, 70–81.
69. Qasim, M. (2017). Sustainability and Wellbeing: A scientometric and bibliometric review of the literature. *Journal of Economic Surveys, 31*(4), 1035–1061.
70. Rungi, A., Morrison, G., & Pammolli, F. (2017). Global ownership and corporate control networks. In *Working Papers 07/2017*. Lucca: IMT Institute for Advanced Studies Lucca.
71. Sacerdote, B. (2011). Peer effects in education: How might they work, how big are they and how much do we know thus far? In *Handbook of the economics of education* (vol. 3, pp. 249–277). New York: Elsevier.
72. Upper, C. (2011). Simulation methods to assess the danger of contagion in interbank markets. *Journal of Financial Stability, 7*, 111–125.
73. Vitali, S., & Battiston, S. (2014). The community structure of the global corporate network. *PloS One, 9*, e104655.
74. Vitali, S., Glattfelder, J. B., & Battiston, S. (2011). The network of global corporate control. *PloS One, 6*, e25995.
75. Wells, S. J. (2004). Financial interlinkages in the United Kingdom's interbank market and the risk of contagion. *Bank of England Working Paper No. 230*.